Homotopy Equivalences of 3-Manifolds and Deformation Theory of Kleinian Groups

[redacted]

MEMOIRS
of the
American Mathematical Society

Number 812

Homotopy Equivalences of 3-Manifolds and Deformation Theory of Kleinian Groups

Richard D. Canary
Darryl McCullough

November 2004 • Volume 172 • Number 812 (first of 4 numbers) • ISSN 0065-9266

American Mathematical Society
Providence, Rhode Island

2000 *Mathematics Subject Classification.*
Primary 57M99; Secondary 20F34, 30F40, 57M50.

Library of Congress Cataloging-in-Publication Data

Canary, Richard Douglas.
Homotopy equivalences of 3-manifolds and deformation theory of Kleinian groups / Richard D. Canary, Darryl McCullough.
p. cm. — (Memoirs of the American Mathematical Society, ISSN 0065-9266 ; no. 812)
"Volume 172, number 812 (first of 4 numbers)."
Includes bibliographical references.
ISBN 0-8218-3549-1 (alk. paper)
1. Three-manifolds (Topology). 2. Homotopy equivalences. 3. Low-dimensional topology. 4. Kleinian groups. I. Title: Homotopy equivalences of three-manifolds and deformation theory of Kleinian groups. II. McCullough, Darryl, 1951– III. Title. IV. Series.
QA3.A57 no. 812
[QA612.14]
510 s–dc22
[514′.3] 2004054528

Memoirs of the American Mathematical Society

This journal is devoted entirely to research in pure and applied mathematics.

Subscription information. The 2004 subscription begins with volume 167 and consists of six mailings, each containing one or more numbers. Subscription prices for 2004 are $583 list, $466 institutional member. A late charge of 10% of the subscription price will be imposed on orders received from nonmembers after January 1 of the subscription year. Subscribers outside the United States and India must pay a postage surcharge of $31; subscribers in India must pay a postage surcharge of $43. Expedited delivery to destinations in North America $35; elsewhere $130. Each number may be ordered separately; *please specify number* when ordering an individual number. For prices and titles of recently released numbers, see the New Publications sections of the *Notices of the American Mathematical Society.*

Back number information. For back issues see the *AMS Catalog of Publications.*

Subscriptions and orders should be addressed to the American Mathematical Society, P. O. Box 845904, Boston, MA 02284-5904, USA. *All orders must be accompanied by payment.* Other correspondence should be addressed to 201 Charles Street, Providence, RI 02904-2294, USA.

Memoirs of the American Mathematical Society is published bimonthly (each volume consisting usually of more than one number) by the American Mathematical Society at 201 Charles Street, Providence, RI 02904-2294, USA. Periodicals postage paid at Providence, RI. Postmaster: Send address changes to Memoirs, American Mathematical Society, 201 Charles Street, Providence, RI 02904-2294, USA.

This publication is indexed in *Science Citation Index*®, *SciSearch*®, *Research Alert*®, *CompuMath Citation Index*®, *Current Contents*®/*Physical, Chemical & Earth Sciences.*
Printed in the United States of America.

∞ The paper used in this book is acid-free and falls within the guidelines established to ensure permanence and durability.
Visit the AMS home page at `http://www.ams.org/`

10 9 8 7 6 5 4 3 2 1 09 08 07 06 05 04

Contents

Abstract

This text investigates a natural question arising in the topological theory of 3-manifolds, and applies the results to give new information about the deformation theory of hyperbolic 3-manifolds. It is well known that some compact 3-manifolds with boundary admit homotopy equivalences that are not homotopic to homeomorphisms. We investigate when the subgroup $\mathcal{R}(M)$ of outer automorphisms of $\pi_1(M)$ which are induced by homeomorphisms of a compact 3-manifold M has finite index in the group $\mathrm{Out}(\pi_1(M))$ of all outer automorphisms. This question is completely resolved for Haken 3-manifolds. It is also resolved for many classes of reducible 3-manifolds and 3-manifolds with boundary patterns, including all pared 3-manifolds.

The components of the interior $\mathrm{GF}(\pi_1(M))$ of the space $\mathrm{AH}(\pi_1(M))$ of all (marked) hyperbolic 3-manifolds homotopy equivalent to M are enumerated by the marked homeomorphism types of manifolds homotopy equivalent to M, so one may apply the topological results above to study the topology of this deformation space. We show that $\mathrm{GF}(\pi_1(M))$ has finitely many components if and only if either M has incompressible boundary, but no "double trouble," or M has compressible boundary and is "small." (A hyperbolizable 3-manifold with incompressible boundary has double trouble if and only if there is a thickened torus component of its characteristic submanifold which intersects the boundary in at least two annuli.) More generally, the deformation theory of hyperbolic structures on pared manifolds is analyzed.

Some expository sections detail Johannson's formulation of the Jaco-Shalen-Johannson characteristic submanifold theory, the topology of pared 3-manifolds, and the deformation theory of hyperbolic 3-manifolds. An epilogue discusses related open problems and recent progress in the deformation theory of hyperbolic 3-manifolds.

Received by the editor February 6, 2001.

2000 *Mathematics Subject Classification.* Primary 57M99; Secondary 20F34, 30F40, 57M50.

Key words and phrases. 3-manifold, hyperbolic, pared, reducible, convex-cocompact, geometrically finite, boundary pattern, Haken, reducible, characteristic submanifold, compression body, homotopy equivalence, homeomorphism, diffeomorphism, automorphism, double trouble.

Supported in part by the National Science Foundation and the Sloan Foundation.

Supported in part by the National Science Foundation.

Preface

This work addresses a question about homotopy equivalences and homeomorphisms of 3-manifolds, and gives an application to the topology of deformation spaces of hyperbolic 3-manifolds. Although the topological question is quite natural, it did not receive much attention until motivated by the geometric application. It is simply this: in the group of homotopy classes of self-homotopy-equivalences of a 3-manifold, when does the subgroup consisting of the classes that contain a homeomorphism have finite index?

Most of our results will apply only to Haken 3-manifolds, although in chapter 12 we develop rather general versions of our theorems for reducible 3-manifolds. For closed Haken manifolds, Waldhausen proved that all homotopy classes contain homeomorphisms, so we need only consider Haken manifolds with nonempty boundary. This case breaks into two very different subcases, the manifolds with incompressible boundary and those with compressible boundary. We resolve the question completely, giving exact conditions for the subgroup of classes realizable by homeomorphisms to have finite index.

In the case when the boundary is incompressible, our principal means to examine homotopy equivalences and homeomorphisms of Haken 3-manifolds is the characteristic submanifold theory due independently to Jaco-Shalen and Johannson. For our geometric application, we need to consider a more general version of the question, in which we work with maps that preserve certain submanifolds of the boundary. The formulation of the characteristic submanifold given by Johannson, with its elegant theory of boundary patterns, provides exactly the kind of control that we need. For this reason, almost all of our work is carried out in the context of manifolds with boundary patterns. A complete resolution of the realization question for Haken manifolds with boundary patterns seems out of reach, and our main results are restricted to the case when the submanifolds forming the boundary pattern are disjoint. But these do include as a very special case the boundary patterns which arise in the study of the deformation theory of hyperbolic 3-manifolds.

The statements of the main results require a number of preliminary concepts. Consequently, they cannot conveniently be given until rather late in the development of our exposition, indeed not until chapter 8. To motivate so much preliminary work, we state and discuss restricted versions of the main results in chapter 1. There we lay out the general program and provide several examples illustrating some of the phenomena that can occur.

Chapter 2 contains an exposition of Johannson's version of the characteristic submanifold theory. We review the concepts which underlie his main results, give a variety of examples of boundary patterns and characteristic submanifolds, and develop some technical results which we need for our later work. All of Johannson's theory takes place under the assumption that the 3-manifold satisfies a certain incompressibility condition on its boundary (namely, that the boundary pattern has the property of being useful). To work with manifolds that do not satisfy this condition (that is, to allow some boundary patterns that are not useful, in the technical sense), we use a second kind of characteristic structure, the characteristic compression body invented by Bonahon. For manifolds with boundary patterns, we need a relative version of his theory, which we develop in chapter 3. (Both the characteristic compression body and our relative version are generalizations of the Abikoff-Maskit decomposition which was developed earlier in the context of Kleinian groups.) With these topological concepts in place, we can prove in chapter 4 some results about homotopy equivalences and homotopy types of 3-manifolds with boundary patterns, which generalize theorems of Johannson and Swarup.

Geometry enters, at least implicitly, in chapter 5, where we introduce the boundary patterns naturally associated to manifolds with hyperbolic structures. These are called pared structures. We review the definitions and some known results, including the strong restrictions on the structure of the characteristic submanifold associated to a pared structure. In chapter 6, we define small 3-manifolds. These arise as exceptional cases for our main topological results, in the case where the manifold has compressible boundary. We determine which of the small manifolds have pared structures, and prove a technical result which lists the homotopy types all of whose homeomorphism classes are either compression bodies or small manifolds.

The actual geometric theory which we use draws on work of many mathematicians, including Ahlfors, Bers, Kra, Marden, Maskit, Sullivan, and Thurston. We consider deformation spaces of well-behaved (i. e. geometrically finite) structures on a 3-manifold, or more generally, on a pared 3-manifold. The Parameterization Theorem given in chapter 7 shows that each of the components of such a deformation space is a manifold parameterized by natural geometric data and that the components are enumerated by topological data. As an application of our topological results we will be able to determine exactly when such deformation spaces have finitely many components. We state our topological and geometric results in full generality, in chapter 8, and show how the hyperbolic results follow from the main topological theorems. The succeeding two chapters complete the proof of those topological theorems.

Chapter 11 is independent; it shows how certain homotopy equivalences that arise in the proofs of the main topological theorems are related to the homotopy equivalences called Dehn flips by Johannson.

In chapter 12, we extend the main theorems to the case when the 3-manifold is reducible, although for simplicity we restrict to the case of trivial boundary pattern. Again, there is a division into the cases when the boundary is incompressible or not. When it is compressible, we again find that the realizable automorphisms have infinite index unless the manifold is "small," as redefined for the reducible case. In the incompressible case, the result is that the realizable automorphisms have finite index if and only if the same is true for each irreducible prime summand of the manifold.

The epilogue, chapter 13, surveys some results and conjectures related to our main topics, and discusses some recent developments.

Not many readers will want to work straight through the details of all of the preliminary material. The shortest route to understanding the statements and underlying ideas of the main theorems will of course depend greatly on the background that the reader brings to the task, but here is one possible approach. One should certainly begin with a study of chapter 1, especially the examples. One might then continue with a quick examination of the first three sections of chapter 2 and a more careful review of sections 2.9 and 2.11 (reading "incompressible boundary" for "useful boundary pattern", if desired). The first three sections of chapter 3 will give the main ideas from the theory of characteristic compression bodies, and from chapter 4 only the statements of the main results are needed. From chapter 5, one needs only the definitions, the description of the characteristic submanifold for the pared case, and the concept of double trouble, and from chapter 6, only the definitions and the statements of the main results. From chapter 7 one needs enough definitions to understand the statement of the Parameterization Theorem, and at this point there should be enough in place to understand the statements of the main results in chapter 8 and the derivation of the main hyperbolic results from the main topological results, as well as to understand the outlines of the proofs given in the succeeding two chapters. Chapter 11 and perhaps chapter 12 are likely to be of interest only to the most topologically-oriented readers, but we hope that chapter 13 will stimulate the further interest of topologists and geometers alike.

We thank the referee for an extraordinarily careful reading of the manuscript, and for pointing out many corrections and improvements.

CHAPTER 1

Introduction

1.1. Motivation

Although most of the work that follows involves results of a purely topological nature, it was motivated by questions arising in the deformation theory of hyperbolic 3-manifolds. We begin by introducing these questions and then proceed to the topological question which will occupy most of our attention.

We will say that a compact, oriented, irreducible 3-manifold M is *hyperbolizable* if its interior admits a (complete) hyperbolic structure. Thurston has shown that a compact, oriented, irreducible 3-manifold with nonempty boundary is hyperbolizable if and only if it is atoroidal. If M has no torus boundary components then there is a convex cocompact uniformization of M, i. e. a discrete faithful representation $\rho\colon \pi_1(M) \to \mathrm{PSL}(2,\mathbb{C})$ such that there exists a orientation-preserving homeomorphism from M to $(\mathbb{H}^3 \cup \Omega(\rho))/\rho(\pi_1(M))$ (where $\Omega(\rho)$ is the domain of discontinuity for the action of $\rho(\pi_1(M))$ on $\overline{\mathbb{C}}$).

We will be interested in studying the space $\mathrm{CC}(M)$ of all convex cocompact uniformizations of M. The work of Ahlfors, Bers, Kra, Marden and Maskit allows one to give an explicit parameterization of $\mathrm{CC}(M)$. Each component of $\mathrm{CC}(M)$ is homeomorphic to $\mathcal{T}(\partial M)/\operatorname{Mod}_0(M)$ where $\mathcal{T}(\partial M)$ is the Teichmüller space of all hyperbolic structures on ∂M and $\operatorname{Mod}_0(M)$ is the group of isotopy classes of orientation-preserving homeomorphisms of M which are homotopic to the identity. Teichmüller space is a finite-dimensional cell and $\operatorname{Mod}_0(M)$ acts discontinuously and freely on $\mathcal{T}(\partial M)$, so each component of $\mathrm{CC}(M)$ is a finite-dimensional manifold. If M has incompressible boundary, then $\operatorname{Mod}_0(M)$ is trivial, so $\mathcal{T}(\partial M)$ is simply a cell.

The most basic form of our motivating question is as follows:

Hyperbolic Question (convex cocompact case): *For which compact hyperbolizable 3-manifolds M without torus boundary components does the space* $\mathrm{CC}(M)$ *of convex cocompact uniformizations of M have finitely many components?*

To give our topological enumeration of the components of $\mathrm{CC}(M)$ we introduce the following notation. The outer automorphism group $\mathrm{Out}(\pi_1(M))$ is equal to the quotient of the group $\mathrm{Aut}(\pi_1(M))$ of automorphisms of $\pi_1(M)$ by the normal subgroup $\mathrm{Inn}(\pi_1(M))$ of inner automorphisms of $\pi_1(M)$. If $h\colon M \to M$ is a homeomorphism, then h determines a well-defined element of $\mathrm{Out}(\pi_1(M))$, although it does not determine a well-defined element of $\mathrm{Aut}(\pi_1(M))$ unless we make a choice of basepoint for M which is preserved by h. In this case we will say that the element of $\mathrm{Out}(\pi_1(M))$ determined by h is *realized* by h. By $\mathcal{R}(M)$ we denote the subgroup of $\mathrm{Out}(\pi_1(M))$ consisting of elements which are realized by homeomorphisms of M, and by $\mathcal{R}_+(M)$ the subgroup realizable by orientation-preserving homeomorphisms.

The work of Ahlfors, Bers, Kra, Marden and Maskit shows that the components of $CC(M)$ are in a one-to-one correspondence with the cosets of $\mathcal{R}_+(M)$ in $\mathrm{Out}(\pi_1(M))$. We will show that if M has no torus boundary components, then $CC(M)$ has finitely many components if and only if either M has incompressible boundary, M is a compression body, or M is obtained from one or two I-bundles (over closed surfaces) by adding a 1-handle. This result is obtained as an immediate corollary of our complete answer to the following topological question.

Finite Index Realization Problem (absolute case): *For which compact, irreducible, orientable 3-manifolds M does the group $\mathcal{R}(M)$ of outer automorphisms which are realizable by homeomorphisms of M have finite index in $\mathrm{Out}(\pi_1(M))$?*

We will answer the topological question by using the characteristic submanifold theory of Johannson and Jaco-Shalen and the compression body decomposition of Bonahon and McCullough-Miller. We will see that $\mathcal{R}(M)$ has finite index in $\mathrm{Out}(\pi_1(M))$ if and only if M is a compression body, M is obtained from one or two I-bundles by adding a 1-handle, or M has incompressible boundary and every Seifert-fibered component of the characteristic submanifold of M is either "not complicated" or disjoint from the boundary of M. A precise statement will be given in section 1.2.

It is also natural to consider the space $CC(\pi_1(M))$ of convex cocompact uniformizations of manifolds homotopy equivalent to M. A result of Swarup guarantees that the set of compact irreducible 3-manifolds homotopy equivalent to M is finite, and Thurston's geometrization theorem implies that each of these is also hyperbolizable. Hence, we need only analyze which 3-manifolds M are homotopy equivalent to compact irreducible 3-manifolds M' such that $\mathcal{R}(M')$ has infinite index in $\mathrm{Out}(\pi_1(M'))$. We will show that if M is hyperbolizable and has no torus boundary components, then $CC(\pi_1(M))$ has finitely many components if and only if either M has incompressible boundary, is a handlebody, or is obtained from one or two I-bundle(s) (over closed surfaces) by adding a 1-handle.

In order to study deformation spaces of hyperbolic structures on hyperbolizable manifolds with torus boundary components we will need to study geometrically finite uniformizations of M. Topologically, we will study manifold pairs (M, P) where M is a compact hyperbolizable 3-manifold and P is a collection of disjoint incompressible annuli and tori in the boundary of M. A geometrically finite uniformization of the pair (M, P) is a discrete faithful representation $\rho\colon \pi_1(M) \to \mathrm{PSL}(2, \mathbb{C})$ for which there is an orientation-preserving homeomorphism from $M - P$ to $(\mathbb{H}^3 \cup \Omega(\rho))/\rho(\pi_1(M))$. We will let $GF(M, P)$ denote the space of geometrically finite uniformizations of (M, P).

Thurston's geometrization theorem asserts that if M has nonempty boundary, then $GF(M, P)$ is nonempty if and only if (M, P) is a pared 3-manifold (see section 5.1 for the definition of a pared manifold). Let $\mathrm{Aut}(\pi_1(M), \pi_1(P))$ be the group of automorphisms of $\pi_1(M)$ which take the fundamental group of any component of P to a subgroup conjugate to the fundamental group of some component of P. Define $\mathrm{Out}(\pi_1(M), \pi_1(P))$ to be the quotient $\mathrm{Aut}(\pi_1(M), \pi_1(P))/\mathrm{Inn}(\pi_1(M))$, $\mathcal{R}(M, P)$ to be the subgroup of $\mathrm{Out}(\pi_1(M), \pi_1(P))$ consisting of elements which are realized by homeomorphisms of the pair (M, P), and $\mathcal{R}_+(M, P)$ to be the orientation-preserving elements of $\mathcal{R}(M, P)$. The work of Ahlfors, Bers, Kra, Marden and Maskit implies that the components of $GF(M, P)$ are in one-to-one correspondence with the cosets of $\mathcal{R}_+(M, P)$ in $\mathrm{Out}(\pi_1(M), \pi_1(P))$.

By $\mathrm{GF}(\pi_1(M), \pi_1(P))$ we denote the set of geometrically finite uniformizations of manifold pairs homotopy equivalent to (M, P). The complete forms of our motivating hyperbolic questions are the following:

Hyperbolic Questions: *For which pared manifolds (M, P) does $\mathrm{GF}(M, P)$ have finitely many components? For which (M, P) does $\mathrm{GF}(\pi_1(M), \pi_1(P))$ have finitely many components?*

A manifold pair (M, P) is a specific case of Johannson's general theory of manifolds with boundary patterns. A boundary pattern $\underline{\underline{m}}$ on M is a collection of submanifolds of ∂M such that any two elements intersect in a (possibly empty) collection of arcs and circles, while any three elements intersect in a finite collection of points. One can define $\mathrm{Out}(\pi_1(M), \pi_1(\underline{\underline{m}}))$ and $\mathcal{R}(M, \underline{\underline{m}})$ much as above. One then asks the general topological question:

Finite Index Realization Problem: *For which compact orientable irreducible 3-manifolds with boundary pattern $(M, \underline{\underline{m}})$ does the subgroup $\mathcal{R}(M, \underline{\underline{m}})$ realizable by homeomorphisms have finite index in $\mathrm{Out}(\pi_1(M), \pi_1(\underline{\underline{m}}))$?*

We will answer this topological question for a variety of boundary patterns, which includes all pared 3-manifolds. Full statements of the results are given in chapter 8.

1.2. The main theorems for Haken 3-manifolds

In this section we will develop the notation necessary to give a complete statement of our results for Haken 3-manifolds with nonempty boundary but empty "boundary pattern". We will also provide outlines of their proofs, which serve as outlines for the proofs of the more general forms of our results.

We first introduce more formally the notation involved in our hyperbolic question. Let M be a compact, oriented, hyperbolizable 3-manifold. Let $\mathcal{D}(\pi_1(M))$ denote the space of discrete, faithful representations of $\pi_1(M)$ into $\mathrm{PSL}(2, \mathbb{C})$ and let $\mathrm{AH}(\pi_1(M))$ denote $\mathcal{D}(\pi_1(M))/\mathrm{PSL}(2, \mathbb{C})$ where $\mathrm{PSL}(2, \mathbb{C})$ acts by conjugation. If $\rho \in \mathrm{AH}(\pi_1(M))$, then let N_ρ denote $\mathbb{H}^3/\rho(\pi_1(M))$. Further denote by $\Omega(\rho)$ the maximal open subset of $\overline{\mathbb{C}}$ on which $\rho(\pi_1(M))$ acts discontinuously, and let $\widehat{N}_\rho = (\mathbb{H}^3 \cup \Omega(\rho))/\rho(\pi_1(M))$. We call $\widehat{N}_\rho$ the *conformal extension* of N_ρ. When $\widehat{N}_\rho$ is compact, ρ is said to be *convex cocompact*. Let $\mathrm{CC}(\pi_1(M))$ denote the space of convex cocompact elements of $\mathrm{AH}(\pi_1(M))$. We further define $\mathrm{CC}(M)$ to consist of all $\rho \in \mathrm{CC}(\pi_1(M))$ such that there exists an orientation-preserving homeomorphism from M to $\widehat{N}_\rho$.

We also consider the space $\mathrm{A}(M)$ of oriented compact irreducible 3-manifolds homotopy equivalent to M. Two elements M_1 and M_2 of $\mathrm{A}(M)$ are regarded as equivalent if there exists an orientation-preserving homeomorphism from M_1 to M_2.

Now we will define the space $\mathcal{A}(M)$ of marked, oriented, compact, irreducible 3-manifolds homotopy equivalent to M. Its basic objects are pairs (M', h') where $M' \in \mathrm{A}(M)$ and $h'\colon M \to M'$ is a homotopy equivalence. Two pairs (M_1, h_1) and (M_2, h_2) are considered equivalent when there exists an orientation-preserving homeomorphism $\phi\colon M_1 \to M_2$ such that $\phi \circ h_1$ is homotopic to h_2. Then $\mathcal{A}(M)$ is the set of all equivalence classes of such pairs.

Since each element M' of $A(M)$ is a $K(\pi, 1)$ every element α of $\mathrm{Out}(\pi_1(M'))$ is realized by a homotopy equivalence $h_\alpha : M' \to M'$. Let $\mathcal{A}_0(M')$ be the set of elements of $\mathcal{A}(M)$ of the form (M', h') for some homotopy equivalence $h' : M \to M'$.

If we fix a homotopy equivalence $h_0 : M \to M'$, then we can define a surjective map $J : \mathrm{Out}(\pi_1(M)) \to \mathcal{A}_0(M)$ by letting $J(\alpha) = (M', h_\alpha \circ h_0)$. Then, by definition, $J(\alpha) = J(\alpha')$ if and only if $\alpha' \circ \alpha^{-1} \in \mathcal{R}_+(M')$. Thus, the elements of $\mathcal{A}_0(M')$ are enumerated by the cosets of $\mathcal{R}_+(M')$ in $\mathrm{Out}(\pi_1(M'))$. Consequently, we may identify $\mathcal{A}(M)$ with

$$\bigsqcup_{M' \in \mathrm{A}(M)} \mathrm{Out}(\pi_1(M'))/\mathcal{R}_+(M').$$

For any element $\rho \in \mathrm{CC}(\pi_1(M))$ there exists a homotopy equivalence $h_\rho \colon M \to \widehat{N}_\rho$ such that $(h_\rho)_\# = \rho$ (where we regard ρ as an identification of $\pi_1(M)$ with the fundamental group $\rho(\pi_1(M))$ of N_ρ). Consequently, there is a well-defined map $\Theta\colon \mathrm{CC}(\pi_1(M)) \to \mathcal{A}(M)$ given by letting $\Theta(\rho) = (\widehat{N}_\rho, h_\rho)$. (In fact, Θ extends to a function defined on $\mathrm{AH}(\pi_1(M))$ simply by letting a compact core for N_ρ play the same role as $\widehat{N}_\rho$.) Thurston's Geometrization Theorem guarantees that Θ is surjective if M is a compact hyperbolizable 3-manifold with no torus boundary components.

For $\rho \in \mathrm{CC}(M)$, results of Ahlfors-Bers [**6**], Bers [**13**], Kra [**66**], and Maskit [**76**] provide a parameterization of the set $\mathrm{QC}(\rho)$ of elements of $\mathrm{CC}(\pi_1(M))$ which are quasiconformally conjugate to ρ. (Two representations $\rho_1, \rho_2 \colon \pi_1(M) \to \mathrm{PSL}(2, \mathbb{C})$ are said to be quasiconformally conjugate when there exists a quasiconformal homeomorphism $\phi\colon \overline{\mathbb{C}} \to \overline{\mathbb{C}}$ such that $\rho_2(g) = \phi \circ \rho_1(g) \circ \phi^{-1}$ for all $g \in \pi_1(M)$.) Let $\mathcal{T}(\partial M)$ denote the Teichmüller space of all (marked) hyperbolic structures on ∂M and let $\mathrm{Mod}_0(M)$ denote the set of isotopy classes of orientation-preserving diffeomorphisms of ∂M which extend to diffeomorphisms of M which are homotopic to the identity. We recall that $\mathcal{T}(\partial M)$ is homeomorphic to a finite-dimensional Euclidean space and that $\mathrm{Mod}_0(M)$ acts properly discontinuously and freely on $\mathcal{T}(\partial M)$. This parameterization is summarized in the following theorem. (See sections 7.1 and 7.2 for more details.)

Quasiconformal Parameterization Theorem: (Ahlfors, Bers, Kra, Maskit) *Let M be a compact 3-manifold with boundary and $\rho \in \mathrm{CC}(M)$. Then $\mathrm{QC}(\rho)$ may be identified with $\mathcal{T}(\partial M)/\mathrm{Mod}_0(M)$.*

Marden's Isomorphism Theorem ([**75**]) implies that if $(M', h') \in \mathcal{A}(M)$, then $\Theta^{-1}(M', h') = \mathrm{QC}(\rho)$ for any $\rho \in \Theta^{-1}(M', h')$. Marden's Stability Theorem ([**75**]) implies that if $\rho \in \mathrm{CC}(\pi_1(M))$, then $\mathrm{QC}(\rho)$ is an open subset of $\mathrm{AH}(\pi_1(M))$. Hence $\mathrm{CC}(\pi_1(M))$ is an open subset of $\mathrm{AH}(\pi_1(M))$. Since Θ is surjective we have the following parameterization of $\mathrm{CC}(\pi_1(M))$. (See section 7.3.)

Parameterization Theorem: *If M is a hyperbolizable compact oriented 3-manifold with no torus boundary components, then $\mathrm{CC}(\pi_1(M))$ is homeomorphic to the disjoint union*

$$\bigsqcup_{(M', h') \in \mathcal{A}(M)} \mathcal{T}(\partial M')/\mathrm{Mod}_0(M')$$

It follows that the components of $CC(\pi_1(M))$ are enumerated by $\mathcal{A}(M)$ which we have identified with

$$\bigsqcup_{M' \in \mathrm{A}(M)} \mathrm{Out}(\pi_1(M'))/\mathcal{R}_+(M').$$

In particular, the components of $CC(M)$ are enumerated by the cosets of $\mathcal{R}_+(M)$ in $\mathrm{Out}(\pi_1(M))$.

The problem of determining the M for which $\mathcal{R}_+(M)$ has finite index in $\mathrm{Out}(\pi_1(M))$ is purely topological. It turns out that the Johannson-Jaco-Shalen characteristic submanifold theory is quite well-adapted to answering this question when M has incompressible boundary. In the discussion of this question it is often simpler to consider $\mathcal{R}(M)$. Note that $\mathcal{R}_+(M)$ has index 1 or 2 in $\mathcal{R}(M)$.

If M is an orientable compact irreducible 3-manifold with incompressible boundary then Johannson [**58**] and Jaco-Shalen [**56**] have exhibited a characteristic submanifold which captures all the "essential fibered submanifolds" of M. An I-bundle V is said to be admissibly imbedded in M if $\partial M \cap V$ is the associated ∂I-bundle. A Seifert fibered space V is said to be admissibly imbedded in M if $V \cap \partial M$ is a union of fibers of V. The characteristic submanifold Σ of M consists of admissibly imbedded I-bundles and Seifert fibered spaces in M and is well-defined up to isotopy. It is characterized by the property that it is the "minimal" such submanifold so that any "essential" map of a Seifert fibered space or I-bundle into M is homotopic to a map whose image lies entirely in Σ. In chapter 2 we give a summary of Johannson's version of this theory, and discuss most of the results from [**58**] that will be used in our work.

Since M is aspherical, every element of $\mathrm{Out}(\pi_1(M))$ is realized by a homotopy equivalence of M. Johannson's Classification Theorem asserts that any homotopy equivalence from M to M is homotopic to one which preserves the characteristic submanifold Σ and is a homeomorphism on its complement. In particular we may assume that it is a homeomorphism on each component of the frontier of Σ in M. (The frontier of Σ in M is the closure of $\partial\Sigma - \partial M$.) Thus we may pass to a subgroup of finite index in $\mathrm{Out}(\pi_1(M))$ such that each element is realized by a homotopy equivalence which preserves each element of the characteristic submanifold and is a homeomorphism on its complement. By this means, the problem quickly reduces to the study of homotopy equivalences of Seifert fiber spaces and I-bundles. The Baer-Nielsen theorem implies that every boundary-preserving homotopy equivalence of a surface is homotopic to a homeomorphism. Hence, perhaps after passing to another finite index subgroup of $\mathrm{Out}(\pi_1(M))$, we may assume that each element is realized by a homotopy equivalence which is also a homeomorphism on each I-bundle component of Σ. Waldhausen's theorem assures us that if a homotopy equivalence preserves the boundary of a Seifert fibered component V of Σ, then it is homotopic to a homeomorphism on V. Hence, we only need worry about Seifert fibered components of Σ which intersect the boundary of M and *typically* these will give rise to infinitely many cosets of nonrealizable outer automorphisms. (See examples 1.4.3 and 1.4.4 in the next section.) The most technical portion of the proof involves analyzing exactly which Seifert fibered components of Σ are "atypical." This completes the outline of the proof of our first topological theorem, which we now state in the "absolute" case (i. e. the case when the boundary pattern is empty).

Main Topological Theorem 2 (absolute case): *Let M be a compact, orientable, irreducible 3-manifold with incompressible and nonempty boundary. Then, $\mathcal{R}(M)$ has finite index in $\mathrm{Out}(\pi_1(M))$ if and only if every Seifert fibered component V of the characteristic submanifold that intersects ∂M satisfies one of the following:*

(1) *V is a solid torus, or*
(2) *V is an S^1-bundle over the Möbius band or annulus and no component of ∂V contains more than one component of $V \cap \partial M$, or*

(3) *V is fibered over the annulus with one exceptional fiber, and no component of $V \cap \partial M$ is an annulus, or*
(4) *V is fibered over the disk with two holes with no exceptional fibers, and $V \cap \partial M$ is one of the boundary tori of V, or*
(5) *$V = M$, and either V is fibered over the disk with two exceptional fibers, V is fibered over the Möbius band with one exceptional fiber, or V is fibered over the torus with one hole with no exceptional fibers, or*
(6) *$V = M$, and V is fibered over the disk with three exceptional fibers each of type $(2,1)$.*

We will now outline the main topological result in the case when M has compressible boundary. An irreducible, orientable, compact connected 3-manifold V is called a *compression body* if it has a boundary component F such that the homomorphism $\pi_1(F) \to \pi_1(V)$ induced by inclusion is surjective (see Bonahon [**16**], McCullough-Miller [**89**], or Brin-Johannson-Scott [**21**]). A compression body has a simple topological structure: it either is a handlebody, or is obtained from a product $(\bigcup_{i=1}^{m} F_i) \times \mathrm{I}$, where each F_i is a closed orientable surface of positive genus, by attaching 1-handles along disks in $\bigcup_{i=1}^{m}(F_i \times \{1\})$. Maskit [**76**] and McCullough-Miller [**89**] proved that $\mathcal{R}(M)$ has finite index in $\mathrm{Out}(\pi_1(M))$ whenever M is a compression body.

If M has a compressible boundary component but is not a compression body, then it contains a loop C which is not freely homotopic into the boundary of M. Suppose further that C is based at a point in a compressing disk D of M, and there exists a regular neighborhood N of D such that C intersects only one component of $N - D$. The regular neighborhood N is homeomorphic to $D^2 \times \mathrm{I}$. Construct a homotopy equivalence h from M to M which is the identity on $M - N$, by cutting M apart along D, dragging one copy of D once around C, and gluing back together. Example 1.4.1 describes this construction in a specific case. Such a homotopy equivalence takes loops in the boundary of M to loops that are not freely homotopic into the boundary, so it cannot be homotopic to a homeomorphism. If there are infinitely many "inequivalent" choices of C, then $\mathcal{R}(\pi_1(M))$ has infinite index in $\mathrm{Out}(\pi_1(M))$. So if M is not a compression body, then *typically* $\mathcal{R}(\pi_1(M))$ has infinite index in $\mathrm{Out}(\pi_1(M))$.

Our main topological result for manifolds with compressible boundary determines the 3-manifolds that are "atypical." A Haken 3-manifold M is *small* if it can be described in one of the following three ways:

I. M is obtained from a twisted I-bundle over a closed surface by gluing on a 1-handle,
II. M is obtained from the boundary connected sum of two product I-bundles over closed surfaces by gluing a twisted I-bundle to one or both of the incompressible boundary components, or
III. M is obtained from the boundary connected sum of two product I-bundles over homeomorphic closed surfaces by gluing the two incompressible boundary components.

Main Topological Theorem 1 (absolute case): *Let M be a compact, orientable, irreducible 3-manifold with compressible boundary. Then, $\mathcal{R}(M)$ has finite index in $\mathrm{Out}(\pi_1(M))$ if and only if M is either small or a compression body.*

We will see in sections 5.2 and 7.1 that if M is a compact hyperbolizable 3-manifold with incompressible boundary and no torus boundary components, then the characteristic submanifold of any element of M consists entirely of I-bundles and solid tori. Hence the following Main Hyperbolic Theorem is an immediate corollary of our Main Topological Theorems and the Parameterization Theorem.

Main Hyperbolic Theorem (convex cocompact case): *Let M be a hyperbolizable, compact 3-manifold with no torus boundary components.*

(1) *If M has compressible boundary, then* $\mathrm{CC}(M)$ *has finitely many components if and only if M is either small or a compression body.*
(2) *If M has incompressible boundary, then* $\mathrm{CC}(M)$ *has finitely many components.*

Swarup [**118**] proved that $\mathrm{A}(M)$ is always finite. We will see, in section 5.2, that if M has incompressible boundary then so does every element of $\mathrm{A}(M)$. In section 6.2 we will determine exactly which compact hyperbolizable 3-manifolds with compressible boundary have the property that every element of $\mathrm{A}(M)$ is either small or a compression body. Given this analysis, the following is an immediate corollary of our Main Hyperbolic Theorem:

Main Hyperbolic Corollary (convex cocompact case): *Let M be a hyperbolizable, compact 3-manifold with no torus boundary components.*

(1) *If M has compressible boundary then* $\mathrm{CC}(\pi_1(M))$ *has finitely many components if and only if $\pi_1(M)$ is either a free group or a free product of two groups, of which one is the fundamental group of a closed surface (orientable or nonorientable) and the other is either infinite cyclic or the fundamental group of a closed surface.*
(2) *If M has incompressible boundary, then* $\mathrm{CC}(\pi_1(M))$ *has finitely many components.*

Remarks: (i) The algebraic condition in part 1 implies that $\pi_1(M)$ is the fundamental group of either a handlebody, a small manifold, or the boundary connected sum of two untwisted I-bundles.

(ii) It is conjectured that $\mathrm{CC}(\pi_1(M))$ is dense in $\mathrm{AH}(\pi_1(M))$. A version of this conjecture first appeared in a paper by Bers [12] (see also Sullivan [**116**] and Thurston [**121**]). Although it has recently (see Anderson and Canary [**8**]) been discovered that closures of components of $\mathrm{CC}(\pi_1(M))$ can intersect, we still conjecture that $\mathrm{AH}(\pi_1(M))$ has finitely many components if and only if $\mathcal{A}(M)$ is finite. Some partial results in the direction of this conjecture will be described in section 13.

We will also prove a version of our Main Hyperbolic Theorem which applies to all pared manifolds. The statement will be quite similar although we will encounter one new phenomenon. If M is hyperbolizable, has incompressible boundary and has torus boundary components, then its characteristic submanifold contains Seifert-fibered components which are homeomorphic to $T^2 \times \mathrm{I}$. If there exists a component V of the characteristic submanifold such that V is homeomorphic to $T^2 \times \mathrm{I}$ and $V \cap \partial M$ contains more than one annulus then $\mathcal{R}(M)$ will have infinite index in $\mathrm{Out}(\pi_1(M))$ (see example 1.4.2). In this case we will say that M has double trouble. We will see that when M is a compact hyperbolizable 3-manifold with incompressible boundary, $\mathrm{GF}(M, P)$ has infinitely many components if and only if M has double trouble.

It is particularly simple to state our results when P is the collection of torus boundary components of M. In this case, we often denote $\mathrm{GF}(M, P)$ simply by $\mathrm{GF}(M)$ and $\mathrm{GF}(\pi_1(M), \pi_1(P))$ by $\mathrm{GF}(\pi_1(M))$. Then $\mathrm{GF}(\pi_1(M))$ is the interior of $\mathrm{AH}(\pi_1(M))$ (as a subset of the appropriate character variety). We will see that $\mathrm{GF}(M)$ has finitely many components if and only if either M has incompressible boundary and does not have double trouble, or M is a compression body, or M is a small manifold. Moreover, $\mathrm{GF}(\pi_1(M))$ has finitely many components if and only if either M has incompressible boundary and does not have double trouble or $\pi_1(M)$ is either a free group, a free product of two (closed, but not necessarily orientable) surface groups, a free product of a (closed, but not necessarily orientable) surface group and an infinite cyclic group, or a free product of a finite number of free abelian groups of rank 2.

1.3. The main theorems for reducible 3-manifolds

In chapter 12, the Main Topological Theorems are extended to reducible 3-manifolds, that is, 3-manifolds which are nontrivial connected sums. The extended results apply also to manifolds which are nonorientable.

The formulation of the results involves the Poincaré associate $P(M)$ of a reducible 3-manifold M. This is the manifold obtained from M by replacing each connected summand of M that is simply-connected by a 3-sphere summand. Put differently, each 2-sphere boundary component of M is filled in with a 3-ball, and any homotopy 3-balls in M not diffeomorphic to the genuine 3-ball are replaced by genuine 3-balls. We prove in proposition 12.1.4 that $\mathcal{R}(M)$ has finite index in $\mathrm{Out}(\pi_1(M))$ if and only if $\mathcal{R}(P(M))$ has finite index in $\mathrm{Out}(\pi_1(P(M)))$. By virtue of this, the extended results need only be proven for 3-manifolds that have no nontrivial simply-connected summands.

As in the irreducible case, the Finite Index Realization Problem breaks into the cases when ∂M is compressible and when it is incompressible. In the former case, there is a direct analogue of Main Topological Theorem 1:

Theorem 12.2.1: *Let M be a compact 3-manifold with compressible boundary, with $P(M)$ reducible. Then $\mathcal{R}(M)$ has finite index in $\mathrm{Out}(\pi_1(M))$ if and only if M is small.*

Here, small means that $P(M)$ is a connected sum $P \# Q$, where P is either a solid torus or a solid Klein bottle, and Q is a 3-manifold with finite fundamental group. The proof of theorem 12.2.1 involves many of the same ideas used for Main Topological Theorem 1.

When ∂M is incompressible, the Finite Index Realization Problem reduces to the case that $P(M)$ is irreducible. It is stated in terms of the decomposition of $P(M)$ as a connected sum of prime 3-manifolds. The irreducible summands in this decomposition are uniquely determined.

Theorem 12.3.1: *Let M be a compact 3-manifold with incompressible boundary. Write $P(M)$ as $(\#_{i=1}^r M_i)\#(\#_{j=1}^s N_j)$ where each M_i is irreducible and each N_j is either $S^2 \times S^1$ or the unique nontrivial S^2-bundle over S^1. Then $\mathcal{R}(M)$ has finite index in $\mathrm{Out}(\pi_1(M))$ if and only if $\mathcal{R}(M_i)$ has finite index in $\mathrm{Out}(\pi_1(M_i))$ for all $1 \leq i \leq r$.*

The proof of theorem 12.3.1 draws on ideas developed in [**86**]. The algebraic underpinnings of these methods are the generators and relations for the automorphism group of a free product of indecomposable groups, which were first given by Fouxe-Rabinovitch [**40, 41**]. These are also used in the proof of Main Topological Theorem 1.

1.4. Examples

In this section we will give explicit examples illustrating some of the topological phenomena which underlie our results. Examples 1.4.1 to 1.4.4 describe phenomena which affect whether the index of $\mathcal{R}(M)$ in $\mathrm{Out}(\pi_1(M))$ is finite. Examples 1.4.5 and 1.4.6 show some ways to produces homotopy equivalent hyperbolizable 3-manifolds which are not homeomorphic.

The first two examples illustrate the two basic types of phenomena which may cause $\mathcal{R}(M)$ to have infinite index in $\mathrm{Out}(\pi_1(M))$ when M is a hyperbolizable manifold. The first can occur only when M has a compressible boundary component, and the second only when M has a torus boundary component.

EXAMPLE 1.4.1. *A 3-manifold with compressible boundary for which the realizable subgroup has infinite index*

When M has a compressible boundary component but is not a compression body the following construction often yields an infinite collection of distinct cosets of $\mathcal{R}(M)$ in $\mathrm{Out}(\pi_1(M))$. Let C be a simple closed curve in M with its basepoint in a 1-handle of M, but which does not pass over the 1-handle. A homotopy equivalence of M can be constructed by taking a map which is the identity off of the 1-handle and wraps the 1-handle around C (and then over the original 1-handle). Below, we describe a specific example and give the resulting automorphism explicitly.

Let S be a surface of genus two and let L be the 3-manifold obtained by taking the boundary connected sum of two copies of $S \times \mathrm{I}$. Form M_1 by gluing two copies of L together along an incompressible boundary component. Then $\pi_1(M_1) \cong \pi_1(S) * \pi_1(S) * \pi_1(S)$ and has a presentation

$$\langle a_1, b_1, a_2, b_2, c_1, d_1, c_2, d_2, e_1, f_1, e_2, f_2 \mid [a_1, b_1] = [a_2, b_2],$$
$$[c_1, d_1] = [c_2, d_2], [e_1, f_1] = [e_2, f_2] \rangle$$

where $\{a_1, b_1, a_2, b_2\}$ and $\{e_1, f_1, e_2, f_2\}$ generate the fundamental groups of the two incompressible boundary components of M_1 and $\{c_1, d_1, c_2, d_2\}$ generates the fundamental group of the surface we glued along to form M_1. The images of the fundamental groups of the two compressible boundary components are generated by $\{a_1, b_1, a_2, b_2, c_1, d_1, c_2, d_2\}$ and $\{c_1, d_1, c_2, d_2, e_1, f_1, e_2, f_2\}$. It is known that M_1 is hyperbolizable.

Define an automorphism ϕ which fixes $a_1, b_1, a_2, b_2, c_1, d_1, c_2$, and d_2 and acts on the remaining generators by

$$e_1 \mapsto a_1 e_1 a_1^{-1}, f_1 \mapsto a_1 f_1 a_1^{-1}, e_2 \mapsto a_1 e_2 a_1^{-1}, f_2 \mapsto a_1 f_2 a_1^{-1}.$$

An element of $\pi_1(M)$ is called *peripheral* if it is conjugate into the image of the fundamental group of a boundary component of M. More topologically, an element of $\pi_1(M)$ is peripheral if the free homotopy class of loops that it determines has a representative lying entirely within the boundary of M. Since homeomorphisms preserve the boundary, a realizable automorphism must take peripheral elements

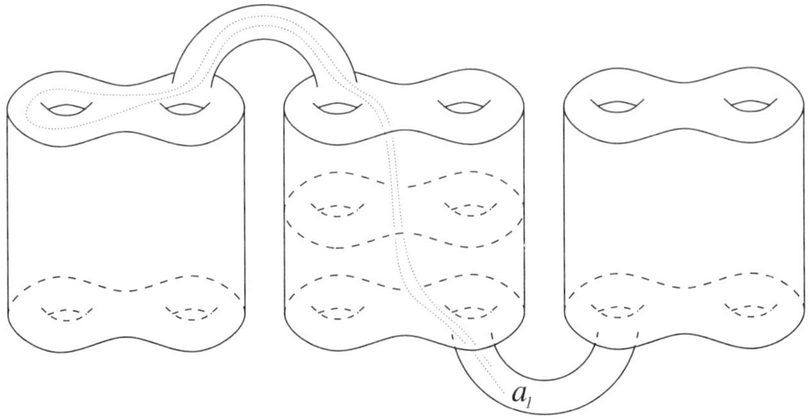

FIGURE 1.1. A 3-manifold with compressible boundary for which the realizable subgroup has infinite index

to peripheral elements. Thus, no nonzero power ϕ^k of ϕ is realizable by a homeomorphism of M_1, since ϕ^k takes the peripheral element c_1e_1 to the nonperipheral element $c_1a_1^ke_1a_1^{-k}$. This automorphism is induced by a homotopy equivalence which is the identity off of the 1-handle in the second copy of L (the copy whose fundamental group is generated by $\{c_1, d_1, c_2, d_2, e_1, f_1, e_2, f_2\}$), and which sends this handle around a loop representing a_1. Figure 1.1 illustrates such a path based at a point in this 1-handle. Since no power of ϕ is realizable, the powers of this homotopy equivalence represent distinct cosets of the realizable subgroup, so $\mathcal{R}(M_1)$ has infinite index in $\mathrm{Out}(\pi_1(M_1))$ and $\mathcal{CC}(M_1)$ has infinitely many components.

EXAMPLE 1.4.2. *A 3-manifold with incompressible boundary for which the realizable subgroup has infinite index*

The following example illustrates the phenomenon caused by double trouble. Again we will first describe the general strategy and then give a specific example (from Thurston [**120**]). Begin with a submanifold V homeomorphic to $T^2 \times \mathrm{I}$ which intersects ∂M in a torus $T^2 \times \{0\}$ and a collection of at least two annuli in $T^2 \times \{1\}$, no pair of which are isotopic in ∂M. The homotopy equivalence is the identity off of a regular neighborhood of one component A of the frontier of V in M, and wraps a collar neighborhood of A once around $T^2 \times \{1\}$. Arcs in ∂M which cross A are carried to arcs in M which travel around $T^2 \times \{1\}$, and loops which cross these annuli in an essential way can be carried to nonperipheral loops.

Let S be a surface of genus two and α a separating curve on S. Let K be the 2-complex (imbedded in 3-space $\mathbb{R}^3$) formed by attaching a torus T to S along the curve α (where α is glued to the longitude of T). Let M_2 be the manifold obtained by taking a regular neighborhood (in $\mathbb{R}^3$) of K. Notice that M_2 is homeomorphic to $S \times \mathrm{I}$ with a tubular neighborhood of $\alpha \times \{\frac{1}{2}\}$ removed. Figure 1.2 illustrates the construction of M_2.

Let h be a homotopy equivalence of $\pi_1(K)$ constructed by fixing T and every point on S except an annulus with one boundary component being α. Then take this annulus and wrap it around the meridian c of T. A presentation for $\pi_1(M_2)$ is

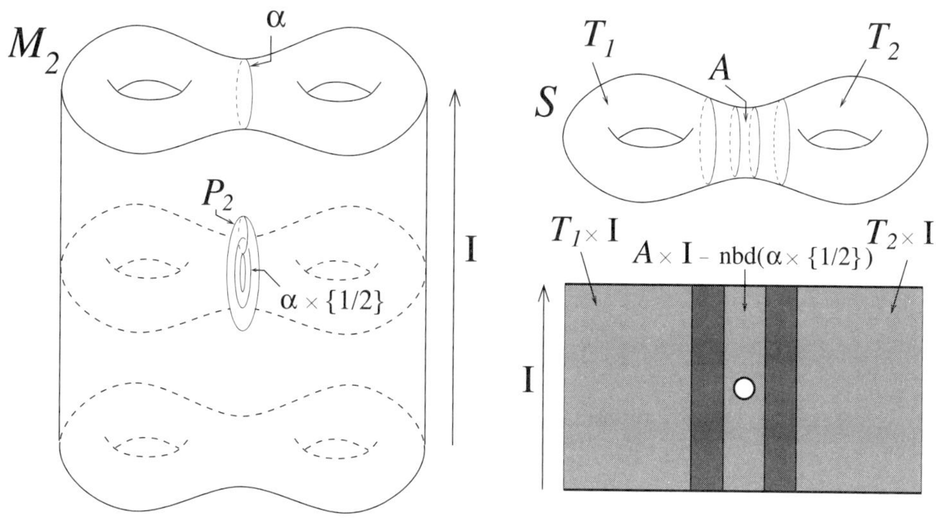

FIGURE 1.2. A 3-manifold for which the realizable subgroup has infinite index

given by

$$\langle a_1, b_1, a_2, b_2, c \mid [a_1, b_1] = [a_2, b_2], [[a_1, b_1], c] = 1\rangle \ .$$

In this presentation, the peripheral subgroups are generated by $\{a_1, b_1, a_2, b_2\}$ and $\{a_1, b_1, c^{-1}a_2c, c^{-1}b_2c\}$. The automorphism ϕ induced by h has the effect

$$a_1 \mapsto a_1, b_1 \mapsto b_1, c \mapsto c, a_2 \mapsto ca_2c^{-1}, b_2 \mapsto cb_2c^{-1} \ .$$

No nonzero power ϕ^k of ϕ is realizable by a homeomorphism. For if $k > 0$, ϕ^k takes the peripheral element a_1a_2 to the nonperipheral element $a_1c^ka_2c^{-k}$, while if $k < 0$ it takes the peripheral element $a_1c^{-1}a_2c$ to $a_1c^{k-1}a_2c^{1-k}$. Hence $\mathcal{R}(M_2)$ has infinite index in $\text{Out}(\pi_1(M_2))$.

The characteristic submanifold Σ of M_2 has three components, as illustrated schematically in figure 1.2. Two of the components are product I-bundles over the tori with one hole, labelled T_1 and T_2, which are the components of the complement in S of a regular neighborhood of α. The other component V is homeomorphic to $T^2 \times \text{I}$. One of the boundary components of V lies in ∂M, and the other meets each of the other two boundary components of M_2 in an annulus whose center circle is homotopic to α. If we let P_2 denote the torus boundary component of M_2, then a geometrically finite uniformization of (M_2, P_2) is explicitly constructed in Kerckhoff-Thurston [**64**]. Since the components of $\text{GF}(M_2, P_2) = \text{GF}(M_2)$ are enumerated by $\text{Out}(\pi_1(M_2))/\mathcal{R}_+(\pi_1(M_2))$, we see that there are infinitely many components of $\text{GF}(M_2)$.

The next two examples illustrate phenomena related to the presence of more complicated Seifert fibered spaces in the characteristic submanifold, which can occur only in nonhyperbolizable examples. Roughly speaking, if Σ has Seifert-fibered components which are complicated and meet the boundary of M, then $\mathcal{R}(M)$ will have infinite index in $\text{Out}(\pi_1(M))$, while if all Seifert-fibered components that meet the boundary are uncomplicated, such as the solid torus, the index can be finite.

However, in the borderline cases the way in which the components meet the boundary can affect the index, as illustrated in examples 1.4.3 and 1.4.4. In both of these examples, Σ is the product of the circle and the disk with two holes, but in example 1.4.3 the index is infinite and in example 1.4.4 it is finite.

EXAMPLE 1.4.3. *A characteristic submanifold which is just complicated enough to make the realizable subgroup have infinite index*

Let F be a disk with two holes, with boundary components C_1, C_2, and C_3, and let $\Sigma = F \times S^1$. Let S be a compact hyperbolizable 3-manifold whose boundary is a single incompressible torus, and form M_3 by identifying the torus boundary component of S with the boundary torus $C_1 \times S^1$ of Σ. In example 2.10.3 below, it will be verified that Σ is the characteristic submanifold of M_3. We first construct a homotopy equivalence h_0 of F which fixes C_1 and is not homotopic to a homeomorphism. Let γ be a properly imbedded arc in F whose endpoints lie in C_2 and C_3 and let N be a collar neighborhood of γ. Let α be a loop in the interior of F which is homotopic to C_1, intersects γ only at its basepoint, and is disjoint from one of the components of $N - \gamma$. Let h_0 be the identity on $F - N$ and let it map N to the union of N and a regular neighborhood of α by wrapping N once about α (and then continuing over N) so that each component of $N \cap \partial F$ is wrapped around α. The map h_0 is an example of a *sweep*; sweeps are described in more detail in lemma 10.2.4 and the discussion preceding it. Define a homotopy equivalence h of M_3 by taking the product of h_0 and the identity on the S^1-factor on Σ, and by taking the identity on S. The peripheral loop in M_3 represented by C_2 is carried by h^k to the element represented by $C_2C_1^k$ in $\pi_1(F) \times \mathbb{Z} = \pi_1(\Sigma) \subset \pi_1(M_3)$. If $|k| \geq 2$, this loop is not homotopic into ∂M_3 so h^k is not homotopic to a homeomorphism. Therefore $\mathcal{R}(M_3)$ has infinite index in $\mathrm{Out}(\pi_1(M_3))$.

EXAMPLE 1.4.4. *A characteristic submanifold which is not quite complicated enough to make the realizable subgroup have infinite index*

Form M_4 from the manifold M_3 in example 1.4.3 by attaching another copy S' of S to Σ along $C_2 \times S^1$. In example 2.10.3 below, it will be verified that the characteristic submanifold Σ of M_4 is also $\Sigma = F \times S^1$. We will show that $\mathcal{R}(M_4)$ has finite index in $\mathrm{Out}(\pi_1(M_4))$.

Any outer automorphism of $\pi_1(M_4)$ can be induced by a homotopy equivalence. By Johannson's Classification Theorem (theorem 2.11.1 below), such a homotopy equivalence is homotopic to a map f which carries $S \cup S'$ to $S \cup S'$ by a homeomorphism and carries Σ to Σ by a homotopy equivalence. By another result of Johannson, theorem 2.11.2 below, the mapping class group of S is finite. It follows that when the automorphism lies in a certain finite-index subgroup of $\mathrm{Out}(\pi_1(M_4))$, the restriction of f to $S \cup S'$ is isotopic to the identity. Then, the restriction of f to Σ is isotopic to the identity on $(C_1 \cup C_2) \times S^1$. Such a homotopy equivalence of Σ is homotopic, relative to $(C_1 \cup C_2) \times S^1$, to a homeomorphism (this can be proven directly, or extracted from proposition 10.2.2 and lemma 10.2.3 below). Consequently, f is homotopic on M_4 to a homeomorphism, and its induced outer automorphism is realizable.

The remaining examples illustrate two ways in which manifolds can be homotopy equivalent but not homeomorphic. In the first case the manifolds have incompressible boundary, while in the second case each has a compressible boundary

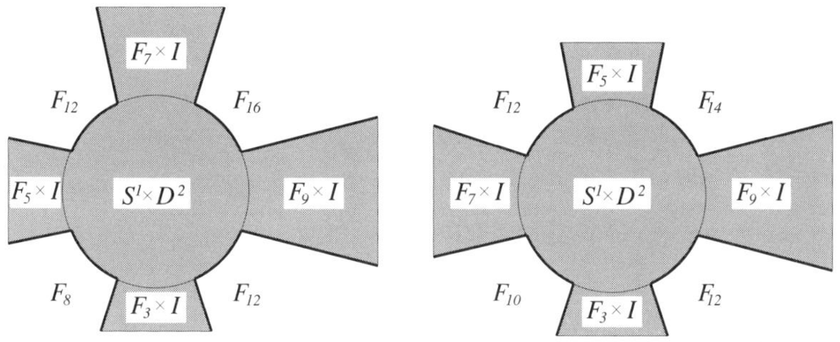

FIGURE 1.3. Homotopy equivalent but not homeomorphic books of I-bundles

component. All four of the manifolds are hyperbolizable, and in fact these examples describe the two basic phenomena that produce hyperbolizable 3-manifolds which are homotopy equivalent but not homeomorphic.

EXAMPLE 1.4.5. *Homotopy equivalent but not homeomorphic 3-manifolds with incompressible boundary*

Let X be a 2-complex built by attaching surfaces of genus 3, 5, 7 and 9, each with one boundary component, along their common boundary circle. One may imbed X in $\mathbb{R}^3$ in two different ways, obtaining image 2-complexes X_1 and X_2 such that in X_1 the surfaces occur in the cyclical order 3, 5, 7, 9, while in X_2 they occur in the cyclical order 3, 7, 5, 9. Let M_i be a regular neighborhood of X_i. Since M_1 and M_2 both have X as a deformation retract, they are homotopy equivalent. But they are not homeomorphic, since M_1 has boundary components of genus 8, 12, 16 and 12, while M_2 has boundary components of genus 10, 12, 14, and 12. Figure 1.3 shows schematic diagrams of M_1 and M_2. These manifolds are examples of books of I-bundles, which will be defined in example 2.10.4.

EXAMPLE 1.4.6. *Homotopy equivalent but not homeomorphic 3-manifolds with compressible boundary*

Let S be a closed surface of genus at least 2. Let M_1 be obtained by attaching a 1-handle to $S \times \mathrm{I}$ such that both ends are attached to $S \times \{1\}$. Let M_2 be obtained by attaching a 1-handle to $S \times \mathrm{I}$ with one end attached to $S \times \{0\}$ and the other attached to $S \times \{1\}$. Both M_1 and M_2 have the one-point union of S with a circle as a deformation retract, so they are homotopy equivalent. But they are not homeomorphic, since they have different numbers of boundary components. The manifold M_1 is an example of a compression body, as defined at the beginning of chapter 3, and M_2 is an example of a small manifold, as defined in section 6.1. Also, these are hyperbolizable 3-manifolds. We saw in the Main Hyperbolic Corollary in section 1.2 that it is of interest to know which homotopy types of hyperbolizable 3-manifolds have the property that each of their homeomorphism classes is either a compression body or a small manifold. Theorem 6.2.1 determines all such homotopy types, and shows that each of them contains either one or two homeomorphism classes. The pair M_1 and M_2 illustrates the homotopy types that contain two classes.

CHAPTER 2

Johannson's Characteristic Submanifold Theory

A remarkable structure theory for Haken manifolds with incompressible boundary was developed independently by W. Jaco and P. Shalen [**56**] and K. Johannson [**58**]. For a Haken manifold M with incompressible boundary, they defined a codimension-zero submanifold Σ, called the characteristic submanifold. It consists of fibered manifolds — I-bundles and Seifert-fibered spaces. It has the enclosing property, which means that every "essential" map of an annulus or torus into M is homotopic into Σ. Essentiality is a strengthening of the condition that the map induce an injection on fundamental groups. Johannson's Classification Theorem shows that any homotopy equivalence $h\colon M \to N$ of Haken manifolds with incompressible boundary is homotopic to one which is a homotopy equivalence on the characteristic submanifolds and a homeomorphism of their complements. The homotopy equivalence of the characteristic submanifolds can usually be assumed to be a fiber-preserving map, and thereby can be effectively analyzed. In this chapter we will review Johannson's version of the characteristic submanifold theory and develop some technical results which will be needed in our work. We include numerous examples, some of which illustrate aspects of our main results, and others which should be of general interest.

A properly imbedded surface S in a 3-manifold M is called *compressible* if either it is a 2-sphere that bounds a ball or a 2-disk that is parallel into ∂M, or there is a 2-disk D in M with $D \cap S = \partial D$ but ∂D not bounding a disk in S. Such a disk D is called a *compression.* If S is not compressible, it is called *incompressible.* When S is connected and two-sided in M, the Loop Theorem shows that S has a compression if and only if the homomorphism $\pi_1(S) \to \pi_1(M)$ induced by inclusion is not injective.

A 3-manifold is said to be *irreducible* if every embedded 2-sphere bounds a ball. A compact, orientable, irreducible 3-manifold is said to be *Haken* if it is the 3-ball or it contains a properly imbedded two-sided incompressible surface. It is well known (see, for example, lemma 6.8 in Hempel [**51**]) that every compact, orientable, irreducible 3-manifold with nonempty boundary is Haken.

We will begin with a review of fibered 3-manifolds in section 2.1. Sections 2.2 through 2.5 contain an introduction to Johannson's theory of boundary patterns, which provides the technical underpinning for his formulation of the characteristic submanifold. A boundary pattern of a 3-manifold is a collection of connected 2-manifolds in its boundary, whose interiors are disjoint. A boundary pattern is called useful when these 2-manifolds are incompressible, and satisfy a certain restriction on their configuration. This is the natural generalization of the property of having an incompressibile boundary to the setting of manifolds with boundary patterns. In Johannson's theory all maps are required to be admissible, which means that they take manifolds of the boundary pattern of the domain into those of the boundary pattern of the range.

After setting up the basic constructs for boundary patterns and admissible maps, we develop the 3-dimensional concepts which are needed to define the characteristic submanifold. The small but important collection of "exceptional" fibered manifolds is examined in section 2.6. These include all the manifolds which admit both an I-fibering and a Seifert fibering, and those that admit nonisotopic fiberings. They are called exceptional because they must be excluded from many of the results that are stated later. Next we examine the surfaces in fibered 3-manifolds which either are unions of fibers or are transverse to all the fibers of a fibered manifold. These vertical and horizontal surfaces are discussed in section 2.7. In section 2.8, we examine the mappings between fibered manifolds. Apart from some fairly obvious exceptions, an essential map from a fibered 3-manifold to itself is homotopic to a map that takes fibers to fibers, and hence induces a map between the quotient 2-manifolds of the fibered 3-manifolds. This allows us later to reduce problems about mappings between fibered manifolds to more accessible questions about the induced map on the quotient 2-manifolds.

Section 2.9 gives the definition and basic properties of the characteristic submanifold, and assembles results from [**58**] in order to give a convenient characterization of the characteristic submanifold. A variety of examples of characteristic submanifolds are given in section 2.10, then in section 2.11 we discuss the central result for homotopy equivalences, Johannson's Classification Theorem. A few topological results, which do not fit very well anywhere else, are collected in section 2.12.

2.1. Fibered 3-manifolds

In this section we review the theory of fibered 3-manifolds. The facts in this section all follow from the elementary theory of fiber bundles and from the general classification of Seifert-fibered 3-manifolds. There are many fine references for Seifert-fibered 3-manifolds, including [**51, 54, 56, 103, 104, 113, 114, 126, 127**]. A nice treatment which includes the nonorientable case is given in sections 1.7, 5.2 and 5.3 of [**103**], and we follow this reference for our description of the fundamental groups.

We first review 3-dimensional I-bundles and S^1-bundles. Let $\mathrm{I}=[0,1]$, and let B be a connected 2-manifold. There is an I-bundle over B having orientable total space. If B is orientable, this bundle is just the product $B\times\mathrm{I}$. If B is nonorientable, it is the quotient of $\widetilde{B}\times\mathrm{I}$ obtained by identifying (x,t) with $(\tau(x),1-t)$ where τ is the nontrivial covering transformation for the orientable double cover $\widetilde{B}$. The boundary of the I-bundle is homeomorphic to $B\times\partial\mathrm{I}\cup\partial B\times\mathrm{I}$ if B is orientable. If B is nonorientable, then the boundary is homeomorphic to $\widetilde{B}\cup\partial B\times\mathrm{I}$, where the two preimage circles of a boundary component C of B are joined by the annulus $C\times\mathrm{I}$.

The Finite Index Theorem (Theorem 10.5 of Hempel [**51**]) provides an extremely useful characterization of I-bundles. Here we give its statement specialized to the irreducible orientable case:

THEOREM 2.1.1. (Finite Index Theorem) *Let M be a compact orientable irreducible 3-manifold, and F ($\neq D^2$ or S^2) a compact, connected, incompressible surface in ∂M. If the index of $\pi_1(F)$ in $\pi_1(M)$ is finite, then either*

(i) *$\pi_1(M)\cong\mathbb{Z}$, F is an annulus, and M is a solid torus, or*
(ii) *$\pi_1(F)=\pi_1(M)$ and $M=F\times I$ with $F=F\times\{0\}$, or*
(iii) *$\pi_1(F)$ has index 2 in $\pi_1(M)$ and M is a twisted I-bundle over a compact surface N with F as the associated ∂I-bundle.*

For S^1-bundles over surfaces with nonempty boundary, the classification is analogous. For each compact 2-manifold B with nonempty boundary, there is an S^1-bundle over B which has orientable total space. It is obtained by taking two copies of the orientable I-bundle over B, and identifying each endpoint of a fiber with the endpoint of the corresponding fiber in the other copy (i. e. it is the double of the I-bundle along the associated ∂I-bundle). This is the unique S^1-bundle over B with orientable total space, where unique means here that any other such bundle is homeomorphic to this one by a homeomorphism that for each $x \in B$ takes the fiber over x in the first bundle to the fiber over x in the second. If B is orientable, the bundle is just the product $B \times S^1$. If B is nonorientable, it is characterized by the property that the preimage of any simple closed loop is a torus if the loop is orientation-preserving and is a Klein bottle if it is orientation-reversing.

When B is closed, the S^1-bundles with orientable total space are classified by an integer b. Orient the total space E, let B_0 result from removing the interior of a small 2-disk from B, and let E_0 be the preimage of B_0. Fix a cross-section of E_0, and let its boundary be a loop c in the torus ∂E_0. Let t be a circle fiber in ∂E_0, oriented so that the pair (c, t) determines the positive orientation on ∂E_0. A cross-sectional disk of the solid torus $\overline{E - E_0}$ then determines an element ct^{-b}, and the integer b is the invariant. There are other choices of cross-section, but all meet the boundary of E_0 in a loop isotopic to c. Reversing the orientation on E replaces b by $-b$.

Since we use only orientable 3-manifolds, we speak of "the" I-bundle over B, and when B is connected with nonempty boundary, we speak of "the" S^1-bundle over B.

Particularly important are the S^1-bundles which are homeomorphic to I-bundles:

(a) The S^1-bundle over the disk is the solid torus. It is the total space of the I-bundles over the annulus and the Möbius band.
(b) The S^1-bundle over the annulus is $S^1 \times S^1 \times I$, which is the I-bundle over the torus.
(c) The S^1-bundle over the Möbius band is the I-bundle over the Klein bottle. One way to see this is to cut it apart along an annulus A which is the preimage of a nonseparating arc in the Möbius band. The result is $[0,1] \times [0,1] \times S^1 = (S^1 \times [0,1]) \times I$, and under the reidentification of the two copies of A, $(S^1 \times [0,1]) \times \{1/2\}$ becomes a Klein bottle which is a cross-section of the I-bundle.

These are the only orientable 3-manifolds which are both an I-bundle and an S^1-bundle. For in an S^1-bundle V, the fundamental group of the fiber determines an infinite cyclic normal subgroup of $\pi_1(V)$. An I-bundle over B has fundamental group $\pi_1(B)$, and the annulus, Möbius band, torus, and Klein bottle are the only 2-manifolds whose fundamental groups have an infinite cyclic normal subgroup. As we will note below, these 3-manifolds are also the only ones with nonempty boundary which admit nonisotopic Seifert-fiberings.

A *Seifert-fibered* 3-manifold V is one which is a union of disjoint fibers which are circles, each having a closed neighborhood which is a fibered solid torus $T_{\mu,\nu}$ for some relatively prime pair μ, ν with $\mu \geq 1$. This means that $T_{\mu,\nu}$ is the quotient space of $D^2 \times I$ that results from identifying $(x, 0)$ with $(\exp(2\pi i\nu/\mu)\, x, 1)$, and the circles are unions of arcs of the form $\{x\} \times I$ in $D^2 \times I$. The fiber which is the quotient of $\{0\} \times \mathrm{I}$ is a core circle of $T_{\mu,\nu}$, while the other fibers represent μ times the core circle. Collapsing each of the circles of $T_{\mu,\nu}$ to a point defines a quotient map from $T_{\mu,\nu}$ to a disk D^2. If $\nu = 0$, then this makes $T_{\mu,\nu}$ an S^1-bundle over D^2, but if $\nu \neq 0$, then $T_{\mu,\nu}$ is an S^1-bundle only over $D^2 - \{0\}$ and the core circle is called an *exceptional fiber.*

There is a homeomorphism from $T_{\mu,\nu+\mu}$ to $T_{\mu,\nu}$ that takes fibers to fibers, so we may always select ν so that $0 \leq \nu < \mu$. If $\nu \neq 0$, define p and q by $p = \mu$, $q\nu = 1 \pmod{\mu}$, and $0 < q < p$. We say that the core circle of $T_{\mu,\nu}$ is an exceptional fiber *of type* (p, q). If m is the boundary of D^2 and t is an S^1-fiber in $\partial T_{\mu,\nu}$, then a cross-section c of the fibering of $\partial T_{\mu,\nu}$ can be selected so that $m = pc + qt$.

The quotient space of V determined by collapsing each fiber to a point is a 2-manifold B, and the quotient map from V to B is called a *Seifert fibering* of V *over* B.

We will now give a general construction of the orientable Seifert-fibered 3-manifolds and their fundamental groups. We start with the case when $\partial V \neq \emptyset$. Let B_0 be a connected surface with nonempty boundary. Denote the boundary components of B_0 by $c_0, \ldots, c_s$, where $s \geq 0$. The fundamental group of B_0 is free, and we select generating sets

$$\{a_1, b_1, \ldots, a_g, b_g, c_1, \ldots, c_s\}, \text{ or}$$
$$\{v_1, \ldots, v_g, c_1, \ldots, c_s\}$$

according as B_0 is orientable or not. Here, $g \geq 0$ if B_0 is orientable and $g \geq 1$ if B_0 is nonorientable. We describe these generators explicitly as follows. Fix a basepoint $x_0 \in c_0$. The generators will be represented by oriented simple loops in B_0 based at x_0. We will use the same letter to denote the loop in B_0 and the element of $\pi_1(B_0, x_0)$ that it represents. If B_0 is orientable, each pair $\{a_i, b_i\}$ determines a torus connected summand of B_0. The loops a_i and b_i intersect at one point other than x_0, and these are the only pairs of representative loops that intersect at any point other than x_0. For suitable orientations on the a_i, b_i, and c_j, the boundary component c_0 represents $\prod_{i=1}^{g}[a_i, b_i] \prod_{j=1}^{s} c_j$. If B_0 is nonorientable, each generating loop v_i passes through a crosscap (i. e. a projective plane connected summand) of B_0. In this case, c_0 represents $\prod_{i=1}^{g} v_i^2 \prod_{j=1}^{s} c_j$.

Now let E_0 be the S^1-bundle over B_0. Orient the fiber of E_0, and use a cross-section of the fibering to regard B_0 as a submanifold of E_0. The fiber represents an element t of $\pi_1(E_0)$, which has presentation

$$\langle a_i, b_i, c_j, t \mid [a_i, t] = [b_i, t] = [c_j, t] = 1 \rangle, \text{ or}$$
$$\langle v_i, c_j, t \mid v_i t v_i^{-1} = t^{-1}, [c_j, t] = 1 \rangle \ .$$

Here and in the remainder of this section, the ranges for the indices in all presentations will be $1 \leq i \leq g$, $1 \leq j \leq s$, and (later) $1 \leq k \leq r$.

Now, to form a Seifert-fibered 3-manifold with nonempty boundary, we fix some r with $0 \leq r \leq s$, and choose relatively prime pairs (p_k, q_k), $1 \leq k \leq r$, with $p_k \geq 2$ and $0 < q_k < p_k$. For each k with $1 \leq k \leq r$, there is a torus boundary component of E_0 whose fundamental group is generated by t and c_k; along each of these boundary components we attach a fibered solid torus T_{μ_k,ν_k} to E_0 so that t is identified with the fiber and c_k with the cross-section c in T_{μ_k,ν_k}, thus the boundary of a meridian disk represents $c_k^{p_k}t^{q_k}$. The resulting manifold V has fundamental group

$$\langle a_i, b_i, c_j, t \mid [a_i,t]=[b_i,t]=[c_j,t]=c_k^{p_k}t^{q_k}=1, \rangle, \text{ or}$$
$$\langle v_i, c_j, t \mid v_i t v_i^{-1}=t^{-1}, [c_j,t]=c_k^{p_k}t^{q_k}=1\rangle \ .$$

Let B be formed from B_0 by filling in each C_k with a disk, for $1 \leq k \leq r$. The bundle fibering of E_0 over B_0 extends to a Seifert fibering of V over B. The core circle of each T_{μ_k,ν_k} is an exceptional fiber of type (p_k, q_k).

Thus far we have constructed a Seifert-fibered 3-manifold with boundary. For the closed case, one has $r = s$, and an additional solid torus with the product fibration $D^2 \times S^1$ is glued to the boundary component containing c_0 so that its fibers agree with the fibers of V and so that the boundary of its meridian disk represents $c_0 t^{-b}$, for some integer b. The effect on the fundamental group is to add one additional relation, obtaining

$$\langle a_i, b_i, c_j, t \mid [a_i,t]=[b_i,t]=[c_j,t]=c_j^{p_j}t^{q_j}=1, \prod [a_i,b_i] \prod c_j = t^b \ \rangle, \text{ or}$$
$$\langle v_i, c_j, t \mid v_i t v_i^{-1}=t^{-1}, [c_j,t]=c_j^{p_j}t^{q_j}=1, \prod v_i^2 \prod c_j = t^b \ \rangle \ .$$

The infinite cyclic subgroup $\langle t \rangle$ generated by the fiber t is normal, and is central if and only if B is orientable. Taking the quotient of $\pi_1(V)$ by this subgroup yields the exact sequences given below for the four cases: $\partial V \neq \emptyset$ and B orientable, $\partial V \neq \emptyset$ and B nonorientable, $\partial V = \emptyset$ and B orientable, $\partial V = \emptyset$ and B nonorientable. For each group, R denotes the set of relations for $\pi_1(V)$ given above:

$$1 \longrightarrow \langle t \rangle \longrightarrow \langle a_i, b_i, c_j, t \mid R \rangle \longrightarrow \langle a_i, b_i, c_j \mid c_k^{p_k} = 1 \rangle \longrightarrow 1,$$
$$1 \longrightarrow \langle t \rangle \longrightarrow \langle v_i, c_j, t \mid R \rangle \longrightarrow \langle v_i, c_j \ldots, c_s \mid c_k^{p_k} = 1 \rangle \longrightarrow 1,$$
$$1 \longrightarrow \langle t \rangle \longrightarrow \langle a_i, b_i, c_j, t \mid R \rangle \longrightarrow$$
$$\langle a_i, b_i, c_j \mid c_1 \cdots c_s \prod [a_i, b_i] = 1, c_j^{p_j} = 1 \rangle \longrightarrow 1, \text{ or}$$
$$1 \longrightarrow \langle t \rangle \longrightarrow \langle v_i, c_j \mid R \rangle \longrightarrow \langle v_i, c_j \mid c_1 \cdots c_s \prod v_i^2 = 1, c_j^{p_j} = 1 \rangle \longrightarrow 1 \ .$$

In the first two of these extensions, corresponding to the cases when $\partial V \neq \emptyset$, the quotient group is a free product of cyclic groups; the a_i, b_i, v_i, and the c_j with $j > r$ are generators of infinite cyclic factors, and the c_k with $k \leq r$ are generators of finite cyclic factors.

Note that in a Seifert-fibered manifold, the preimage in V of a simple loop in B which is disjoint from the images of the exceptional fibers is a torus if the loop is orientation-preserving and is a Klein bottle if the loop is orientation-reversing.

Some of the treatments of Seifert fiberings in the literature use the language of orbifolds to describe the quotient object of a Seifert-fibered 3-manifold (see [**113**], or for a much more general context [**19**]). The quotient orbifold is obtained from the quotient surface B by declaring each point corresponding to an exceptional fiber of type (p,q) to be an order p cone point. The quotient group in the extensions given above is the orbifold fundamental group of this quotient orbifold. We will not need the orbifold viewpoint in our work, so we do not discuss it further.

A map between fibered manifolds is called *fiber-preserving* if the image of each fiber of the domain lies in a fiber of the range. Two fibered structures on a manifold V are called *isotopic* if there is a fiber-preserving homeomorphism between them which is isotopic to the identity.

Here are some of the fibered manifolds that occur most frequently. As shown in the Unique Fibering Theorem 2.8.1 below, they are the only cases (with nonempty boundary) which admit nonisotopic Seifert fiberings.

EXAMPLE 2.1.2. *The solid torus.*

The solid torus $V = D^2 \times S^1$ is the I-bundle over the annulus and also over the Möbius band; in the latter case the annulus $\widetilde{B}$ in ∂V wraps twice in the longitudinal direction, and (for some choice of longitude in the boundary) once in the meridianal direction. It also admits infinitely many Seifert fiberings. One is the product fibering, the others each have exactly one exceptional fiber. The latter are distinct up to isotopy, and distinct up to fiber-preserving homeomorphism except that the fibering with exceptional fiber of type (p, q) is homeomorphic to the one with exceptional fiber of type $(p, p-q)$, by a homeomorphism which (up to isotopy) is the identity on the S^1-factor and is complex conjugation in the D^2-factor.

EXAMPLE 2.1.3. *The I-bundle over the torus.*

The I-bundle over the torus is $S^1 \times S^1 \times I$. It admits infinitely many nonisotopic Seifert fiberings, one for each pair $\{(p, q), (-p, -q)\}$ of pairs of relatively prime integers. These are all product fiberings, in which the fiber represents $\pm(p, q) \in \pi_1(V) \cong \mathbb{Z} \times \mathbb{Z}$. Any two of these fiberings are equivalent by a fiber-preserving homeomorphism.

EXAMPLE 2.1.4. *The I-bundle over the Klein bottle.*

The I-bundle over the Klein bottle admits two Seifert fiberings. One is as the S^1-bundle over the Möbius band, as discussed in example (c) above. The other Seifert fibering has quotient surface a disk and two exceptional fibers of type $(2, 1)$. To see this alternate fibration, take an arc that cuts the disk into two disks, each containing one exceptional point. The preimage of each half disk is a solid torus, and the two solid tori intersect in an annulus whose core curve wraps once around the meridian and twice arounds the longitude of each solid torus. This core curve is the boundary of a Möbius band in each solid torus. In fact this Möbius band is the zero section of a twisted I-bundle structure on each solid torus. When the two solid tori are glued together so that the two copies of the annulus are identified, the two Möbius bands form a Klein bottle, which is the zero section of the I-bundle structure on the union. As detailed in lemma 2.8.5 below, these are the only two Seifert fiberings up to isotopy.

2.2. Boundary patterns

In this section, we introduce Johannson's theory of boundary patterns. A boundary pattern for a 3-manifold is a set $\underline{\underline{m}}$ of 2-manifolds in its boundary, which satisfy certain conditions. In particular, they meet only in their boundaries. Roughly speaking, the part of the boundary that lies in the elements of $\underline{\underline{m}}$ is no longer "free", and in the context of Johannson's theory the manifold behaves in some respects as though its only boundary were the part that does not lie in elements of $\underline{\underline{m}}$. In particular, if all of its boundary lies in elements of $\underline{\underline{m}}$, then the 3-manifold enjoys some of the properties of closed 3-manifolds. For example, if the boundary pattern is also "useful," there is an analogue of Waldhausen's fundamental theorem for closed manifolds which assures that all homotopy equivalences are homotopic to homeomorphisms (see theorem 2.5.6 below). In this section we will give many examples of boundary patterns, most of which will be used in our later work.

Precisely, a *boundary pattern* $\underline{\underline{m}}$ for an n-manifold M is a finite set of compact, connected $(n-1)$-manifolds in ∂M, such that the intersection of any i of them is empty or consists of $(n-i)$-manifolds. Thus when $n=3$, the components of the intersections of pairs of elements of the boundary pattern are arcs or circles, and if three elements meet, their intersection consists of a finite collection of points at which three intersection arcs meet.

On a 2-manifold, a boundary pattern is simply a collection of arcs and circles in the boundary, which are disjoint except that two arcs may meet in an endpoint, or in both endpoints. In particular, an *i-faced disk* is a 2-disk whose boundary pattern has i elements, such that every point in the boundary lies in some element of the boundary pattern. For an i-faced disk (but not for any other manifolds), the elements of the boundary pattern are called *faces.* A 4-faced disk is called a *square.* An i-faced disk with $i \leq 3$ is called a *small-faced* disk. Squares and small-faced disks play important roles in Johannson's theory. A 2-manifold with boundary pattern is called an *annulus* if it is $\mathrm{S}^1 \times \mathrm{I}$ with boundary pattern consisting of the components of its boundary. This boundary pattern is often denoted by $\overline{\underline{\underline{\emptyset}}}$ (see example 2.2.1 below). To indicate a 2-manifold homeomorphic to $\mathrm{S}^1 \times \mathrm{I}$, but not necessarily having the boundary pattern $\overline{\underline{\underline{\emptyset}}}$, we use the term *topological annulus.* Similarly, a 2-manifold with boundary pattern is called a *Möbius band* if it is a topological Möbius band and carries the boundary pattern $\overline{\underline{\underline{\emptyset}}}$ having the entire boundary as its only element.

For a boundary pattern $\underline{\underline{m}}$, we denote by $J(\underline{\underline{m}})$ the union of ∂F for all $F \in \underline{\underline{m}}$. When M is 2-dimensional, $J(\underline{\underline{m}})$ is a finite collection of points, and when M is 3-dimensional, it is a union of circles and trivalent graphs.

The symbol $|\underline{\underline{m}}|$ will mean the set of points of ∂M that lie in some element of $\underline{\underline{m}}$. It is important in arguments to distinguish between the elements of $\underline{\underline{m}}$ and the points of M which lie in these elements, and we will always be precise in this distinction. The elements of $\underline{\underline{m}}$ are called *bound sides,* and the closures of the components of $\partial M - |\underline{\underline{m}}|$ are called *free sides.* When $|\underline{\underline{m}}| = \partial M$, $\underline{\underline{m}}$ is said to be *complete.* Provided that ∂M is compact, we define the *completion* of $\underline{\underline{m}}$ to be the complete boundary pattern $\overline{\underline{\underline{m}}}$ which is the union of $\underline{\underline{m}}$ and the collection of free sides. Note that $J(\underline{\underline{m}}) = J(\overline{\underline{\underline{m}}})$.

Boundary patterns arise naturally in various ways. Here are some examples.

EXAMPLE 2.2.1. *Trivial (but important) examples*

For any manifold M, one has the empty boundary pattern $\underline{\underline{\emptyset}}$. Its completion $\overline{\underline{\underline{\emptyset}}}$ is the set of boundary components of M. When ∂M is empty, $\underline{\underline{\emptyset}}$ is the only possible boundary pattern on M.

EXAMPLE 2.2.2. *A boundary pattern containing a square*

Here is an explicit example of a 3-manifold S with boundary pattern $\underline{\underline{s}}$ such that one of the elements of $\underline{\underline{s}}$ is a square. Its other two elements are topological annuli, although they are not annuli as 2-manifolds with boundary pattern (that is, they do not carry the boundary pattern $\overline{\underline{\underline{\emptyset}}}$). This boundary pattern will be examined further in examples 2.4.7 and 2.10.10, and will be used in example 2.10.11 to construct characteristic submanifolds with certain properties.

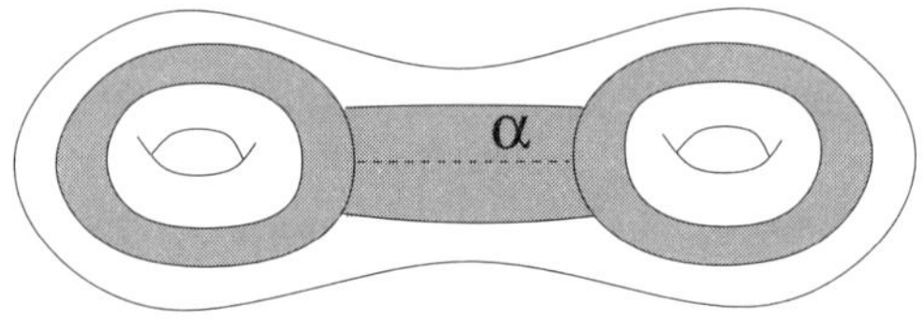

FIGURE 2.1. A boundary pattern containing a square

Let S be a 3-manifold with boundary. Let A_0 and A_1 be disjoint annuli in a component of ∂S, and fix an arc $\alpha = [0,1]$ in ∂S running from A_0 to A_1 and meeting them only in its endpoints. Let $N = [0,1] \times [-1,1] \subset \partial S$ so that $\alpha = [0,1] \times \{0\}$ and $N \cap (A_0 \cup A_1) = \partial\alpha \times [-1,1]$, i. e. N is a bicollar neighborhood of the arc α in the manifold $\overline{\partial S - (A_0 \cup A_1)}$. Putting $\underline{\underline{s}} = \{A_0, N, A_1\}$ gives a boundary pattern as described above. Figure 2.1 shows a picture of such a boundary pattern.

EXAMPLE 2.2.3. *Seifert-fibered 3-manifolds*

Some information on Seifert-fibered 3-manifolds was given in section 2.1. As used in Johannson's theory, they always have a boundary pattern which is adapted to the fibered structure. A Seifert fibering on a 3-manifold $(V, \underline{\underline{v}})$ with boundary pattern is called an *admissible Seifert fibering* when the elements of $\underline{\underline{v}}$ are the preimages of the elements of a boundary pattern of the quotient surface. Equivalently, the elements of $\underline{\underline{v}}$ are tori or fibered annuli.

EXAMPLE 2.2.4. *I-bundles*

There are nonhomeomorphic I-bundle structures on a 3-dimensional orientable handlebody, and for a given homeomorphism class there are many nonisotopic structures. These can be distinguished in a natural way using boundary patterns. Assume that a handlebody V carries a fixed structure as an I-bundle over B. Each component of the associated ∂I-bundle is a 2-manifold in ∂V, called a *lid* of the I-bundle. There are two lids when the bundle is a product, and one when it is twisted. Let $\underline{\underline{b}}$ be a boundary pattern on B. The preimages of the elements of $\underline{\underline{b}}$ form a collection of squares and annuli in ∂V, called the *sides* of the I-bundle. The lid or lids, together with the sides, if any, form a boundary pattern $\underline{\underline{v}}$ on V (in this somewhat unfortunate terminology, a side is always a bound side, and a free side is never a side). When V carries this boundary pattern, the fibering is called an *admissible I-fibering* of $(V, \underline{\underline{v}})$ over $(B, \underline{\underline{b}})$. Note that an I-bundle over a closed 2-manifold is admissibly I-fibered if and only if its boundary pattern is $\overline{\underline{\underline{\emptyset}}}$. We emphasize that for an admissible I-fibering, *the lids are always elements of the boundary pattern*. Consequently, the bundle projection map from V to B is never an admissible map.

EXAMPLE 2.2.5. *The submanifold boundary pattern*

Let X be a submanifold of positive codimension of a manifold with boundary pattern $(M, \underline{\underline{m}})$, such that $X \cap \partial M = \partial X$. Define $\underline{\underline{x}}$ by

$$\underline{\underline{x}} = \{\text{components of } X \cap G \mid G \in \underline{\underline{m}}\} \ .$$

Provided that ∂X meets each intersection of a collection of elements of $\overline{\underline{\underline{m}}}$ transversely, $\underline{\underline{x}}$ is a boundary pattern on X.

Suppose now that X is a codimension-zero submanifold of M. Then $\underline{\underline{x}}$ will be a boundary pattern provided that $\mathrm{Fr}(X) \cap \partial M$ meets each intersection of elements of $\overline{\underline{\underline{m}}}$ transversely, where $\mathrm{Fr}(X)$ is the frontier (that is, the topological boundary) of X.

Unless otherwise stated, it will be assumed that all submanifolds of M meet these transversality conditions, and carry the boundary pattern $\underline{\underline{x}}$. To emphasize this, we sometimes say that $(X, \underline{\underline{x}})$ is a submanifold of $(M, \underline{\underline{m}})$.

EXAMPLE 2.2.6. *The proper boundary pattern on a codimension-zero submanifold*

For codimension-zero submanifolds, a boundary pattern other than the submanifold boundary pattern of the previous example is sometimes used. Let $\underline{\underline{x}}$ denote the submanifold boundary pattern on X, and let $\underline{\underline{x}}'$ denote the collection of components of $\mathrm{Fr}(X)$. The boundary pattern $\underline{\underline{x}} \cup \underline{\underline{x}}'$ is called the *proper* boundary pattern on X.

EXAMPLE 2.2.7. *Product boundary patterns*

If $(M, \underline{\underline{m}})$ and $(N, \underline{\underline{n}})$ are compact connected manifolds with boundary patterns, then the product $(M, \underline{\underline{m}}) \times (N, \underline{\underline{n}})$ is defined to be $M \times N$ with the boundary pattern

$$\{F \times N \mid F \in \underline{\underline{m}}\} \cup \{M \times G \mid G \in \underline{\underline{n}}\} \ .$$

To verify the boundary pattern condition, we note that if $\dim(M) = m$ and $\dim(N) = n$, then $F_1 \times N \cap \cdots \cap F_i \times N \cap M \times G_1 \cap \cdots \cap M \times G_j = (F_1 \cap \cdots \cap F_i) \times (G_1 \cap \cdots \cap G_j)$, which is a manifold of dimension $(m-i)+(n-j) = (m+n)-(i+j)$, verifying the boundary pattern condition. The main cases of interest for us are

$$\begin{aligned}
(M, \underline{\underline{m}}) \times (\mathrm{I}, \underline{\underline{\emptyset}}) &= (M \times \mathrm{I}, \{F \times \mathrm{I} \mid F \in \underline{\underline{m}}\}) \\
(M, \underline{\underline{m}}) \times (\mathrm{I}, \{0\}) &= (M \times \mathrm{I}, \{F \times \mathrm{I} \mid F \in \underline{\underline{m}}\} \cup \{F \times \{0\}\}) \\
(M, \underline{\underline{m}}) \times (\mathrm{I}, \overline{\underline{\underline{\emptyset}}}) &= (M \times \mathrm{I}, \{F \times \mathrm{I} \mid F \in \underline{\underline{m}}\} \cup \{M \times \{0\}\} \cup \{M \times \{1\}\})
\end{aligned}$$

In particular, the admissibly fibered product I-bundle over $(F, \underline{\underline{f}})$ is $(F, \underline{\underline{f}}) \times (\mathrm{I}, \overline{\underline{\underline{\emptyset}}})$.

EXAMPLE 2.2.8. *Splitting* $(M, \underline{\underline{m}})$ *along a submanifold*

Let F be a properly imbedded two-sided 2-manifold in $(M, \underline{\underline{m}})$, with the submanifold boundary pattern $\underline{\underline{f}}$. A sufficiently small product neighborhood $F \times [-1, 1]$ of F with its submanifold boundary pattern will be $(F, \underline{\underline{f}}) \times ([-1, 1], \underline{\underline{\emptyset}})$. The manifold obtained from $(M, \underline{\underline{m}})$ by *splitting along* F is $M - (F \times (-1, 1))$ with its *proper* boundary pattern. In this way, boundary patterns can be used to "remember" a sequence of splittings of a 3-manifold along two-sided surfaces. Among the powerful technical tools used in [**58**] are hierarchies for 3-manifolds, remembered by the resulting boundary patterns, and satisfying certain strong incompressibility assumptions.

In the next section we will give another example of a boundary pattern.

2.3. Admissible maps and mapping class groups

For manifolds with boundary patterns, one usually restricts attention to maps which take manifolds of the boundary pattern of the domain into those of the boundary pattern of the range; such maps are called admissible. We will define admissible homotopy equivalences, and three groups of automorphisms $\mathcal{H}(M,\underline{\underline{m}})$, $\operatorname{Out}(\pi_1(M),\pi_1(\underline{\underline{m}}))$, and $\mathcal{R}(M,\underline{\underline{m}})$. Obtaining information about these groups of automorphisms is the ultimate objective of most of the topological work in this paper.

Precisely, a map f from $(M,\underline{\underline{m}})$ to $(N,\underline{\underline{n}})$ is called *admissible* when $\underline{\underline{m}}$ is the disjoint union

$$\underline{\underline{m}} = \bigsqcup_{G\in\underline{\underline{n}}} \{\text{components of } f^{-1}(G)\}\ .$$

Notice that the requirement that the union be disjoint implies that for each element F of $\underline{\underline{m}}$, there is exactly one element of $\underline{\underline{n}}$ that contains the entire image of F. For elements of $\underline{\underline{m}}$ that meet, the elements of $\underline{\underline{n}}$ that contain their images must be distinct. Thus, two neighboring bound sides F_1 and F_2 must be mapped to neighboring bound sides G_1 and G_2, in such a way that $F_1\cap F_2$ consists of some components of the preimage of $G_1\cap G_2$ in $F_1\cup F_2$. Moreover, the full preimage of the intersection of two elements of $\underline{\underline{n}}$ must consist of components of intersections of pairs of elements of $\underline{\underline{m}}$.

When $(X,\underline{\underline{x}})$ is a submanifold of $(M,\underline{\underline{m}})$, the inclusion map of X is admissible. If additionally $(X,\underline{\underline{x}})$ is admissibly imbedded in $(M,\overline{\underline{\underline{m}}})$, then X cannot meet the closure of $\partial M - |\underline{\underline{m}}|$. In particular, an element of $\underline{\underline{x}}$ which does not meet any other element of $\underline{\underline{x}}$ must be imbedded in the manifold interior of some element of $\underline{\underline{m}}$.

An *admissible homotopy* between maps from $(M,\underline{\underline{m}})$ to $(N,\underline{\underline{n}})$ is a homotopy which is admissible as a map from $(M,\underline{\underline{m}})\times(\mathrm{I},\underline{\underline{\emptyset}})$ to $(N,\underline{\underline{n}})$. An *admissible isotopy* is an isotopy which is an admissible homotopy. An admissible map $f\colon(M,\underline{\underline{m}})\to(N,\underline{\underline{n}})$ is called an *admissible homotopy equivalence* if there is an admissible map $g\colon(N,\underline{\underline{n}})\to(M,\underline{\underline{m}})$ such that gf and fg are admissibly homotopic to the identity maps. When the elements of $\underline{\underline{m}}$ (and consequently of $\underline{\underline{n}}$) are pairwise disjoint, this simply says that $f\colon(M,|\underline{\underline{m}}|)\to(N,|\underline{\underline{n}}|)$ is a homotopy equivalence of pairs.

The group of admissible isotopy classes of admissible homeomorphisms from $(M,\underline{\underline{m}})$ to $(M,\underline{\underline{m}})$ is denoted by $\mathcal{H}(M,\underline{\underline{m}})$. Suppose that $\langle h\rangle\in\mathcal{H}(M,\underline{\underline{m}})$. Since $h^{-1}(|\underline{\underline{m}}|)=|\underline{\underline{m}}|$, h must carry each free side of $(M,\underline{\underline{m}})$ homeomorphically to a free side of $(M,\underline{\underline{m}})$. Therefore h is also admissible for $(M,\overline{\underline{\underline{m}}})$. That is, $\mathcal{H}(M,\underline{\underline{m}}) = \mathcal{H}(M,\overline{\underline{\underline{m}}})$.

If $\underline{\underline{m}}$ is a boundary pattern for M, for which the elements of $\underline{\underline{m}}$ are incompressible, define $\operatorname{Out}(\pi_1(M),\pi_1(\underline{\underline{m}}))$ to be the group of outer automorphisms $[\phi]$ of $\pi_1(M)$ such that for each $F\in\underline{\underline{m}}$, $\phi_\#(\pi_1(F))$ is conjugate in $\pi_1(M)$ to $\pi_1(G)$ for some $G\in\underline{\underline{m}}$. Sending h to its induced outer automorphism defines a homomorphism $\mathcal{H}(M,\underline{\underline{m}})\to\operatorname{Out}(\pi_1(M),\pi_1(\underline{\underline{m}}))$, and we define $\mathcal{R}(M,\underline{\underline{m}})$ to be the image of this homomorphism. Elements of $\mathcal{R}(M,\underline{\underline{m}})$ are said to be *realizable* and we call $\mathcal{R}(M,\underline{\underline{m}})$ the *realizable subgroup* of $\operatorname{Out}(\pi_1(M),\pi_1(\underline{\underline{m}}))$. The subgroup of $\mathcal{H}(M,\underline{\underline{m}})$ consisting of orientation-preserving elements is denoted by $\mathcal{H}_+(M,\underline{\underline{m}})$, and its image in $\operatorname{Out}(\pi_1(M),\pi_1(\underline{\underline{m}}))$ is denoted by $\mathcal{R}_+(M,\underline{\underline{m}})$. Of course, these have index at most 2 in $\mathcal{H}(M,\underline{\underline{m}})$ and $\mathcal{R}(M,\underline{\underline{m}})$. Our main topological theorem characterizes, in a fairly general setting, when $\mathcal{R}(M,\underline{\underline{m}})$ has finite index in $\operatorname{Out}(\pi_1(M),\pi_1(\underline{\underline{m}}))$.

We will close this section with another example of a boundary pattern. Although it will not be used in any of our geometric applications, we include it in order to illustrate another way to use boundary patterns to exert control on mapping classes.

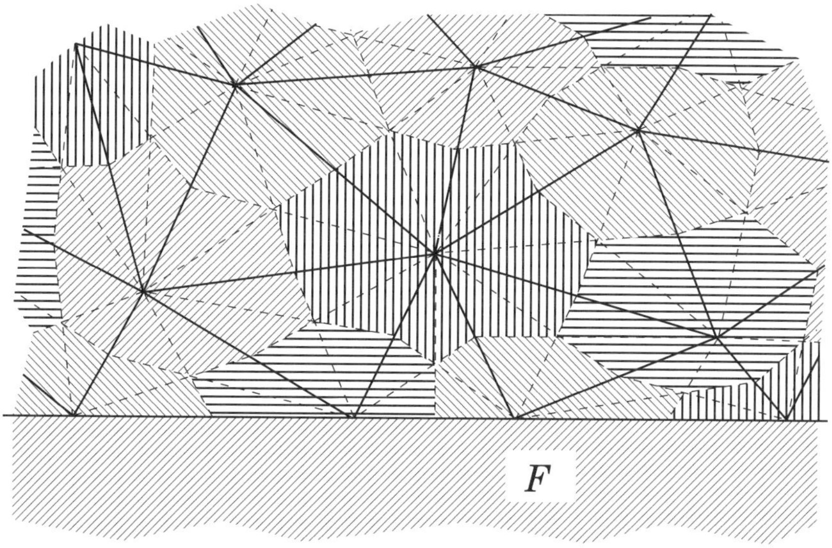

FIGURE 2.2. The cell complex dual to a triangulation of $\overline{\partial M - F}$

EXAMPLE 2.3.1. *Tiling a submanifold of* ∂M

Let F be a 2-manifold in ∂M, and let $G = \overline{M - F}$. We will construct a complete boundary pattern $\underline{\underline{m}}$ on M that contains the components of F among its elements, and such that $\mathcal{H}(M, \underline{\underline{m}})$ contains a subgroup of finite index isomorphic to $\mathcal{H}(M \text{ rel } G)$ (the relative mapping class group, consisting of the path components of the space of homeomorphisms of M that fix each point of G). After giving this construction, we show how to modify it to obtain a boundary pattern for which $\mathcal{H}(M, \underline{\underline{m}})$ is itself isomorphic to $\mathcal{H}(M \text{ rel } G)$. In example 2.4.10 we will verify that when F is incompressible, these boundary patterns satisfy the condition of being "useful", which is needed to apply the full strength of Johannson's theory. Thus, the mapping class groups $\mathcal{H}(M, \underline{\underline{m}})$ of manifolds with useful boundary pattern include as special cases all relative mapping class groups $\mathcal{H}(M \text{ rel } G)$ for which $\overline{\partial M - G}$ is incompressible.

Let T be a triangulation of G, and let T' be its first barycentric subdivision. Figure 2.2 illustrates the cell complex structure $\mathcal{C}$ on G dual to T: each 2-cell is the closed star in T' of a vertex of T, each 1-cell is the intersection of two 2-cells, and each vertex is either the intersection of three 2-cells (if it was the barycenter of a 2-simplex of T) or is the intersection of two 2-cells with ∂G (if it was the barycenter of a 1-simplex of ∂G). Note that the intersection of any two 2-cells is empty or is a single arc, and the intersection of any three is empty or is a single point. Consequently the set whose elements are the components of F and the 2-cells of $\mathcal{C}$ is a complete boundary pattern $\underline{\underline{m}}$. We call it a boundary pattern obtained by *tiling* G.

Each class in $\mathcal{H}(M, \underline{\underline{m}})$ induces a well-defined permutation on the elements of $\underline{\underline{m}}$; let $\mathcal{H}_0(M, \underline{\underline{m}})$ be the subgroup of finite index that permutes the 2-cells of $\mathcal{C}$ trivially. We will show that the inclusion of groups of homeomorphisms induces an isomorphism $\mathcal{H}(M \text{ rel } G) \to \mathcal{H}_0(M, \underline{\underline{m}})$. Let $\langle h \rangle \in \mathcal{H}_0(M, \underline{\underline{m}})$. Since any two 2-cells

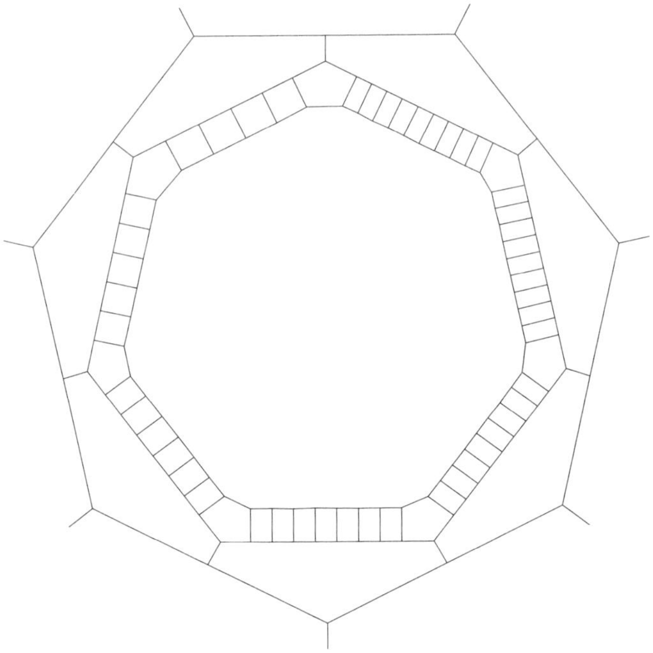

FIGURE 2.3. The subdivision of a tiling cell

of $\mathcal{C}$ intersect in at most one 1-cell, and any three 3-cells intersect in at most one 0-cell, h must also preserve each 1-cell and 0-cell of $\mathcal{C}$. It fixes both endpoints of each 1-cell, so by admissible isotopy we may assume that h is the identity on each 1-cell of $\mathcal{C}$. By the Alexander trick on each 2-cell of $\mathcal{C}$, we may change h by admissible isotopy to be the identity on all of G. This shows that $\mathcal{H}(M \text{ rel } G) \to \mathcal{H}_0(M, \underline{\underline{m}})$ is surjective. By a similar argument, using the cells of $\mathcal{C} \times \mathrm{I}$, one may prove that it is injective.

In fact, one can even select $\underline{\underline{m}}$ so that $\mathcal{H}(M, \underline{\underline{m}})$ equals $\mathcal{H}_0(M, \underline{\underline{m}})$, hence is isomorphic to $\mathcal{H}(M \text{ rel } G)$. The idea is to break each 2-cell c of $\mathcal{C}$ into smaller 2-cells. Figure 2.3 illustrates this subdivision for a 7-faced 2-cell. An annular regular neighborhood A of ∂c in an n-faced 2-cell c is broken into n 5-faced 2-cells $c_1, \ldots, c_n$, by introducing n 1-cells, each running from an interior point of one of the faces of c to the inner circle of A. Each of these 1-cells will be a common face of two adjacent c_i's. Each c_i will have two faces on ∂c, but the arc of intersection of c_i with the inside circle of A will be subdivided into a very large number of faces, to be determined later.

Let B be a smaller annulus in c whose outer circle is the inner circle of A. It is subdivided into a circle of 4- and 5-faced 2-cells: for each adjacent pair of c_i's there is one 5-faced cell containing the two faces that meet their intersection arc, and all the rest of the cells are 4-faced. Adjacent 2-cells of B meet in a single face. Each 5-faced cell of B has two faces on the inner circle of A, and each 4-faced cell of B has one face on the inner circle of A. The remainder of c is a central 2-cell d whose boundary is the inner circle of B. We repeat the process with each 2-cell of $\mathcal{C}$. The arcs that subdivide the first annulus A in each 2-cell are chosen to meet the 1-cell faces of c in points different from any corresponding arcs that meet the 1-cell faces

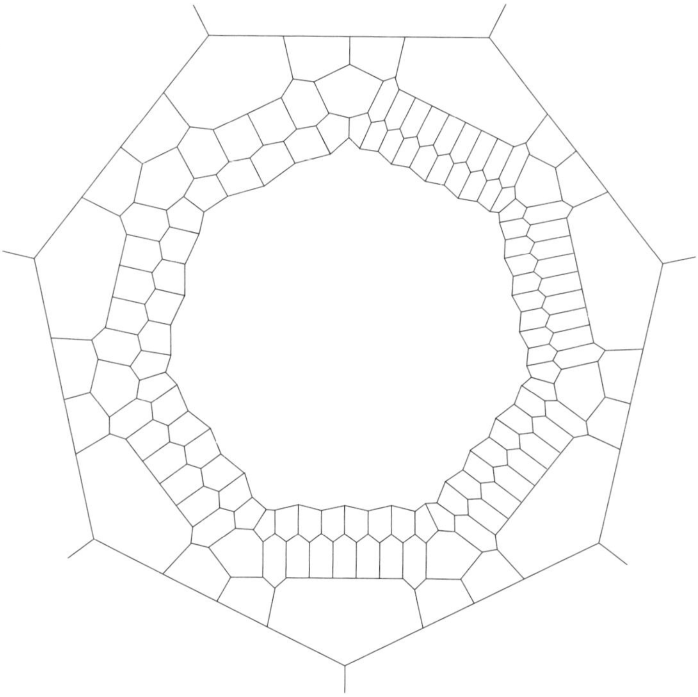

FIGURE 2.4. The modified subdivision of a tiling cell

and lie in 2-cells adjacent to c; this ensures that the graph formed by the 1-cells in the subdivision has valence 3 at each vertex, and thus the 2-cells of the subdivision will form a boundary pattern.

The number of faces of each c_i can be chosen arbitrarily. Therefore, we can perform the construction in such a way that all the c_i and all the central d produced from 2-cells of $\mathcal{C}$ have different numbers of faces. Then, any homeomorphism admissible for $(M, \underline{\underline{m}})$ must preserve each c_i and d, since all have different numbers of faces. This forces the intermediary 4- and 5-faced cells each to be preserved as well. Therefore, $\mathcal{H}_0(M, \underline{\underline{m}}) = \mathcal{H}(M, \underline{\underline{m}})$.

This version of the construction produces numerous essential squares (defined in section 2.4 below). For example, the frontier of a regular neighborhood of any of the 4-faced 2-cells that lies in B is an essential square. Moreover, if the original cell c was 4-faced, then a loop parallel to the boundary of c and slightly inside c will bound a square (and if c was 3-faced, such a loop will bound a 3-faced disk that gives a violation of usefulness of the boundary pattern, as defined in section 2.4 below). To eliminate these, we modify the construction as shown in Figure 2.4. We begin as before, but this time the annulus B is subdivided into a circle of 5- and 6-faced 2-cells. Each has two faces on the inner circle of B. Let C be an even smaller annulus whose outer circle is the inner circle of B. It is subdivided into a circle of 5-faced cells, each meeting one face from each of two adjacent 5- or 6-faced cells in B, and having one face on the inner circle of C. As before, the remainder of c is a central 2-cell d whose boundary is the inner circle of C. This eliminates all the essential squares except possibly one near the boundary of c, in case c was

4-faced. To eliminate such squares, we add two more arcs near each of the arcs that runs from an interior point of a face of c, and two 5-faced 2-cells in A as seen in Figure 2.4. This eliminates these remaining squares (as well as 3-faced disks violating usefulness, in case c was 3-faced).

We will see in example 2.4.10 below that under minimal assumptions on the triangulation, both of these retilings produces a boundary pattern that satisfies the condition of usefulness. For the second one, the characteristic submanifold will be disjoint from G.

2.4. Essential maps and useful boundary patterns

In this section we are going to introduce two of the important concepts in Johannson's theory: essential maps and useful boundary patterns. To motivate these, we first recall a bit of the history of the theory of 3-manifolds, focusing on some of its classical arguments. Essentiality and usefulness are the conditions that allow these arguments to be carried out in the more general context of 3-manifolds with boundary patterns.

One of the great advances in the theory of 3-manifolds was the Loop Theorem proved by C. D. Papakyriakopoulos in 1957, a generalization of a result formulated many years earlier by M. Dehn, known as Dehn's Lemma. Along with the Sphere Theorem, also proved by Papakyriakopoulos, the Loop Theorem and Dehn's Lemma allow singular maps to be replaced by imbeddings, giving a means to utilize homotopy-theoretic or algebraic information about M to obtain topological results. In particular, when F is a two-sided imbedded surface in a 3-manifold M and $\pi_1(F) \to \pi_1(M)$ is not injective (for some choice of basepoint), the existence of a "compressing" disk is guaranteed; this is an imbedded disk D in M with $D \cap F = \partial D$ and ∂D a loop which is not contractible in F. Such a disk allows a surgery process called compression to be performed on F that produces a lower-genus surface in M. Precisely, one fixes a product region $D \times [-1,1]$ with $D = D \times \{0\}$ and $D \times [-1,1] \cap F = \partial D \times [-1,1]$, and replaces F by $(F - \partial D \times (-1,1)) \cup D \times \{-1,1\}$. When this process cannot be performed further, the resulting surface is still two-sided, and the inclusion induces an injection on the fundamental group (for every choice of basepoint).

Suppose that M and N are closed irreducible 3-manifolds and $f\colon M \to N$ is a map which is transverse to a fixed two-sided imbedded incompressible surface G in N. If the preimage surface $f^{-1}(G)$ admits compressing disks, then f can be changed by homotopy to achieve a compression of $f^{-1}(G)$. The construction of this homotopy is rather complicated; lemma 6.5 of [**51**] contains a detailed description. When this simplification is repeated as many times as possible, the result is a new map f for which $f^{-1}(G)$ has no compressing disks. Some of its components may be 2-spheres. Since M is irreducible, these are compressible (bound 3-balls in M), and there is a second homotopy process for changing f to remove any such 2-spheres from the preimage of G. If these two homotopy processes are carried out as far as possible, the result is a map homotopic to f for which $f^{-1}(G)$ is incompressible. This is a key step in Waldhausen's method for proving that homotopy equivalences between closed Haken 3-manifolds are homotopic to homeomorphisms [**128**].

For irreducible manifolds with boundary, one often needs a stronger condition on imbedded surfaces than just being incompressible. Additionally, there should be no boundary-compressing disks. A *boundary-compressing disk* for a (properly) imbedded surface F has boundary consisting of an arc in ∂M and an arc in F which is not properly homotopic in F into ∂F; a compression along such a disk creates a further simplification of F. Again, when F is a preimage surface $f^{-1}(G)$, a homotopy process allows these simplifications to be accomplished by homotopy of f. Also, any components $f^{-1}(G)$ that are boundary-parallel disks may be removed by homotopy, resulting in a preimage surface which is incompressible and boundary-incompressible.

In Johannson's extension of these ideas to the context of irreducible 3-manifolds with boundary pattern, the role of compressing disks is played by admissibly imbedded small-faced disks. Incompressibility and boundary-incompressibility become the condition of essentiality: there are no small-faced compressing disks with one face an essential loop or arc in F and (when they are 2- or 3-faced) the other faces in elements of $\underline{\underline{m}}$. Essentiality is actually defined for admissible maps rather than just for submanifolds, and for admissible maps between manifolds of any dimensions. The Loop Theorem extends to show that if a surface imbedded in M has singular compressing disks, in this more general sense, then it has an imbedded one.

A boundary pattern will be called useful if every imbedded small-faced disk admits an admissible isotopy that shrinks it down to a point. This property is somewhat analogous to the property that an irreducible 3-manifold has incompressible boundary. A more direct analogue of incompressible boundary is the property that the completed boundary pattern $\overline{\underline{\underline{m}}}$ is useful, and this is a frequent hypothesis for the stronger results in Johannson's theory. In section 4.1, we will observe that admissible homotopy equivalences preserve usefulness of boundary patterns, and prove that they preserve usefulness of completed boundary patterns.

Following Johannson, we now give the precise definition of essential maps and essential submanifolds. An admissible loop or path is a map $h\colon (K, \underline{\underline{k}}) \to (X, \underline{\underline{x}})$, where in the case of a loop $(K, \underline{\underline{k}})$ is $(\mathrm{S}^1, \underline{\underline{\emptyset}})$, and for a path it is $(\mathrm{I}, \overline{\underline{\underline{\emptyset}}})$, and where $(X, \underline{\underline{x}})$ is a 2- or 3-manifold. The loop or path is called *inessential* if it is homotopic through admissible maps to a constant map. That is, the final map of the homotopy, the constant map, might not be admissible, but all earlier maps are admissible. If it is not inessential, it is called *essential.* A circle or arc admissibly imbedded in $(X, \underline{\underline{x}})$ is called *essential* if its inclusion map is essential.

A map $f\colon (X, \underline{\underline{x}}) \to (Y, \underline{\underline{y}})$ between 2- or 3-manifolds (not necessarily of the same dimension) is called *essential* when for any essential loop or path $h\colon (K, \underline{\underline{k}}) \to (X, \underline{\underline{x}})$, the composition $fh\colon (K, \underline{\underline{k}}) \to (Y, \underline{\underline{y}})$ is essential. It is immediate that essential maps always induce injections on fundamental groups, for any choice of basepoints.

Note that if the domain of f contains no essential loop or arc, then f is automatically essential. In particular, any admissible map with domain a small-faced disk is essential. It may be philosophically disquieting that all such maps are essential, but as noted in the first paragraph on p. 32 of [**58**], this convention seems to lead to fewer technical complications (see for example the remark after theorem 2.5.4 below). In fact, there are seven (compact, connected) 2-manifolds $(F, \underline{\underline{f}})$ that contain no essential loop or arc. Since F must be simply-connected, it is either S^2 or D^2. In the latter case, any two elements of $\underline{\underline{f}}$ must meet, since otherwise there is an essential arc connecting them, so $(F, \overline{\underline{\underline{f}}})$ is a small-faced disk. The possible

$(F,\underline{\underline{f}})$ are: S^2, $(D^2,\underline{\underline{\emptyset}})$, $(D^2,\{k\})$ where k is an arc, $(D^2,\{k,\ell\})$ where k and ℓ are arcs meeting in an endpoint, and the three small-faced disks.

We will now show that for maps between most surfaces whose boundary patterns consist of their boundary components, there is a relatively simple test for essentiality: it is equivalent to π_1-injectivity. The cases when F is an annulus or Möbius band must be excluded, since then there can be an admissible map $(F,\overline{\underline{\underline{\emptyset}}}) \to (G,\overline{\underline{\underline{\emptyset}}})$, taking F into a small neighborhood of a boundary component of G, which is injective on fundamental groups but is not essential.

LEMMA 2.4.1. *Let $f\colon (F,\overline{\underline{\underline{\emptyset}}}) \to (G,\overline{\underline{\underline{\emptyset}}})$ be an admissible map between connected 2-manifolds which is injective on fundamental groups. If F is not an annulus or Möbius band, then f is essential.*

PROOF. We may assume F is not S^2 or D^2, since then f is essential by definition. Since f is injective on fundamental groups, we need only prove it takes essential paths to essential paths. Let α be an essential path in F connecting boundary components β and γ (possibly $\beta=\gamma$), and suppose the restriction of f to α is inessential. Then the images under f of the loops β and $\alpha\gamma\overline{\alpha}$ are homotopic preserving basepoints into the same boundary component of G. Since $f_\#$ is injective, there are nonzero powers m and n so that β^m and $\alpha\gamma^n\alpha^{-1}$ represent the same element of $\pi_1(F)$. Let $\widetilde{F}$ be the covering corresponding to this element. It has cyclic fundamental group, and since β^m and γ^n both lift to boundary components of $\widetilde{F}$, it is either of the form $S^1\times I$ or $S^1\times I - C$ for some closed subset $C\subset S^1\times\{1\}$. If the latter occurs, then since both β^m and $\alpha\gamma^n\overline{\alpha}$ have lifts which are closed loops, α must have a lift which has both endpoints in $S^1\times\{0\}$. This lift is admissibly homotopic into $S^1\times\{0\}$, so α is inessential in $(F,\overline{\underline{\underline{\emptyset}}})$, contrary to its selection. Therefore $\widetilde{F}$ is homeomorphic to $S^1\times I$ and is a finite cover of F, so F is either an annulus or a Möbius band. □

A 2-dimensional submanifold $(F,\underline{\underline{f}})$ of $(M,\underline{\underline{m}})$ is called *essential* if and only if the inclusion map of $(F,\underline{\underline{f}})$ into $(M,\underline{\underline{m}})$ is essential. For codimension-zero submanifolds, this condition is not quite strong enough. For example, if W is a regular neighborhood of an essential loop in the interior of M, with empty boundary pattern, or a regular neighborhood in M of an arc or circle of intersection of two elements of $\underline{\underline{m}}$, with the submanifold boundary pattern, then the inclusion of $(W,\underline{\underline{w}})$ into $(M,\underline{\underline{m}})$ is an essential map. So as in **[58]** we define a codimension-zero submanifold $(W,\underline{\underline{w}})$ of $(M,\underline{\underline{m}})$ to be *essential* when its frontier is an essential 2-manifold in $(M,\underline{\underline{m}})$. This implies that the inclusion of $(W,\underline{\underline{w}})$ into $(M,\underline{\underline{m}})$ is an essential map. To see this, denote the frontier of $(W,\underline{\underline{w}})$ by $(F,\underline{\underline{f}})$, and suppose there is an essential loop or path $h\colon (K,\underline{\underline{k}}) \to (W,\underline{\underline{w}})$ which is inessential as a map into $(M,\underline{\underline{m}})$. Then there is a map $g\colon (D,\underline{\underline{d}}) \to (M,\underline{\underline{m}})$ with $(D,\overline{\underline{\underline{d}}})$ a small-faced disk whose restriction to $\partial D - |\underline{\underline{d}}|$ is h. We may assume that g is transverse to F. The restriction of g to each preimage arc of F is inessential in $(M,\underline{\underline{m}})$. If all of these arcs were inessential in F, then by doing cutting and pasting we could assume that this preimage is empty, so that h would be inessential in $(W,\underline{\underline{w}})$. Therefore some preimage arc must be essential in F, showing that F is not essential in $(M,\underline{\underline{m}})$.

We will now define useful boundary patterns for 3-manifolds. First, note that each element of $\underline{\underline{m}}$ is incompressible if and only if whenever D is an admissibly imbedded 1-faced disk in $(M,\underline{\underline{m}})$, ∂D bounds a disk in $|\underline{\underline{m}}|$ which does not meet

$J(\underline{\underline{m}})$ (where as defined in section 2.2, $J(\underline{\underline{m}}) = \bigcup_{F\in\underline{\underline{m}}} \partial F$). We say that $\underline{\underline{m}}$ is *useful* when the boundary of every admissibly imbedded small-faced disk in $(M, \underline{\underline{m}})$ bounds a disk D in ∂M such that $D \cap J(\underline{\underline{m}})$ is the cone on $\partial D \cap J(\underline{\underline{m}})$. This cone is empty if $(D, \underline{\underline{d}})$ is 1-faced, is an arc if it is 2-faced, and is a cone on three points if it is 3-faced.

Here are some examples of useful boundary patterns.

EXAMPLE 2.4.2. *Trivial (but important) examples*

The empty boundary pattern is always useful, and the boundary pattern $\overline{\underline{\underline{\emptyset}}}$ is useful if and only ∂M is incompressible.

EXAMPLE 2.4.3. *Fibered 3-manifolds*

The product of a small-faced disk with S^1 yields a boundary pattern on a product fibered torus which is not useful. With very few exceptions, however, the boundary patterns on admissibly fibered 3-manifolds are useful and even have useful completion; lemma 2.6.1 below details the failures of usefulness in the fibered case.

EXAMPLE 2.4.4. *3-manifolds containing no essential circles or arcs*

There are eight irreducible 3-manifolds $(M, \underline{\underline{m}})$ with nonempty boundary and useful boundary pattern that contain no essential arc or loop. Although these will not be needed for our work, we determine them here to provide another example. Suppose that $(M, \underline{\underline{m}})$ is a connected irreducible 3-manifold with useful boundary pattern which contains no essential path or loop. Then M is simply-connected, so is a 3-ball, and either $\underline{\underline{m}} = \{\partial M\}$ or each element of $\underline{\underline{m}}$ is a 2-disk. In the latter case, any two elements must meet, since otherwise there is an essential path connecting them, and must meet in a single arc or a circle, since otherwise there will be a 2-faced disk in $(M, \underline{\underline{m}})$ which violates the definition of usefulness. There cannot be more than four disks. For if there are three whose union is not all of ∂M, they must be configured as three of the faces of the tetrahedron, by usefulness, and since there are no violations of usefulness by 3-faced disks, a fourth disk that meets all three of them, each in an arc, can only be the other face. Thus there are only eight possibilities: (1) $(D^3, \underline{\underline{\emptyset}})$, (2) the suspensions of the small-faced disks, with boundary pattern consisting of the suspensions of their faces, and (3) $(T, \underline{\underline{t}})$, where T is a tetrahedron and $\underline{\underline{t}}$ is some nonempty subset of the set of faces of T.

EXAMPLE 2.4.5. *Boundary patterns with disjoint elements*

Much of our later work concerns 3-manifolds with boundary patterns $\underline{\underline{m}}$ which consist of disjoint incompressible submanifolds. Any such boundary pattern is useful, since the incompressibility prevents violations of usefulness by 1-faced disks, and the disjointness of the elements ensures that there are no admissibly imbedded 2- or 3-faced disks at all. If ∂M and $\overline{\partial M - |\underline{\underline{m}}|}$ are incompressible, then the completed boundary pattern $\overline{\underline{\underline{m}}}$ is also useful. To see this we first note that there are still no admissibly imbedded 3-faced disks, and that the incompressibility assumptions ensure that there are no violations of usefulness by 1-faced disks. Suppose that D is a 2-faced disk admissibly imbedded in $(M, \overline{\underline{\underline{m}}})$. Since ∂M is incompressible, ∂D bounds a disk D_0 in ∂M for which $\partial D_0 \cap J(\overline{\underline{\underline{m}}})$ consists of two points. Now D_0 cannot contain any circles of $J(\overline{\underline{\underline{m}}})$ since this would lead to a violation of incompressibility of the elements of $\overline{\underline{\underline{m}}}$. Since $J(\overline{\underline{\underline{m}}})$ consists of circles, $D_0 \cap J(\overline{\underline{\underline{m}}})$ must consist of a single arc, and D does not violate the usefulness condition.

When ∂M is compressible, violations of usefulness by 2-faced disks occur readily. For example if F is a disk with g holes, $g \geq 1$, and $(M, \underline{\underline{m}}) = (F \times \mathrm{I}, \{F \times \{1\}\})$, then there are numerous nonseparating admissible 2-faced disks in $(M, \overline{\underline{\underline{m}}})$. Some of our later work, however, will concern boundary patterns $\underline{\underline{a}}$ for which each element is either closed or is an annulus. In this case, unless $(M, \overline{\underline{\underline{a}}})$ is the product of a 2-faced disk with S^1, such a boundary pattern will have useful completion if and only if each free side is incompressible. For later reference we state this as a lemma and prove it here. It will apply to the boundary patterns associated to pared manifolds, which will be introduced in chapter 5.

LEMMA 2.4.6. *Let $(M, \underline{\underline{a}})$ be an irreducible 3-manifold such that the elements of $\underline{\underline{a}}$ are disjoint, incompressible, and each element either is closed or is an annulus. Assume that $(M, \overline{\underline{\underline{a}}})$ is not the product of a 2-faced disk with S^1. Then $\overline{\underline{\underline{a}}}$ is useful if and only if $(M, \underline{\underline{a}})$ has no compressible free side.*

PROOF. If some free side is compressible, then $\overline{\underline{\underline{a}}}$ is not useful. Conversely, suppose each free side is incompressible. Then each 1-faced admissible disk is parallel into an element of $\overline{\underline{\underline{a}}}$, so we need only check that there are no nontrivial 2- or 3-faced disks. Since the elements of $\underline{\underline{a}}$ are disjoint, the bound sides of any surface admissibly imbedded in $(M, \overline{\underline{\underline{a}}})$ must alternate between those contained in elements of $\underline{\underline{a}}$ and those in free sides of $(M, \underline{\underline{a}})$. Consequently, there are no admissibly imbedded 3-faced disks.

If D is an admissible 2-faced disk then its boundary is a compressible loop intersecting some annulus A of $\underline{\underline{a}}$ in a single arc α. If α begins and ends on the same boundary component of A, then α is parallel, in A, to an arc in ∂A. Since $\partial M - |\underline{\underline{a}}|$ is incompressible, $\partial D - \alpha$ must be parallel, in $\partial M - |\underline{\underline{m}}|$, to the same arc in ∂A, showing that D does not give a violation of usefulness. So we may assume that α joins distinct boundary components of A. Let C be the boundary of a regular neighborhood of $A \cup \partial D$ in ∂M. It is null-homotopic in M, since it is homotopic into $A - \partial D$. Also, it lies in a free side W of $(M, \underline{\underline{a}})$; since W is incompressible, C bounds a disk in W. It follows that W is an annulus, so $W \cup A$ is a compressible torus boundary component of M. Since M is irreducible, it must be a solid torus. Since D meets each of the annuli of $\overline{\underline{\underline{a}}}$ in a single arc, $(M, \overline{\underline{\underline{a}}})$ is the product of a 2-faced disk with S^1, contrary to hypothesis. □

EXAMPLE 2.4.7. *A useful boundary pattern containing a square*

In example 2.2.2 we gave an explicit example of a 3-manifold with boundary pattern $(S, \underline{\underline{s}})$ such that one of the elements of $\underline{\underline{s}}$ was a square N and the other two A_0 and A_1 were topological annuli. Using the notation of that example, we will now verify that if ∂S, A_0, and A_1 are incompressible, then $\underline{\underline{s}}$ is useful. This fact will be used in example 2.10.11 when we use $(S, \underline{\underline{s}})$ to construct characteristic submanifolds with certain properties.

Recall that α is an arc in ∂S running from A_0 to A_1 and meeting them only in its endpoints, so that $N = [0, 1] \times [-1, 1] \subset \partial S$ with $\alpha = [0, 1] \times \{0\}$ and $N \cap (A_0 \cup A_1) = \partial\alpha \times [-1, 1]$. Observe that $J(\underline{\underline{s}})$ consists of the boundary components C_0 and C_1 of A_0 and A_1 that do not meet N, together with the graph G obtained from the four sides of N by adding as edges the remainders of the two boundary components B_j of the A_j that do meet N. The circles B_0, B_1, C_0, and C_1 are essential in S.

Suppose that D is an admissibly imbedded small-faced disk in $(S, \underline{\underline{s}})$ which gives a violation of usefulness. Since ∂S is incompressible, ∂D bounds a disk D_0 in ∂S. If D is 1-faced, then ∂D must be disjoint from $J(\underline{\underline{s}})$. Since each component of $J(\underline{\underline{s}})$ contains an essential circle, $J(\underline{\underline{s}}) \cap D_0$ is empty and D cannot violate usefulness. Suppose that D is 2-faced, so that $D \cap J(\underline{\underline{s}})$ consists of two points a and b contained in edges of $J(\underline{\underline{s}})$. Since no component of $J(\underline{\underline{s}})$ can be separated by a single point, a and b are both contained in a single component of $J(\underline{\underline{s}})$. If they were contained in C_0 or C_1, then they would be connected by a single arc in $J(\underline{\underline{s}}) \cap D_0$ and D could not violate usefulness. So a and b both lie in G. Since C_0 and C_1 are essential, they cannot be contained in D_0, so $J(\underline{\underline{s}}) \cap D_0$ is contained in G. Since a and b together separate G, and $J(\underline{\underline{s}}) \cap D_0$ is not just a single arc, each of the edges $\alpha \times \{\pm 1\}$ contains one of a or b. But then, D_0 contains either B_0 or B_1, contradicting their essentiality. Finally, suppose that D is 3-faced, so that $\partial D \cap J(\underline{\underline{s}})$ contains three points. Every simple loop in $J(\underline{\underline{s}})$ that contains any of the points must contain exactly two of them (since the loop must enter and leave D_0 exactly once). The only such three-point subsets consist of a point in one of the edges $\alpha \times \{-1\}$ or $\alpha \times \{1\}$ together with points in the each of the two edges that form B_0 or in each of the two edges that form B_1. These are joined in G by a cone on three points, and this must be $J(\underline{\underline{s}}) \cap D_0$ since otherwise D_0 would contain one of the B_i. Therefore, no small-faced disk in $(S, \underline{\underline{s}})$ can give a violation of usefulness, so $(S, \underline{\underline{s}})$ is useful.

EXAMPLE 2.4.8. *Gluing 3-manifolds with boundary patterns*

Let $(M, \underline{\underline{m}})$ be a 3-manifold with boundary pattern. In this rare instance, we do not assume that M is connected, because we want to consider either gluing two manifolds together along elements of their boundary patterns, or identifying two elements of the boundary pattern of a single manifold. Suppose that F_1 and F_2 are two elements of $\underline{\underline{m}}$, with $F_1 \cap F_2 = \emptyset$. For each i, let $\underline{\underline{f_i}}$ be the boundary pattern consisting of the components of $F_i \cap G$ as G ranges over the elements of $\underline{\underline{m}} - \{F_1, F_2\}$. Let $h\colon (F_1, \underline{\underline{f_1}}) \to (F_2, \underline{\underline{f_2}})$ be an admissible homeomorphism, and form the quotient manifold N from M by identifying each x with $h(x)$.

Each time an element k_1 of $\underline{\underline{f_1}}$ is identified to an element k_2 of $\underline{\underline{f_2}}$, there are elements G_1 and G_2 in $\underline{\underline{m}}$ (possibly $G_1 = G_2$) such that k_i is a component of $G_i \cap F_i$. Define a boundary pattern $\underline{\underline{n}}$ on N as follows. Start with the disjoint union of the elements of $\underline{\underline{m}} - \{F_1, F_2\}$, and if x lies in a component of such a $G_1 \cap F_1$, identify x with the copy of $h(x)$ that lies in the component G_2 that contains the image of $G_1 \cap F_1$. The collection of components of the quotient space is $\underline{\underline{n}}$. In particular, the common image of k_1 and k_2 is properly imbedded in the component that contains the images of G_1 and G_2. Therefore the surface $(F, \underline{\underline{f}})$ in $(N, \underline{\underline{n}})$ which is the image of $(F_1, \underline{\underline{f_1}})$ and $(F_2, \underline{\underline{f_2}})$ is properly and admissibly imbedded.

In this construction, usefulness is preserved. Since this provides a general method for constructing new useful boundary patterns from old, we prove this fact as a lemma. It will be applied in example 2.10.11 below to give examples illustrating some cases of Main Topological Theorem 2.

LEMMA 2.4.9. *Let $(N, \underline{\underline{n}})$ be formed from $(M, \underline{\underline{m}})$ by identifying faces $(F_1, \underline{\underline{f_1}})$ and $(F_2, \underline{\underline{f_2}})$ as described above. If $\underline{\underline{m}}$ is useful, then $\underline{\underline{n}}$ is useful.*

PROOF. Let $(D, \underline{\underline{d}})$ be an admissible small-faced disk imbedded in $(N, \underline{\underline{n}})$. We must show that ∂D bounds a disk D_0 in ∂N, such that $D_0 \cap J(\underline{\underline{n}})$ is the cone on

$\partial D_0 \cap J(\underline{\underline{n}})$. The existence of such a D_0 is not altered by performing surgery on D along circles in the interior of D, nor by admissible isotopy of D. So when we perform these operations, we can call the result D as well.

By admissible isotopy, put D transverse to F. If there are circle intersections, perform surgery on D, starting from an innermost intersection circle, to assume that D intersects F only in arcs.

If D is not disjoint from F, consider an intersection arc k of $D \cap F$ which is outermost on D. It is the frontier of a disk E in D which meets F only in k. With the boundary pattern consisting of k and the components of the intersections of E with elements of $\underline{\underline{n}}$, E is admissibly imbedded in $(M, \underline{\underline{m}})$. It will be a small-faced, except in the case when $(D, \underline{\underline{d}})$ was 3-faced, k connects two different faces of $(D, \underline{\underline{d}})$, and E contains the third face of $(D, \underline{\underline{d}})$ so is a square. If this happens, then $\overline{D - E}$ either meets F only in k, in which case we take it as E, or contains more intersection arcs, in which case we use an outermost disk cut off by one of those arcs as E. So we can always choose E to be small-faced.

Since $\underline{\underline{m}}$ was useful, ∂E bounds a disk E_0 in ∂M, such that $E_0 \cap J(\underline{\underline{m}})$ is the cone on $\partial E_0 \cap J(\underline{\underline{m}})$. There is an admissible isotopy of D that pushes $\partial D \cap E$ across $E_0 \cap \partial N$. It changes k, and any other intersections of D with $F \cap E_0$, into intersection that are circles. After performing surgery to remove these circle intersections, the number of intersections of D with F has been reduced. Eventually, we make D disjoint from F.

Since $\underline{\underline{m}}$ was useful, ∂D bounds a disk D_0 in ∂M, such that $D_0 \cap J(\underline{\underline{m}})$ is the cone on $\partial D_0 \cap J(\underline{\underline{m}})$. The latter condition ensures that D_0 does not meet either of the F_i, so D_0 lies in ∂N after the identifications. □

The converse of lemma 2.4.9 is false. An extreme example arises from taking two handlebodies with boundary pattern $\underline{\underline{\bar{\emptyset}}}$ and identifying their boundaries to form a closed 3-manifold. A more amusing example is to stack two cubes one on top of the other, where the bottom cube has complete boundary pattern consisting of the top, the left- and right-hand faces, and the remainder, and the top cube has complete boundary pattern consisting of the bottom, the front and back faces, and the remainder. The boundary patterns on the original cubes are not useful, but $\underline{\underline{n}}$ is the useful boundary pattern consisting of two disks with a common boundary circle that looks like the seam of a baseball. Examples resulting in more complicated useful boundary patterns can be constructed by gluing together two non-useful boundary patterns on two 3-balls.

EXAMPLE 2.4.10. *Tiling a submanifold of ∂M*

Recall the boundary pattern $\underline{\underline{m}}$ obtained by tiling a submanifold $G = \overline{\partial M - F}$, using the dual cell-complex $\mathcal{C}$ of a triangulation T of G as in example 2.3.1. It will be useful provided that

(1) F is incompressible,
(2) the intersection of each 2-simplex of T with ∂G is a vertex or a 1-simplex, and
(3) any three vertices such that any two bound a 1-simplex of T are the vertices of a 2-simplex of T.

Many triangulations satisfy conditions (2) and (3), for example any triangulation that is the barycentric subdivision of another triangulation.

To see that $\underline{\underline{m}}$ is useful if conditions (1), (2), and (3) are satisfied, we note first that by condition (1) all elements of $\underline{\underline{m}}$ are incompressible, so there are no violations of usefulness by 1-faced disks. If two elements of $\underline{\underline{m}}$ meet, then at least one is a disk, and their intersection is a single arc, so again using condition (1) there are no violations of usefulness by 2-faced disks. Consider an admissibly imbedded 3-faced disk E. Suppose first that its faces lie in three 2-cells of $\mathcal{C}$. Regard each of the 2-cells as having a boundary pattern consisting of the 1-cells of $\mathcal{C}$ in its boundary. The face of E in each 2-cell is admissibly isotopic in that 2-cell to a union of two arcs each of which is the portion of a 1-simplex of T that runs from its barycenter to one of its endpoints. By condition (3), the union of these six arcs is the boundary of a 2-simplex D of T. It is possible that D meets ∂G; if so, replace it by a slightly smaller disk that does not. Then D meets $J(\underline{\underline{m}})$ in a cone on three points, and since E is admissibly isotopic so that $\partial E = \partial D$, E does not give a violation of usefulness. Suppose now that one of the faces of E lies in F. Then the other two lie in two 2-cells C_1 and C_2 of $\mathcal{C}$. Now C_1 and C_2 both meet ∂G, and $C_1 \cap C_2$ is a single 1-cell α, since it is the dual 1-cell to a 1-simplex of the triangulation, and two such intersection cells would correspond to distinct 1-simplices with common endpoints. At least one endpoint of α is the barycenter of a 2-simplex of the triangulation, so lies in the interior of G. The other endpoint of α must lie in ∂G, for if it were in the interior, then the 1-simplex of the triangulation dual to α would meet ∂G in its endpoints, (since these are the common intersection point of the simplices of the subdivision that form C_1 and C_2), giving a violation of condition (2). Since α has an endpoint in ∂G, the union of the two faces of E in G is parallel in $C_1 \cup C_2$ into ∂G. Now, condition (1) shows that ∂E bounds a disk D in ∂M, and the face of E in F is parallel in D into ∂F. The intersection of $\alpha \cup \partial F$ with D is a cone on three points, so again E does not violate usefulness.

Note that if one now subdivides $\mathcal{C}$ as in the final two constructions of example 2.3.1 of section 2.3, the resulting cell complex is also dual to a triangulation satisfying conditions (1), (2), and (3), so it yields a useful boundary pattern $\underline{\underline{m}}$ for which $\mathcal{H}(M, \underline{\underline{m}}) \cong \mathcal{H}(M \text{ rel } F)$.

2.5. The classical theorems

We will now examine how some of the fundamental results of the theory of 3-manifolds extend to manifolds with boundary pattern. The Loop Theorem of Papakyriakopoulos can be formulated as saying that ∂M is incompressible if and only if for every admissible map $f\colon (D, \underline{\underline{d}}) \to (M, \underline{\underline{\emptyset}})$ where $(D, \underline{\underline{d}})$ is a 1-faced disk, there is a map $g\colon D \to \partial M$ such that $g|_{\partial D} = f|_{\partial D}$. Notice that f need not be an embedding. That usefulness is a natural generalization of boundary irreducibility is seen in the following version of the Loop Theorem. It is given as proposition 2.1 in [**58**].

THEOREM 2.5.1. (Loop Theorem) *Let $(M, \underline{\underline{m}})$ be a 3-manifold with boundary pattern. Then $\underline{\underline{m}}$ is useful if and only if for any admissible map $f\colon (D, \underline{\underline{d}}) \to (M, \underline{\underline{m}})$, with $(D, \underline{\underline{d}})$ a small-faced disk, there exists a map $g\colon D \to \partial M$ so that $g|_{\partial D} = f|_{\partial D}$ and $g^{-1}(J(\underline{\underline{m}}))$ is the cone on $g^{-1}(J(\underline{\underline{m}})) \cap \partial D$.*

Using the Loop Theorem, we can reformulate the definition of usefulness in terms of maps of small-faced disks.

PROPOSITION 2.5.2. *Let* $\underline{\underline{m}}$ *be a boundary pattern on an irreducible* 3*-manifold* M. *Then* $\underline{\underline{m}}$ *is useful if and only if every admissible map of a small-faced disk into* $(M, \underline{\underline{m}})$ *is admissibly homotopic to a constant map.*

PROOF. Suppose that $(M, \underline{\underline{m}})$ is useful, and let $f\colon (D, \underline{\underline{d}}) \to (M, \underline{\underline{m}})$ be an admissible map of a small-faced disk. By the Loop Theorem 2.5.1, there exists a map $g\colon D \to \partial M$ so that $g|_{\partial D} = f|_{\partial D}$, and $g^{-1}(J(\underline{\underline{m}}))$ is the cone on $g^{-1}(J(\underline{\underline{m}})) \cap \partial D$. Regard the 3-ball as the cone on D, that is, the quotient space $D \times \mathrm{I}/(x, 1) \sim (y, 1)$ for all $x, y \in D$. Define a map H from the cone on D into M as follows. On $D \times \{0\}$, it is f. Using g, extend this to the cone on ∂D in such a way that the preimage of $J(\underline{\underline{m}})$ is the cone on $f^{-1}(J(\underline{\underline{m}})) \cap \partial D$. Since M is irreducible, $\pi_3(M) = 0$, so the map may be extended to the cone on D; this may be done in such a way that the preimage of ∂M is exactly the cone on ∂D. Now, let q be the quotient map from $D \times \mathrm{I}$ to the cone on D, then Hq is a homotopy from f to a constant map through admissible maps.

For the converse, we will show that if $\underline{\underline{m}}$ is not useful, then there exists an admissible map of a small-faced disk into $(M, \underline{\underline{m}})$ which is not admissibly homotopic to a constant map. When $\underline{\underline{m}}$ is not useful, the Loop Theorem 2.5.1 yields an admissible map $f\colon (D, \underline{\underline{d}}) \to (M, \underline{\underline{m}})$, where $(D, \underline{\underline{d}})$ is a small-faced disk, for which there is no map $g\colon D \to \partial M$ so that $g|_{\partial D} = f|_{\partial D}$ and $g^{-1}(J(\underline{\underline{m}}))$ is the cone on $g^{-1}(J(\underline{\underline{m}})) \cap \partial D$. Suppose for contradiction that f is admissibly homotopic to a constant map. Then there would be a homotopy of f through admissible maps to a constant map, say $K\colon D \times \mathrm{I} \to M$. It induces a map H from the cone on D, and the restriction of H to the cone on ∂D would be a map g of the kind already excluded. □

As discussed above, the Loop Theorem of Papakyriakopoulos is often used in procedures where the preimage of a surface under a map between 3-manifolds is being simplified. To extend these to 3-manifolds with boundary patterns, the following analogue of Dehn's Lemma is needed. It is given as lemma 4.2 of [**58**].

LEMMA 2.5.3. (Compression Lemma) *Let* $(M, \underline{\underline{m}})$ *be a* 3*-manifold with useful boundary pattern, and let* $(F, \underline{\underline{f}})$ *be an admissibly imbedded surface in* M *with* $F \cap \partial M = \partial F$, *none of whose components is a small-faced disk. Then* $(F, \underline{\underline{f}})$ *is inessential in* $(M, \underline{\underline{m}})$ *if and only if there is an admissibly imbedded disk* $(D, \underline{\underline{d}})$ *in* $(M, \underline{\underline{m}})$ *such that* $(D, \underline{\underline{d}})$ *is a small-faced disk and* $D \cap F$ *is a face of* $(D, \underline{\underline{d}})$ *which is an essential arc or circle in* F.

Using the Compression Lemma, Johannson deduces the following general theorem for simplifying the preimage of an essential surface. It is lemma 4.4 of [**58**].

THEOREM 2.5.4. (Essential Preimage Theorem) *Let* $(M_i, \underline{\underline{m_i}})$ *be irreducible and aspherical* 3*-manifolds with useful boundary patterns. Let* $(F, \underline{\underline{f}})$ *be an essential surface imbedded in* $(M_2, \underline{\underline{m_2}})$, *with* $F \cap \partial M_2 = \partial F$, *such that no component of* $(F, \underline{\underline{f}})$ *is a* 2*-sphere or a small-faced disk. Then any admissible map* $f\colon (M_1, \underline{\underline{m_1}}) \to (M_2, \underline{\underline{m_2}})$ *is admissibly homotopic to a map* g *such that* $g^{-1}(F)$ *is an essential surface in* $(M_1, \underline{\underline{m_1}})$, *no component of which is a* 2*-sphere or a small-faced disk. Moreover:*

(i) *If in addition* $\overline{\underline{\underline{m_1}}}$ *is useful, then* g *may be chosen so that* $g^{-1}(F)$ *is essential in* $(M_1, \overline{\underline{\underline{m_1}}})$*, and no component of* $g^{-1}(F)$ *is a 2-sphere or a small-faced disk in* $(M_1, \overline{\underline{\underline{m_1}}})$*.*

(ii) *Suppose that* $f^{-1}(F)$ *is already an admissible surface in* $(M_1, \overline{\underline{\underline{m_1}}})$*,* $\overline{\underline{\underline{m_1}}}$ *is useful,* $(N, \underline{\underline{n}})$ *is an essential codimension-zero submanifold in* $(M_1, \overline{\underline{\underline{m_1}}})$ *such that* $f^{-1}(F)$ *is disjoint from* N*, and no component of the frontier of* N *is a 2-sphere or a small-faced disk. Then the homotopy of* f *may be chosen to be constant on* N*.*

An example here will illustrate one of the reasons that all maps whose domains have no essential 1-submanifolds are considered to be essential. Let $(S, \underline{\underline{s}})$ be a square, and let $(M_2, \underline{\underline{m_2}})$ be $(S, \underline{\underline{s}}) \times \mathrm{S}^1$. Let D be a small disk that meets one of the sides of S in an arc α in the interior of the side, so that $\underline{\underline{d}} = \{\alpha\}$ is a boundary pattern on D. Put $(M_1, \underline{\underline{m_1}}) = (D, \underline{\underline{d}}) \times \mathrm{S}^1$. The inclusion map $i\colon (M_1, \underline{\underline{m_1}}) \to (M_2, \underline{\underline{m_2}})$ is admissible. Let $F = (S, \underline{\underline{s}}) \times \{p\}$ for some $p \in \mathrm{S}^1$, an essential surface in $(M_2, \underline{\underline{m_2}})$. Then $i^{-1}(F) = D \times \{p\}$, and this preimage surface cannot be eliminated or simplified by admissible homotopy of i. So to make the Essential Preimage Theorem true as stated here, this copy of $(D, \underline{\underline{d}})$ is allowed as an essential surface in $(M_1, \underline{\underline{m_1}})$. Note that according to statement (i) in the Essential Preimage Theorem, such examples cannot arise when $\underline{\underline{m_1}}$ has useful completion.

The classical Baer-Nielsen theorem asserts that if $f\colon F \to G$ is a homotopy equivalence between two compact surfaces (other than disks) such that $f(\partial F) = \partial G$, then f is properly homotopic to a homeomorphism. We will make frequent use of the following generalization of the Baer-Nielsen theorem (see proposition 3.3 of [**58**]).

THEOREM 2.5.5. (Baer-Nielsen Theorem) *Let* $(F, \underline{\underline{f}})$ *and* $(G, \underline{\underline{g}})$ *be connected surfaces with complete boundary patterns. Suppose that* $(F, \underline{\underline{f}})$ *is not a 1-faced disk or the 2-sphere, and that* G *is not the projective plane. Then any essential map* $f\colon (F, \underline{\underline{f}}) \to (G, \underline{\underline{g}})$ *is admissibly homotopic to a covering map. If the restriction of* f *to* ∂F *is a local homeomorphism, then the homotopy may be chosen to be constant on* ∂F*.*

Waldhausen's generalization of the Baer-Nielsen theorem to dimension 3 extends to manifolds with complete and useful boundary patterns, as given in proposition 3.4 of [**58**]:

THEOREM 2.5.6. (Waldhausen's Theorem) *Let* $(M, \underline{\underline{m}})$ *and* $(N, \underline{\underline{n}})$ *be connected irreducible 3-manifolds with complete and useful boundary patterns. Suppose that* M *has nonempty boundary and* $(M, \underline{\underline{m}})$ *is not a 3-ball with one or two bound sides. Then any essential map* $f\colon (M, \underline{\underline{m}}) \to (N, \underline{\underline{n}})$ *is admissibly homotopic to a covering map. If the restriction of* f *to* ∂M *is a covering map, then the homotopy may be chosen to be constant on* ∂M*.*

A strong property of essential surfaces is given in the next result. Two disjoint admissibly imbedded surfaces $(F, \underline{\underline{f}})$ and $(G, \underline{\underline{g}})$ in a 3-manifold $(M, \underline{\underline{m}})$ are *admissibly parallel* if there is a component of the complement of $F \cup G$ whose closure, with the proper boundary pattern, is a product I-bundle whose lids are F and G. The next result is proposition 19.1 of [**58**]. It is stated there for complete boundary pattern, but since we assume that $\underline{\underline{f}}$ and $\underline{\underline{g}}$ are complete, one can apply it to $(M, \overline{\underline{\underline{m}}})$ to obtain the version stated here.

THEOREM 2.5.7. (Parallel Surfaces Theorem) *Let $(M, \underline{\underline{m}})$ be an irreducible 3-manifold with boundary pattern whose completion is useful, and let $(F, \underline{\underline{f}})$ and $(G, \underline{\underline{g}})$ be connected essential surfaces, with complete boundary patterns, admissibly and essentially imbedded in $(M, \underline{\underline{m}})$, and properly imbedded in M. Assume that $(G, \underline{\underline{g}})$ is admissibly homotopic into $(F, \underline{\underline{f}})$. Then $(G, \underline{\underline{g}})$ is admissibly isotopic into $(F, \underline{\underline{f}})$. Moreover, if F and G are disjoint, then $(G, \underline{\underline{g}})$ is admissibly parallel to $(F, \underline{\underline{f}})$.*

2.6. Exceptional fibered 3-manifolds

We gave a brief description of fibered 3-manifolds in section 2.1, and in examples 2.2.3 and 2.2.4 of section 2.2, we defined their admissible fiberings. In this section, we will discuss the exceptional fibered manifolds. As detailed in lemma 2.6.1, the cases listed below as (EF1)-(EF5) include all the admissibly fibered manifolds with complete boundary pattern which are not irreducible or whose boundary patterns are not useful. The remaining cases, (EIB) and (ESF), are those which contain a horizontal square, annulus, or torus (that is, an imbedded square, annulus, or torus which meets all fibers transversely).

Many of the exceptional fibered manifolds admit more than one isotopy class of fibering, in fact some have both I- and Seifert fiberings. By the Unique Fibering Theorem 2.8.1 below, the exceptional manifolds include all manifolds with complete boundary pattern which admit more than one isotopy class of admissible fibering.

All of the manifolds in (EF4), (EF5), and (ESF) are closed, and consequently are not relevant for most of our later work, but we include them here in order to give a fuller exposition of Johannson's theory. Recall that the terms annulus and Möbius band refer to 3-manifolds which carry the boundary pattern $\overline{\emptyset}$.

Exceptional Fibered Manifolds: An orientable 3-manifold $(V, \underline{\underline{v}})$ with an admissible fibering as an I-bundle or Seifert-fibered space is called an *exceptional fibered manifold* if its completion $(V, \overline{\underline{\underline{v}}})$ is one of the following:

(EF1) An I-bundle over a small-faced disk.

(EF2) The S^1-bundle over an i-faced disk, $i = 2, 3$, or a Seifert-fibered space over a 1-faced disk with at most one exceptional fiber.

(EF3) An I-bundle over the 2-sphere or projective plane.

(EF4) A Seifert-fibered space with the 2-sphere as quotient surface and at most three exceptional fibers.

(EF5) A Seifert-fibered space with the projective plane as quotient surface and at most one exceptional fiber.

Certain other manifolds are frequently exceptional cases because they admit a horizontal square, annulus, or torus. These are:

(EIB) A manifold $(V, \underline{\underline{v}})$ such that $(V, \overline{\underline{\underline{v}}})$ can be admissibly fibered as an I-bundle over the square, annulus, Möbius band, torus, or Klein bottle.

(ESF) A closed 3-manifold which can be obtained by gluing two I-bundles over the torus or Klein bottle together along their boundaries.

Note that some of the manifolds in (EIB) may be Seifert-fibered; the condition in (EIB) only says that the completed boundary pattern *can be given* some admissible fibering as an I-bundle. If $(V, \overline{\underline{\underline{v}}})$ admits an I-fibering over the square, annulus,

Möbius band, torus, or Klein bottle, then $(V, \underline{\underline{v}})$ admits a (possibly nonisotopic) one over the same surface, unless V fibers over the Klein bottle and $\underline{\underline{v}} = \emptyset$, or V fibers over the torus and either $\underline{\underline{v}} = \emptyset$ or $\underline{\underline{v}}$ is a single torus boundary component. To see this, one notes that $\underline{\underline{v}}$ is obtained from $\overline{\underline{\underline{v}}}$ by removing a collection of elements that are pairwise disjoint subsets of ∂V, and considers the five bundles case-by-case.

The manifolds in cases (EF1) and (EF2) have boundary patterns that are not useful, and those in case (EF3) and some of those in cases (EF4) and (EF5) are not irreducible. These are the only such fibered manifolds, as we now check.

LEMMA 2.6.1. *Let $(V, \underline{\underline{v}})$ be an admissibly fibered I-bundle or Seifert-fibered 3-manifold.*

(i) *If $(V, \underline{\underline{v}})$ is not an exceptional case (EF1) or (EF2), then $\overline{\underline{\underline{v}}}$ is useful.*
(ii) *If $(V, \underline{\underline{v}})$ is not an exceptional case (EF3), (EF4), or (EF5), then V is Haken.*

PROOF. Without loss of generality we may assume that $\underline{\underline{v}}$ is complete. Assume first that $\underline{\underline{v}}$ is not useful. The Loop Theorem 2.5.1 yields an imbedded small-faced disk D whose boundary does not bound a disk D_0 in ∂V whose intersection with $J(\underline{\underline{v}})$ is the cone on $\partial D_0 \cap J(\underline{\underline{v}})$.

Suppose that $(V, \underline{\underline{v}})$ is an I-bundle. If D is 1-faced, then ∂D lies in a side of $(V, \underline{\underline{v}})$, since the lids are incompressible. Since ∂D does not bound a disc in the side, the side must be a compressible annulus, so V is an I-bundle over the 1-faced disk as in (EF1).

Suppose D is 2-faced. If both faces lie in sides, then V is an I-bundle over a 2-faced disk as in (EF1). Suppose one face lies in a lid. If the bundle is a product, then the face lying in a side G of V is parallel into the lid (in a square, every arc with both endpoints in one edge is parallel into that edge, and similarly for an arc in an annulus with both endpoints in one boundary component), and again using incompressibility of the lid, D_0 must exist. Suppose the bundle is twisted. Lift D to a 2-faced admissible disk $\widetilde{D}$ in the product I-bundle $\widetilde{V}$ which double covers V. This cover is not an I-bundle over a disk, since then $\pi_1(V)$ would be of order 2, so by the product case already completed, $\widetilde{D}$ bounds a disk $\widetilde{D}_0$ in $\partial\widetilde{V}$ whose intersection with $J(\widetilde{\underline{\underline{v}}})$ is the cone on $J(\widetilde{\underline{\underline{v}}}) \cap \partial\widetilde{D}_0$. Now $\widetilde{D}_0$ cannot meet its translate under the covering transformation, since its boundary does not meet the boundary of the translate and if one were contained in the other, the covering transformation would have a fixed point. So $\widetilde{D}_0$ descends to D_0 in ∂V, showing that D did not violate usefulness.

Suppose that D is 3-faced. If it has a face in a lid, it meets two sides and D_0 exists, much as for the 2-faced case. If it meets three sides of V, then V must be the I-bundle over the triangle as in (EF1).

Assume now that V is Seifert-fibered. If ∂D is not essential in ∂V, then it bounds a disk D_0 in ∂V. Since D has at most three faces, and $J(\underline{\underline{v}})$ consists of parallel circles in tori, it follows that $D_0 \cap J(\underline{\underline{v}})$ is the cone on $\partial D_0 \cap J(\underline{\underline{v}})$. So ∂D is essential in ∂V. Therefore ∂V is compressible, so V is a solid torus. Cutting V along D yields a product $(D, \underline{\underline{d}}) \times (\mathrm{I}, \emptyset)$, where $\underline{\underline{d}}$ is the boundary pattern on D. If D is 1-faced, then V is as in (EF2). If $i = 2$ or $i = 3$, then the identifications of ends producing V from $D \times \mathrm{I}$ must preserve the faces of D, since otherwise there would be adjacent faces of D lying in the same element of $\underline{\underline{v}}$. Again, $(V, \underline{\underline{v}})$ is as in (EF2).

Assume now that $(V, \underline{\underline{v}})$ is not Haken. If $(V, \underline{\underline{v}})$ is Seifert-fibered, then from [**126**] (see the discussion in chapter 8 of [**103**], or Lemma VI.15 and Remark VI.16 of [**54**]), V must be either of type (EF4) or type (EF5). If $(V, \underline{\underline{v}})$ is an I-bundle with $\pi_2(V) \neq 0$, it must be an I-bundle over a 2-sphere or a projective plane, as in (EF3). If $\pi_2(V)=0$, then $(V, \underline{\underline{v}})$ is irreducible and has nonempty boundary, so it is Haken. □

2.7. Vertical and horizontal surfaces and maps

In the study of fibered manifolds, an important role is played by surfaces which either are unions of fibers or are transverse to all the fibers of a fibered manifold. Assume that V carries a fixed structure as an I-bundle or Seifert-fibered space, with quotient surface B. Let $p\colon V \to B$ be the quotient map, and let G be a manifold. A map $g\colon G \to V$ is called *vertical* if its image is a union of nonexceptional fibers. It is called *horizontal* if $g^{-1}(\partial V) = \partial G$ and g is transverse to the fibers. When g is horizontal, pg is a branched covering map. Branch points can occur only if V is Seifert-fibered, and then they lie over the exceptional points of B. A submanifold of V is called vertical or horizontal when its inclusion map is vertical or horizontal. In a 3-manifold, a vertical submanifold can be 1-, 2-, or 3-dimensional, but a horizontal submanifold can only be 2-dimensional.

In general, an essential surface in a fibered manifold is isotopic to one which is horizontal or vertical. The next result, proposition 5.6 in [**58**], makes this precise.

THEOREM 2.7.1. (Vertical-horizontal Theorem) *Let $(M, \underline{\underline{m}})$ be an I-bundle or Seifert-fibered space, with fixed admissible fibration, but not one of the exceptional fibered manifolds (EF1)-(EF5). Let $(G, \underline{\underline{g}})$ be an essential surface in $(M, \underline{\underline{m}})$ with complete boundary pattern, none of whose components is a 2-sphere or a small-faced disk. Then G is admissibly isotopic to a vertical surface or to a horizontal surface. If, in addition, B is any element of $\underline{\underline{m}}$ which is not a lid of $(M, \underline{\underline{m}})$, such that $B \cap G$ is either horizontal or vertical, then the admissible isotopy of G may be chosen to be constant on $B \cap G$.*

As a first application of the Vertical-horizontal Theorem, we classify the essential surfaces with complete boundary pattern in I-bundles. Let $(V, \underline{\underline{v}})$ be an admissibly fibered I-bundle over $(B, \underline{\underline{b}})$, such that $(V, \underline{\underline{v}})$ is not an exceptional case (EF1)-(EF5). Let $(F, \underline{\underline{f}})$ be a surface with complete boundary pattern, admissibly and essentially imbedded in $(V, \underline{\underline{v}})$, and which is not a 2-sphere or a small-faced disk. By the Vertical-horizontal Theorem 2.7.1, $(F, \underline{\underline{f}})$ is admissibly isotopic to a vertical or horizontal surface. If F meets a lid of $(V, \underline{\underline{v}})$, then this surface must be vertical, an annulus or Möbius band. If F is disjoint from the lids of $(V, \underline{\underline{v}})$, then the surface must be horizontal. Suppose first that $(V, \underline{\underline{v}})$ is a product bundle $(B, \underline{\underline{b}}) \times (\mathrm{I}, \underline{\underline{\emptyset}})$. Then $(F, \underline{\underline{f}})$ is admissibly homotopic into $B \times \{1/2\}$. By the Parallel Surfaces Theorem 2.5.7, $(F, \underline{\underline{f}})$ is admissibly isotopic to $B \times \{1/2\}$. Suppose now that $(V, \underline{\underline{v}})$ is a twisted I-bundle. If F is orientable, then it lifts to the 2-fold covering of $(V, \underline{\underline{v}})$ by a product bundle, where it is admissibly homotopic to a lid. This admissible homotopy projects to an admissible homotopy of $(F, \underline{\underline{f}})$ into the lid of $(V, \underline{\underline{v}})$, so by the Parallel Surfaces Theorem 2.5.7, it is admissibly parallel to the lid. Suppose F is nonorientable. Choose a small regular neighborhood of F that meets each I-fiber J in segments which are neighborhoods of the points of $F \cap J$. The frontier $\widetilde{F}$ of this regular neighborhood is orientable and essential, so is parallel to the lid and its projection to the standard cross-section of the I-fibering is a double covering of B. Therefore F meets each fiber in a single point, so is isotopic to the standard cross-section.

For squares, annuli, and tori which are not imbedded, propositions 5.10 and 5.13 (and the remark following it) of [**58**] give the following verticalization result for maps:

THEOREM 2.7.2. (Essential Singular Annulus and Torus Theorem) *Let $(M, \underline{\underline{m}})$ be an I-bundle or Seifert-fibered space with a fixed admissible fibration, but not one of the exceptional fibered manifolds (EF1)-(EF2). Suppose $f\colon (T, \underline{t}) \to (M, \underline{\underline{m}})$ is an essential map of a square or annulus into $(M, \underline{\underline{m}})$. Then either*

(i) *there exists an admissible homotopy which makes f vertical, or*
(ii) *$(M, \underline{\underline{m}})$ is one of the exceptions (EIB).*

Moreover,

(iii) *if k is any bound side of $(T, \underline{t})$ which is mapped by f into a lid of $(M, \underline{\underline{m}})$, then f is admissibly homotopic to a vertical map by a homotopy which is constant on k.*

Suppose $f\colon T \to M$ is an essential map of a torus into $(M, \underline{\underline{m}})$, and that $(M, \underline{\underline{m}})$ is not one of the exceptional fibered manifolds (EF4). Then

(iv) *there exists an admissible Seifert fibration of $(M, \underline{\underline{m}})$, possibly not isotopic to the original one, and an admissible homotopy which makes f vertical with respect to this fibration.*

2.8. Fiber-preserving maps

The most powerful tool for analyzing admissible homotopy equivalences between 3-manifolds with useful boundary patterns is Johannson's Classification Theorem 2.11.1, which we will state in section 2.9. Roughly speaking, it says that an admissible homotopy equivalence is admissibly homotopic to one which fails to be a homeomorphism only on certain fibered submanifolds. The results of the present section then allow one, apart from exceptional cases, to deform the map to be fiber-preserving on these fibered submanifolds. A fiber-preserving map between fibered 3-manifolds is very closely related to the map it induces between the quotient 2-manifolds, thus some questions about homotopy equivalences between 3-manifolds can eventually be reduced to 2-dimensional problems which are much easier to analyze. This will be the philosophy used in chapter 10 below. In preparation for this and other parts of our work, we present in this section the necessary results on fiber-preserving maps.

We will first give a couple of uniqueness results that ensure that apart from exceptional cases, an admissible homeomorphism between admissibly fibered manifolds is admissibly isotopic to a fiber-preserving homeomorphism. More difficult are results which guarantee that essential maps are admissibly homotopic to fiber-preserving maps. We give several such results, whose ultimate objective is the Fiber-preserving Self-map Theorem 2.8.6. It says that if $f\colon (V, \underline{v}) \to (V, \underline{v})$ is an essential self-map of an admissibly fibered irreducible I-bundle or Seifert-fibered space, such that $\overline{\underline{v}}$ is useful and nonempty, then under minimal hypotheses on f and $(V, \underline{v})$, f is admissibly homotopic to a map which is fiber-preserving with respect to the given fibering.

The first result shows that the fibering of a fibered manifold is usually unique up to isotopy. It follows from corollary 5.9 of [**58**].

THEOREM 2.8.1. (Unique Fibering Theorem) *Suppose that* $(M_1, \underline{\underline{m_1}})$ *and* $(M_2, \underline{\underline{m_2}})$ *are each an I-bundle or Seifert-fibered space with a fixed admissible fibration, but neither is a solid torus with* $\underline{\underline{\overline{m_i}}} = \underline{\underline{\overline{\emptyset}}}$*, or an exceptional fibered manifold (EF3)-(EF5), (EIB), or (ESF). Then every admissible homeomorphism* $h\colon (M_1, \underline{\underline{m_1}}) \to (M_2, \underline{\underline{m_2}})$ *is admissibly isotopic to a fiber-preserving homeomorphism. Moreover,*

(i) *the conclusion holds if the* $(M_i, \underline{\underline{m_i}})$ *are each admissibly I-fibered as exceptional cases (EIB), provided that* h *maps lids to lids, and*
(ii) *if* $(M_1, \underline{\underline{m_1}})$ *is an I-bundle and* $h\colon (M_1, \underline{\underline{m_1}}) \to (M_1, \underline{\underline{m_1}})$ *is the identity on one lid, then the isotopy may be chosen to be constant on this lid.*

Applied to the identity map, the Unique Fibering Theorem 2.8.1 shows that apart from exceptional cases any two Seifert fiberings on $(M, \underline{\underline{m}})$ are admissibly isotopic. We will now give another convenient criterion, which provides a relative version of unique fibering. Also, when one of $\underline{\underline{m_i}}$ (hence also the other) contains an annulus, any admissible homeomorphism will satisfy the hypothesis of theorem 2.8.2. In particular, it applies to some of the exceptional cases excluded in the Unique Fibering Theorem 2.8.1, such as the Seifert manifolds which can be admissibly I-fibered over the annulus or Möbius band.

THEOREM 2.8.2. (Seifert Fibering Isotopy Criterion) *Suppose that each of* $(M_1, \underline{\underline{m_1}})$ *and* $(M_2, \underline{\underline{m_2}})$ *is a Seifert-fibered space with nonempty boundary and with fixed admissible fibration, but that neither* $(M_i, \underline{\underline{\overline{m_i}}})$ *is a solid torus with* $\underline{\underline{\overline{m_i}}} = \underline{\underline{\overline{\emptyset}}}$*. Let* $f\colon (M_1, \underline{\underline{m_1}}) \to (M_2, \underline{\underline{m_2}})$ *be an admissible homeomorphism, and suppose that for some nonexceptional fiber* τ *in* M_1*,* $f(\tau)$ *is homotopic in* M_2 *to a nonexceptional fiber. Then* f *is admissibly isotopic to a fiber-preserving homeomorphism. If* f *is already fiber-preserving on some union* F *of elements of* $\underline{\underline{m_1}}$*, then the isotopy may be chosen to be relative to* F.

Theorem 2.8.2 holds true for almost all closed manifolds (for example, all those to which the Unique Fibering Theorem 2.8.1 applies). It fails in the case of S^3, which has nonisomorphic fiberings with isotopic fibers. An extensive analysis of the question of deforming homeomorphisms of Seifert-fibered 3-manifolds to fiber-preserving homeomorphisms is given in [**57**].

PROOF. We will first show that after admissible isotopy of f there is a fiber τ' in ∂M_1 for which $f(\tau')$ is a fiber of M_2. If F is nonempty, this is immediate. If $\underline{\underline{m_1}}$ contains an annulus, let τ' be a boundary component of this annulus. Then $f(\tau')$ is a boundary component of an annulus of $\underline{\underline{m_2}}$, so is a fiber of M_2. If F is empty and $\underline{\underline{m_1}}$ and $\underline{\underline{m_2}}$ contain no annuli, then the hypothesized fiber τ is homotopic to a fiber τ' in ∂M_1 (since any two nonexceptional fibers are homotopic). Since $f(\tau')$ is homotopic to $f(\tau)$, it is homotopic in M_2 to a fiber σ in the boundary component of M_2 that contains $f(\tau')$. This says that the element of $\pi_1(M_2)$ represented by $f(\tau')$ is conjugate to the element represented by σ (after selecting suitable orientations). Since σ is a fiber, the element it represents generates a cyclic normal subgroup of $\pi_1(M_2)$, so $f(\tau')$ and σ represent the same element of $\pi_1(M_2)$. So if $f(\tau')$ is not homotopic to σ in ∂M_2, then the boundary torus that contains them is compressible, showing that M_2 is a solid torus. Since $\underline{\underline{m_2}}$ contains no annuli, this case is excluded by hypothesis. We conclude that $f(\tau')$ and σ are homotopic, and hence isotopic, in ∂M_2. After changing f by admissible isotopy, we may assume that there is a fiber τ in ∂M_1 for which $f(\tau)$ is a fiber of M_2.

If F has components that are annuli, then f is isotopic relative to F to a homeomorphism that is fiber-preserving on the boundary tori of M_1 that contain these annuli. Replacing the annuli of F with these tori, we may assume that F is a union of boundary components of M_1.

We now apply Lemma VI.19 of [**54**], which says that if τ' is in ∂M_1 and $f(\tau')$ is a fiber of M_2, then f is isotopic relative to τ' to a fiber-preserving homeomorphism. The proof of Lemma VI.19 in [**54**] proceeds by an induction in which the inductive step produces the isotopy relative to boundary components on which the homeomorphism is already fiber-preserving, so the isotopy may be taken relative to F.

It remains to show that this isotopy of f to a fiber-preserving homeomorphism can be chosen to be admissible. In the special case when $\underline{\underline{m_1}}$ and $\underline{\underline{m_2}}$ do not contain annuli, all isotopies are admissible and the proof is complete. Suppose now that $\underline{\underline{m_1}}$ and $\underline{\underline{m_2}}$ contain annuli. Let G be the union of the components of ∂M_1 that contain annuli of $\underline{\underline{m_1}}$. Since f takes annuli of $\underline{\underline{m_1}}$ homeomorphically to annuli of $\underline{\underline{m_2}}$, we can change f, by an admissible isotopy, to assume that it is fiber-preserving on each component of G. Since our method allows the isotopy making f fiber-preserving to be selected to be relative to any boundary components on which f is already fiber-preserving, there is such an isotopy that is relative to $F \cup G$. This isotopy is relative to the annuli of $\underline{\underline{m_1}}$, so is admissible. □

A more difficult issue is when a *map* between two different fibered manifolds is homotopic to a fiber-preserving one. Sufficient conditions are given in proposition 28.4 of [**58**]:

THEOREM 2.8.3. (Fiber-preserving Map Theorem) *Let each of* $(M_1, \underline{\underline{m_1}})$ *and* $(M_2, \underline{\underline{m_2}})$ *be an I-bundle or Seifert-fibered space with a fixed admissible fibering. Assume that*

(i) *neither of the* $(M_i, \underline{\underline{m_i}})$ *is a solid torus with* $\overline{m_i} = \underline{\underline{\emptyset}}$*, or one of the exceptional fibered manifolds (EF1)-(EF5) or (ESF),*
(ii) *neither* $(M_i, \underline{\underline{m_i}})$ *is one of the exceptional manifolds (EIB), and*
(iii) *if* $(M_2, \underline{\underline{m_2}})$ *is an I-bundle, then* M_1 *is neither a ball nor a solid torus.*

Then every essential map $f\colon (M_1, \underline{\underline{m_1}}) \to (M_2, \underline{\underline{m_2}})$ *is admissibly homotopic to a map which is fiber-preserving with respect to the given fiberings.*

We will need a version of the Fiber-preserving Map Theorem that applies to the exceptional cases (EIB).

THEOREM 2.8.4. (I-bundle Mapping Theorem) *Suppose that each of* $(M_1, \underline{\underline{m_1}})$ *and* $(M_2, \underline{\underline{m_2}})$ *is an irreducible I-bundle with a fixed admissible fibration, and that each* $\underline{\underline{m_i}}$ *has useful completion. Then every essential map* $f\colon (M_1, \underline{\underline{m_1}}) \to (M_2, \underline{\underline{m_2}})$ *that takes lids to lids is admissibly homotopic to a fiber-preserving map.*

PROOF. We will follow the general approach of the proof of proposition 28.4 of [**58**]. Each of (EF1)-(EF5) consists of manifolds $(V, \underline{\underline{v}})$ for which either V is closed, or V is not irreducible, or $\overline{\underline{\underline{v}}}$ is not useful, so neither of $(M_i, \underline{\underline{m_i}})$ can be one of these exceptional cases.

Consider first the case when $\underline{\underline{m_1}}$ is complete. Then f is also admissible and essential as a map to $(M_2, \overline{\underline{\underline{m_2}}})$. By Waldhausen's Theorem 2.5.6, $f\colon (M_1, \underline{\underline{m_1}}) \to (M_2, \overline{\underline{\underline{m_2}}})$ is admissibly homotopic to a covering map. Use this map to lift the admissible fibration of $(M_2, \underline{\underline{m_2}})$ to a fibration of $(M_1, \underline{\underline{m_1}})$. Since f carries lids to lids, statement (i) of the Unique Fibering Theorem 2.8.1, applied to the identity map of $(M_1, \underline{\underline{m_1}})$, shows that this fibering is admissibly isotopic to the original fibering of $(M_1, \overline{\underline{\underline{m_1}}})$, and it follows that f is admissibly homotopic to a fiber-preserving map. From now on, assume that $\underline{\underline{m_1}}$ is not complete.

Case 1: $(M_2, \underline{\underline{m_2}})$ is an I-bundle over a disk.

Since f is essential, $f_\#\colon \pi_1(M_1) \to \pi_1(M_2)$ is injective, and therefore M_1 is also an I-bundle over the disk. Since $\underline{\underline{m_1}}$ has useful completion, $(M_1, \underline{\underline{m_1}})$ has at least one bound side which is not a lid. Since f maps lids to lids, it maps at least one side to a side. The statement called Case 1 in the proof of proposition 28.4 of [**58**] now applies directly.

Case 2: $(M_2, \underline{\underline{m_2}})$ is an I-bundle over a nonclosed surface other than a disk.

Let A be a system of disjoint vertical squares, essentially imbedded in $(M_2, \overline{\underline{\underline{m_2}}})$, which cuts M_2 into balls. Using the Essential Preimage Theorem 2.5.4, f can be changed by admissible homotopy so that $f^{-1}(A)$ is essential in $(M_1, \overline{\underline{\underline{m_1}}})$, and has no component which is a 2-sphere or a small-faced disk. By the Vertical-horizontal Theorem 2.7.1, it may be assumed that each component of $f^{-1}(A)$ is horizontal or vertical.

Since $f_\#\colon \pi_1(M_1) \to \pi_1(M_2)$ is injective, the components of $f^{-1}(A)$ must be disks. If each component of $f^{-1}(A)$ is vertical, and hence is a square, then cutting M_1 and M_2 along A and $f^{-1}(A)$ and applying Case 1 completes the argument. Some care is needed to ensure that the deformations fit together to give a deformation on M_1. The key facts are that any two fiber-preserving maps between squares are homotopic, and that any two homotopies on a square S (or annulus or torus) which have the same starting map and same ending map are homotopic (as maps from $S \times \mathrm{I}$ to S, relative to $S \times \partial\mathrm{I}$). So, once one has the deformation from f to a fiber-preserving map on the cut-apart M_1, one may first deform it to agree on the two copies of $f^{-1}(A)$, then may assume that the homotopies from f to the fiber-preserving map agree on each copy. This way, the homotopy on the cut-apart M_1 may be pieced together to give a homotopy on M_1.

Suppose that some component of $f^{-1}(A)$ is horizontal. Any horizontal surface in an I-bundle is a covering space of the base surface, and consequently must meet every fiber. Since the components of $f^{-1}(A)$ are disks, it follows that $(M_1, \underline{\underline{m_1}})$ is an I-bundle over the disk. So when $(M_1, \underline{\underline{m_1}})$ is not an I-bundle over the disk, the proof of Case 2 is complete.

When $(M_1, \underline{\underline{m_1}})$ is an I-bundle over the disk, it is possible for the components of $f^{-1}(A)$ to be horizontal. Figure 2.5 illustrates how this can occur: Start with

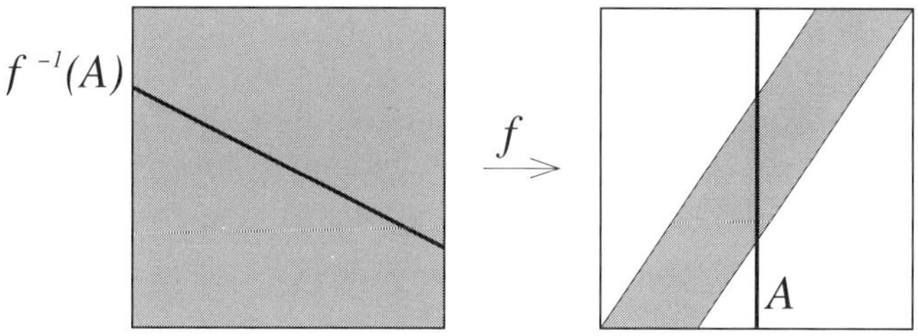

FIGURE 2.5. A horizontal square in $f^{-1}(A)$

a linear imbedding of a square into itself that preserves the lower-left and upper-right corners, takes the lower-right corner to a point on the bottom edge near the lower-left corner, and takes the upper-left corner to a point on the upper edge near the upper-right corner. In coordinates (x,t) on the square $[-1,1]\times[0,1]$, such an imbedding can be defined by $j(x,t)=(-1+3t/2+(x+1)/4,t)$. The preimage of the vertical line $\{0\}\times[0,1]$ is the straight line connecting $(-1,2/3)$ to $(1,1/3)$, that is, the points of the form $(x,1/2-x/6)$. For a three-dimensional example, take the product of this example with the interval $[-1,1]$. Precisely, let $J=[-1,1]$ and $\mathrm{I}=[0,1]$, and let $V=J\times J\times \mathrm{I}$ be the I-bundle over the square $J\times J$. Give V the boundary pattern $\underline{\underline{v}}=\{J\times\{-1\}\times \mathrm{I}, J\times\{1\}\times \mathrm{I}, J\times J\times\{0\}, J\times J\times\{1\}\}$. An admissible imbedding taking lids to lids can be defined from $(V,\underline{\underline{v}})$ to $(V,\underline{\underline{v}})$ by taking (x,y,t) to $(-1+3t/2+(x+1)/4,y,t)$. For this imbedding, the preimage of the vertical square $\{0\}\times J\times \mathrm{I}$ is the horizontal square consisting of all points of the form $(x,y,1/2-x/6)$. The range V may be replaced by a larger I-bundle by adding other I-bundles to it along its free sides $\{-1\}\times J\times \mathrm{I}$ and $\{1\}\times J\times \mathrm{I}$.

Suppose, then, that $(M_1,\underline{\underline{m_1}})$ is an I-bundle over the disk and some component of $f^{-1}(A)$ is horizontal. The sides of $(M_1,\underline{\underline{m_1}})$ must be pairwise disjoint. For since f maps lids to lids, it must map the intersection of two sides to the intersection of two sides of $(M_2,\underline{\underline{m_2}})$. The latter intersection does not meet A, so the preimage intersection would be a fiber disjoint from the preimage of A, a contradiction. Applying statement (iii) of the Essential Singular Annulus and Torus Theorem 2.7.2 to the free sides of $(M_1,\underline{\underline{m_1}})$, there is an admissible homotopy of f which makes the restriction of f to the free sides of $(M_1,\underline{\underline{m_1}})$ vertical. Since f maps lids to lids, we may assume it is fiber-preserving on these free sides. Now each side has top and bottom edges mapping to the lids, and its other edges mapping to fibers in a side of $(M_2,\underline{\underline{m_2}})$, so f can be changed to be fiber-preserving on the sides as well. The argument is now completed as in Case I of Proposition 28.4 of [**58**]; details are given in the following paragraph.

Let L_1 and L_2 be the lids of $(M_1,\underline{\underline{m_1}})$, and for $i=1,2$ let τ_i be the involution of $(M_i,\underline{\underline{m_i}})$ that reflects each fiber across its intersection with a 0-section. Since f is fiber-preserving on the sides of $(M_1,\underline{\underline{m_1}})$, we have $\tau_2 f=f\tau_1$ on $\partial L_1\cup\partial L_2$. Define $g\colon \partial M_1\to M_2$ by $g=f$ on $L_1\cup \overline{\partial L_1\times \mathrm{I}}$ and $g=\tau_2 f\tau_1$ on L_2. On L_2, f and g are maps from a disk into a lid of $(M_2,\underline{\underline{m_2}})$ which agree on ∂L_2, so they are homotopic (the lid is aspherical since $(M_2,\underline{\underline{m_2}})$ is not an exception (EF3)). So we may change f by admissible homotopy so that it agrees with g on all of ∂M_1. Now g already takes the endpoints of every fiber of M_1 to the endpoints of a fiber of M_2, so it extends to a fiber-preserving map $G\colon (M_1,\underline{\underline{m_1}})\to(M_2,\underline{\underline{m_2}})$ (one way to do this is to fix an I-structure on $(M_2, \underline{\underline{m_2}})$, for which by homotopy we could assume that f and therefore g are either the identity or reflection on each fiber in $\partial L_1 \times \mathrm{I}$, and extend g using the identity or reflection from each fiber in $L_1 \times \mathrm{I}$). Since G and f agree on ∂M_1, and $\pi_3(M_2) = 0$, there is a homotopy relative to ∂M_1 that changes f to the fiber-preserving map G.

Case 3: $(M_2, \underline{\underline{m_2}})$ is an I-bundle over a closed surface.

Since no two elements of $\underline{\underline{m_2}}$ and hence no two elements of $\underline{\underline{m_1}}$ can meet, $\underline{\underline{m_1}}$ must consist only of lids. Let A be a vertical essential annulus in $(M_2, \underline{\underline{m_2}})$. As in Case 2, we may assume that $f^{-1}(A)$ is essential in $(M_1, \overline{\underline{\underline{m_1}}})$, and that each component of $f^{-1}(A)$ is horizontal or vertical. If each component of $f^{-1}(A)$ is vertical, we can cut M_1 along $f^{-1}(A)$ and M_2 along A, and apply Case 2 to complete the proof. Again some care is needed to ensure that deformations on the pieces fit together to give a deformation of M_1. This time, components of $f^{-1}(A)$ can be annuli as well as squares. On an annulus, it is not true that there is a deformation between any two homotopies between two given maps, since the components of the space of admissible homotopy equivalence of the annulus are not simply-connected. Indeed, they have infinite cyclic fundamental group, generated by composition of the basepoint homotopy equivalence and the isotopy from the identity to the identity that rotates in the S^1-factor. However, taking the fibering to be the product fibering $\mathrm{S}^1 \times \mathrm{I}$, this isotopy from the identity to the identity is fiber-preserving at each stage, and this allows one to adjust the fiber-preserving map on the cut-apart M_1 in a neighborhood in one of the copies of each annulus of $f^{-1}(A)$ so that the homotopies between the starting and ending maps can be made to agree on both copies of the annulus. Then, they define the desired homotopy on M_1.

Again, though, it can happen that the components of $f^{-1}(A)$ are horizontal. For example, start with the imbedding j of the square $J \times \mathrm{I}$ into itself in Case 2 above, and take $j \times id_{\mathrm{S}^1} \colon J \times \mathrm{I} \times \mathrm{S}^1 \to J \times \mathrm{I} \times \mathrm{S}^1$. The preimage of the vertical annulus $\{0\} \times \mathrm{I} \times \mathrm{S}^1$ will be a horizontal annulus in $J \times \mathrm{I} \times \mathrm{S}^1$.

Suppose some component of $f^{-1}(A)$ is horizontal. Since horizontal surfaces meet every fiber, no component of $f^{-1}(A)$ can be vertical. Since $f_\# \colon \pi_1(M_1) \to \pi_1(M_2)$ is injective, and $f^{-1}(A)$ is two-sided in M_1, the horizontal component must be a disk or annulus. Disks are impossible since $\underline{\underline{m_1}}$ can consist only of lids, and $\underline{\underline{m_1}}$ has useful completion. So $f^{-1}(A)$ contains a horizontal annulus. Since a horizontal surface is a covering space of the base surface of the I-bundle, $(M_1, \underline{\underline{m_1}})$ is an I-bundle over an annulus or Möbius band.

Let T be a free side of $(M_1, \underline{\underline{m_1}})$. Since $\underline{\underline{m_1}}$ consists of lids, T is an annulus. Let k be a boundary component of T. Note that k is disjoint from $f^{-1}(A)$, since otherwise there would be a vertical component of $f^{-1}(A)$. By statement (iii) of the Essential Singular Annulus and Torus Theorem 2.7.2, f is admissibly homotopic to a map g, which agrees with f on k, such that the restriction of g to T vertical. Since g agrees with f on k, T is disjoint from $g^{-1}(A)$. Applying statement (ii) of the Essential Preimage Theorem 2.5.4, with N a regular neighborhood of T, we may further change g by admissible homotopy relative to T to ensure that $g^{-1}(A)$ is essential in $(M_1, \overline{\underline{\underline{m_1}}})$, and each component of $g^{-1}(A)$ is horizontal or vertical. But since g is unchanged on T, $g^{-1}(A)$ is disjoint from T. Therefore each component of $g^{-1}(A)$ is vertical. One may now complete the proof by arguing exactly as above when $f^{-1}(A)$ was vertical. □

The twisted I-bundle over the Klein bottle has unique fibering properties, which we detail in the next lemma.

LEMMA 2.8.5. *Let W be the twisted I-bundle over the Klein bottle, with fundamental group presented as* $\langle a, b \mid bab^{-1} = a^{-1} \rangle$.

(i) *There are two isotopy classes of Seifert fiberings, one as the S^1-bundle over the Möbius band, and one which has quotient space the disk and has two exceptional fibers of type* $(2,1)$. *For the first, the fibers represent the element a, and for the second, the nonexceptional fibers represent b^2.*

(ii) *Suppose that $(W, \underline{\underline{w}})$ has a fixed admissible Seifert fibering. Then any essential map $f\colon (W, \underline{\underline{w}}) \to (W, \underline{\underline{w}})$ is admissibly homotopic to a fiber-preserving map. If $\overline{\underline{\underline{w}}} = \overline{\underline{\underline{\emptyset}}}$, then f is admissibly homotopic to a fiber-preserving covering map.*

PROOF. Observe that $\pi_1(\partial W)$ is the subgroup generated by a and b^2. The fibers of the S^1-bundle structure represents a, and the nonexceptional fibers of the fibering over the disk represent b^2. Moreover, since $b(a^i b^{2j})b^{-1} = a^{-i}b^{2j}$, the only cyclic normal subgroups generated by a simple closed curve in ∂W are those generated by a and b^2. These are the only possible nonexceptional fibers for Seifert fiberings, since the fiber generates an infinite cyclic normal subgoup. So given an arbitrary fibering, a principal fiber in the boundary must be homotopic, hence isotopic, to either a or b^2. By the Seifert Fibering Isotopy Criterion 2.8.2, the fiberings must be isotopic to one of the two standard fiberings.

We now prove (ii). If $\overline{\underline{\underline{w}}} \neq \overline{\underline{\underline{\emptyset}}}$ (equivalently, $\underline{\underline{w}}$ contains an annulus), then the Fiber-preserving Map Theorem 2.8.3 applies. So we may assume that either $\underline{\underline{w}} = \overline{\underline{\underline{\emptyset}}}$ or $\underline{\underline{w}} = \underline{\underline{\emptyset}}$.

We will first show that f is admissibly homotopic to a covering map. If $\underline{\underline{w}} = \overline{\underline{\underline{\emptyset}}}$, then the result follows from Waldhausen's Theorem 2.5.6. Suppose $\underline{\underline{w}} = \underline{\underline{\emptyset}}$. Let $p\colon W \to K$ be the I-fibering over the Klein bottle. The restriction of f to ∂W is an essential map (since it is injective on fundamental groups), so by the Baer-Nielsen Theorem 2.5.5, $p \circ f|_{\partial W}$ is homotopic to a covering map. Therefore the image of $(p \circ f|_{\partial W})_\# : \pi_1(\partial W) \to \pi_1(K)$ lies in the orientation-preserving subgroup, which is the subgroup generated by a and b^2. This implies that f is homotopic to a map taking ∂W to ∂W. Again Waldhausen's Theorem shows that f is homotopic to a covering map. So we may assume that f is a covering map.

Since W is compact, f must be a finite-sheeted covering, and if $\widetilde{W}$ is the covering space of W corresponding to the image of $f_\#$ then f lifts to a homeomorphism $\widetilde{f}\colon W \to \widetilde{W}$. Regarding $\pi_1(\widetilde{W})$ as a subgroup of $\pi_1(W)$, write $\widetilde{f}_\#(a) = a^i b^j$ and $\widetilde{f}_\#(b) = a^k b^\ell$ for some i, j, k, and ℓ. Since a is homotopic into ∂W and b is not, we must have j even and ℓ odd. This implies that $\widetilde{f}_\#(bab^{-1}) = a^{-i}b^j$, and since this must equal $\widetilde{f}_\#(a^{-1}) = a^{-i}b^{-j}$, we must have $j = 0$. Therefore, $\widetilde{f}_\#(a) = a^i$ and $\widetilde{f}_\#(b^2) = a^k b^\ell a^k b^\ell = a^k a^{-k} b^{2\ell} = b^{2\ell}$, and these two elements generate $\pi_1(\partial\widetilde{W})$.

Suppose first that W has the fibering with principal fiber a. Then (since we know that a^i and $b^{2\ell}$ generate $\pi_1(\partial\widetilde{W})$) the fiber of the lifted fibering of $\widetilde{W}$ is a^i. Since $\widetilde{f}_\#(a) = a^i$, the Seifert Fibering Isotopy Criterion 2.8.2 shows that $\widetilde{f}$ is isotopic to a fiber-preserving homeomorphism and consequently f is homotopic to a fiber-preserving covering map. If W has the fibering with principal fiber b^2, then the fiber of $\widetilde{W}$ is $b^{2\ell} = \widetilde{f}_\#(b^2)$, and again the conclusion follows. □

For maps whose domain and range are the same manifold with a fixed fibering, we can combine the previous three results into the following statement, which is convenient for our purposes.

THEOREM 2.8.6. (Fiber-preserving Self-map Theorem) *Let* $(V, \underline{\underline{v}})$ *be an irreducible I-bundle or Seifert-fibered space with a fixed admissible fibering, such that* $\overline{\underline{\underline{v}}}$ *is useful, but not one of the exceptional cases (EF4), (EF5), or (ESF). Let* $f\colon (V, \underline{\underline{v}}) \to (V, \underline{\underline{v}})$ *be an essential map. Assume that*

(i) *if the admissible fibering of* $(V, \underline{\underline{v}})$ *is as an I-bundle, then* f *takes lids to lids, and*

(ii) *if the admissible fibering of* $(V, \underline{\underline{v}})$ *is as a Seifert-fibered space, then* $(V, \overline{\underline{\underline{v}}}) \neq (S^1 \times S^1 \times I, \overline{\underline{\underline{\emptyset}}})$.

Then f *is admissibly homotopic to a map which is fiber-preserving with respect to the given fibering.*

PROOF. Since $\overline{\underline{\underline{v}}}$ is useful, $(V, \underline{\underline{v}})$ is not one of the exceptional cases (EF1) or (EF2). Since V is irreducible, $(V, \underline{\underline{v}})$ is not one of the exceptional cases (EF3). Exceptional cases (EF4), (EF5), and (ESF) are ruled out by hypothesis.

If $(V, \underline{\underline{v}})$ is fibered as an I-bundle, then using hypothesis (i), the I-bundle Mapping Theorem 2.8.4 applies directly to give the result. So we may assume that $(V, \underline{\underline{v}})$ is Seifert-fibered.

Since $\overline{\underline{\underline{v}}}$ is useful, $(V, \underline{\underline{v}})$ is not a solid torus with $\overline{\underline{\underline{v}}} = \overline{\underline{\underline{\emptyset}}}$, so hypothesis (i) of the Fiber-preserving Map Theorem 2.8.3 is satisfied. Its hypothesis (iii) holds vacuously for Seifert-fibered spaces, so if $(V, \overline{\underline{\underline{v}}})$ does not admit an I-fibering which makes it an exceptional case (EIB), then the Fiber-preserving Map Theorem 2.8.3 applies to complete the proof.

In the remaining cases, $(V, \overline{\underline{\underline{v}}})$ can be fibered as an I-bundle over the square, annulus, Möbius band, torus or Klein bottle. An I-bundle over the square cannot be Seifert-fibered, so this case does not occur. Hypothesis (ii) implies that $(V, \overline{\underline{\underline{v}}})$ cannot be fibered as an I-bundle over the torus. If $(V, \overline{\underline{\underline{v}}})$ can be fibered as the I-bundle over the Klein bottle, then lemma 2.8.5 completes the proof.

Suppose $(V, \overline{\underline{\underline{v}}})$ can be fibered as an I-bundle over the annulus or Möbius band. In these cases, $(V, \overline{\underline{\underline{v}}})$ is a solid torus with useful boundary pattern. For the I-bundle over the annulus the Seifert fibering is nonsingular and for the I-bundle over the Möbius band there is one exceptional fiber of type $(2, 1)$. If $\underline{\underline{v}}$ is not complete, then lemma 28.1 of [**58**] applies to yield the conclusion. Suppose $\underline{\underline{v}}$ is complete. Applying Waldhausen's Theorem 2.5.6, we may change f by admissible homotopy to be a covering map. There are two Seifert fiberings of $(V, \overline{\underline{\underline{v}}})$: the given fibering and the lift of the given fibering using f. By the Seifert Fibering Isotopy Criterion 2.8.2, these two fiberings are isotopic, and it follows that f is admissibly homotopic to a fiber-preserving covering map. □

2.9. The characteristic submanifold

We are now ready for the central concept in Johannson's theory, the characteristic submanifold. This is an essential codimension-zero submanifold $(\Sigma, \underline{\underline{\sigma}})$ in $(M, \underline{\underline{m}})$ whose components are admissibly fibered. In particular, any components that are I-fibered must have their lids contained in elements of $\underline{\underline{m}}$. Whenever we discuss the characteristic submanifold of a 3-manifold, it is to be assumed that the boundary pattern is useful. Then, the existence and uniqueness of the characteristic submanifold are guaranteed. Also, we work only with Haken manifolds.

The characteristic submanifold can be characterized in several ways. The way we choose to define it is as a maximal ("full") essential admissibly fibered submanifold $(\Sigma, \underline{\underline{\sigma}})$ of $(M, \underline{\underline{m}})$ such that every essential fibered submanifold of $(M, \underline{\underline{m}})$ is admissibly isotopic into Σ (the "engulfing" property), and every essential map of a square, annulus, or torus into $(M, \underline{\underline{m}})$ is admissibly homotopic into Σ (the "enclosing" property).

When $(V, \underline{\underline{v}})$ is admissibly fibered, the characteristic submanifold of $(V, \overline{\underline{\underline{v}}})$ is $(V, \overline{\underline{\underline{v}}})$ itself. At the other extreme are the manifolds $(W, \underline{\underline{w}})$ for which the characteristic submanifold of $(W, \overline{\underline{\underline{w}}})$ is as small as possible. For any square, annulus, or torus F of $\overline{\underline{\underline{w}}}$, a regular neighborhood of F in W is an essential admissibly fibered submanifold of $(W, \overline{\underline{\underline{w}}})$, so by the engulfing property must be isotopic into the characteristic submanifold of $(W, \overline{\underline{\underline{w}}})$. If every component of the characteristic submanifold of $(W, \overline{\underline{\underline{w}}})$ is just a regular neighborhood of a square, annulus, or torus of $\overline{\underline{\underline{w}}}$, then $(W, \underline{\underline{w}})$ is said to be simple.

After giving some formal definitions and stating the existence and uniqueness, we will combine some results from [**58**] to give a convenient characterization of the characteristic submanifold of a 3-manifold with complete and useful boundary pattern. Roughly speaking, $(\Sigma, \underline{\underline{\sigma}})$ is the characteristic submanifold if and only if it is a maximal essential fibered submanifold and each of its complementary pieces in M is either simple or can be admissibly fibered as an I- or S^1-bundle over the square or annulus. In section 2.10, we will give examples of 3-manifolds and their characteristic submanifolds.

Recall that unless otherwise stated, codimension-zero submanifolds $(X, \underline{\underline{x}})$ of $(M, \underline{\underline{m}})$ carry the submanifold boundary pattern, which ensures that the inclusion $(X, \underline{\underline{x}})$ to $(M, \underline{\underline{m}})$ is admissible. An admissibly fibered I-bundle or Seifert-fibered space $(X, \underline{\underline{x}})$ in $(M, \underline{\underline{m}})$ is said to be *admissible* in $(M, \underline{\underline{m}})$ when the inclusion of $(X, \underline{\underline{x}})$ into $(M, \overline{\underline{\underline{m}}})$ is also admissible. As mentioned in section 2.2, this implies that X does not meet $\overline{\partial M - |\underline{\underline{m}}|}$.

Let $(M, \underline{\underline{m}})$ be a Haken 3-manifold, possibly closed, with useful boundary pattern. A disjoint collection $(\Sigma, \underline{\underline{\sigma}})$ of essential admissible I-bundles and Seifert-fibered spaces is a *characteristic submanifold* for $(M, \underline{\underline{m}})$ if

(1) $(\Sigma, \underline{\underline{\sigma}})$ is full, i. e. the union of Σ with any of the complementary components of M cannot be fibered as a disjoint union of essential admissible I-bundles and Seifert-fibered spaces,
(2) (Engulfing Property) every essential admissible I-bundle or Seifert-fibered space $(X, \underline{\underline{x}})$ in $(M, \underline{\underline{m}})$ is admissibly isotopic into Σ, and
(3) (Enclosing Property) every essential map $f\colon (T, \underline{\underline{t}}) \to (M, \underline{\underline{m}})$ of a square, annulus, or torus into $(M, \underline{\underline{m}})$ is admissibly homotopic to a map with image in Σ.

Combining corollaries 10.9 and 10.10 and theorem 12.5 of [**58**], we obtain the following fundamental existence and uniqueness result.

THEOREM 2.9.1. (Existence and Uniqueness) *Let $(M, \underline{\underline{m}})$ be a Haken 3-manifold with useful boundary pattern. Then there exists a characteristic submanifold in $(M, \underline{\underline{m}})$. Any two characteristic submanifolds in $(M, \underline{\underline{m}})$ are admissibly isotopic.*

The Engulfing Property implies that every essentially imbedded square, annulus, or torus in $(M, \underline{\underline{m}})$ is admissibly isotopic into Σ, since such a surface can be thickened to a fibered manifold. Also, if $(M, \underline{\underline{m}})$ itself can be admissibly fibered as an I-bundle or Seifert fibered space, then the Engulfing Property and the fullness condition imply that $(M, \underline{\underline{m}}) = (\Sigma, \underline{\underline{\sigma}})$, as mentioned in the introduction to this section.

There is a more general version of the Enclosing Property given as proposition 13.1 of [**58**]:

THEOREM 2.9.2. (Extended Enclosing Theorem) *Let $(M, \underline{\underline{m}})$ be a Haken 3-manifold with useful boundary pattern, and let $(\Sigma, \underline{\underline{\sigma}})$ be its characteristic submanifold. Let $(X, \underline{\underline{x}})$ be an admissibly fibered I-bundle or Seifert fibered space whose completed boundary pattern is useful. Suppose that $(X, \underline{\underline{x}})$ is not one of the exceptional cases (EF1)-(EF5). Then every essential map $f\colon (X, \underline{\underline{x}}) \to (M, \underline{\underline{m}})$ is admissibly homotopic into Σ.*

Simple 3-manifolds are the analogues of acylindrical 3-manifolds in the setting of manifolds with boundary pattern. They arise naturally in the characteristic submanifold theory, since the complement S of the characteristic submanifold is "relatively acylindrical," i. e. every essential annulus or torus is homotopic into the frontier of S in M.

Precisely, a 3-manifold with useful boundary pattern $(W, \underline{\underline{w}})$ is called *simple* if $\overline{\underline{\underline{w}}}$ is useful and every component of the characteristic submanifold of $(W, \overline{\underline{\underline{w}}})$ is a regular neighborhood $(F, \underline{\underline{f}}) \times (\mathrm{I}, \{0\})$ of a square, annulus, or torus F of $\overline{\underline{\underline{w}}}$. As usual, "square" and "annulus" here mean that if F is given the boundary pattern consisting of the components of its intersections with the other elements of $\overline{\underline{\underline{w}}}$, then it either is a square or is an annulus with boundary pattern $\overline{\underline{\underline{\emptyset}}}$. We remark that the manifolds of the form $(F, \underline{\underline{f}}) \times (\mathrm{I}, \underline{\underline{k}})$, where $(F, \underline{\underline{f}})$ is a square, annulus, or torus, and $\underline{\underline{k}}$ is either $\{0\}$ or $\overline{\underline{\underline{\emptyset}}}$ are not simple; if $(W, \underline{\underline{w}})$ is one of these manifolds, then the characteristic submanifold of $(W, \overline{\underline{\underline{w}}})$ is all of W. This is not a regular neighborhood of an element of $\overline{\underline{\underline{w}}}$, since it has no admissible deformation retraction to an element of $\overline{\underline{\underline{w}}}$.

We now obtain a convenient characterization of the characteristic submanifold. Recall from example 2.2.6 that the proper boundary pattern on a submanifold consists of the intersections of the submanifold with the boundary pattern of the ambient manifold, together with the components of the frontier of the submanifold.

THEOREM 2.9.3. *Let $(M, \underline{\underline{m}})$ be a Haken 3-manifold with complete and useful boundary pattern, let $(\Sigma, \underline{\underline{\sigma}})$ be an essential admissible fibered submanifold of $(M, \underline{\underline{m}})$. Let $(S, \underline{\underline{s}})$ be $\overline{M - \Sigma}$ with its proper boundary pattern. Then $(\Sigma, \underline{\underline{\sigma}})$ is characteristic for $(M, \underline{\underline{m}})$ if and only if*

(i) *$(\Sigma, \underline{\underline{\sigma}})$ is full,*
(ii) *$(\Sigma, \underline{\underline{\sigma}})$ contains a regular neighborhood of every element of $\underline{\underline{m}}$ that is a square, annulus, or torus, and*
(iii) *every component of $(S, \underline{\underline{s}})$ is either simple or can be admissibly fibered as an I-bundle or S^1-bundle over a square or annulus.*

PROOF. Assume that $(\Sigma, \underline{\underline{\sigma}})$ is characteristic. By definition, it is full. Let F be an element of $\underline{\underline{m}}$ which is a square, annulus, or torus. There is a collar neighborhood of F which is an admissibly imbedded copy of $(F, \underline{\underline{f}}) \times (\mathrm{I}, \{0\})$ with $F = F \times \{0\}$.

This neighborhood can be admissibly Seifert-fibered if F is an annulus or torus, and can be admissibly I-fibered if F is a an annulus or a square. Moreover, it is essential. For suppose that $(D, \underline{d})$ is a small-faced disk in $(M, \underline{\underline{m}})$ with one face k essential in $F \times \{1\}$. Notice that $D \cap F \times \mathrm{I} = k$, since $F \times \{1\}$ is essential in $F \times I$. Adding $k \times \mathrm{I} \subset F \times \mathrm{I}$ to D would produce a small-faced disk imbedded in $(M, \underline{\underline{m}})$ that shows that $\underline{\underline{m}}$ is not useful. By the Engulfing Property, $F \times \mathrm{I}$ is admissibly isotopic into $(\Sigma, \underline{\underline{\sigma}})$, so Σ contains a regular neighborhood of F. The assertion about the complementary components is remark 3 on p. 159 of [**58**].

Assume now that (i), (ii), and (iii) hold. We will verify that $(\Sigma, \underline{\underline{\sigma}})$ is "complete", as defined on p. 90 of [**58**]. That is, if $(T, \underline{t})$ is any essential admissible square, annulus, or torus in $(S, \underline{\underline{s}})$, then either

(1) $T \cap \Sigma$ is nonempty, and the component of $(S, \underline{\underline{s}})$ that contains T can be admissibly fibered as an I- or S^1-bundle over a square or annulus, or
(2) $T \cap \Sigma$ is empty, and T is admissibly parallel in $(S, \underline{\underline{s}})$ to a component of $S \cap \Sigma$.

This will finish the proof, since by (i), $(\Sigma, \underline{\underline{\sigma}})$ is full, and by corollary 10.10 of [**58**], a full, complete, essential fibered submanifold of $(M, \underline{\underline{m}})$ must be characteristic.

Let $(W, \underline{\underline{w}})$ be the component of $(S, \underline{\underline{s}})$ that contains $(T, \underline{t})$. By hypothesis (iii), the boundary pattern $\underline{\underline{w}}$ is complete and useful. Let $(\Gamma, \underline{\underline{\gamma}})$ be the characteristic submanifold of $(W, \underline{\underline{w}})$. By the Engulfing Property, we may assume that $(T, \underline{t})$ lies in $(\Gamma, \underline{\underline{\gamma}})$.

Suppose first that $T \cap \Sigma$ is nonempty, so T meets some component F of $S \cap \Sigma$. Note that T must be a square or an annulus, since if it were a torus it would not be admissibly imbedded in $(W, \underline{\underline{w}})$. We must show that $(W, \underline{\underline{w}})$ can be fibered as an I-bundle or S^1-bundle over a square or annulus. Suppose not, then by assumption (iii), $(W, \underline{\underline{w}})$ is simple. Since $(\Sigma, \underline{\underline{\sigma}})$ is admissibly fibered, F must be a square, annulus, or torus, so from the direction of the theorem already proven, Γ contains a regular neighborhood $F \times \mathrm{I}$ of F. Since $(W, \underline{\underline{w}})$ is simple, the component of $(\Gamma, \underline{\underline{\gamma}})$ that contains F must be a regular neighborhood $F \times \mathrm{I}$ of F, with $F = F \times \{0\}$. Since $(\Gamma, \underline{\underline{\gamma}}) \neq (W, \underline{\underline{w}})$, $F \times \{1\}$ is a component of the frontier of $F \times \mathrm{I}$, so ∂T lies in $F \times \{0\} \cup \partial F \times \mathrm{I}$. In this case, T would be inessential, which is a contradiction.

Suppose now that T is disjoint from Σ. If $(W, \underline{\underline{w}})$ can be admissibly fibered as an I- or S^1-bundle over the square or annulus, then since T is essential and disjoint from Σ, the Vertical-horizontal Theorem 2.7.1 shows that T is parallel in $(W, \underline{\underline{w}})$ to a component of $\Sigma \cap W$. So assume that $(W, \underline{\underline{w}})$ is simple. As in the previous case, T lies in a component of Γ which is a regular neighborhood of an element F of $\underline{\underline{w}}$. Since $(W, \underline{\underline{w}})$ cannot be admissibly fibered as an I- or S^1-bundle over the square or annulus, the argument of the previous paragraph shows that $(T, \underline{t})$ cannot meet F. So $(T, \underline{t})$ is admissibly parallel to F. Since by (ii), every square, annulus, and torus of $\underline{\underline{m}}$ must be contained in the topological interior of Σ, F cannot be in $\underline{\underline{m}}$, so F is a component of $S \cap \Sigma$. We have verified the completeness condition, and as already mentioned, this together with assumption (i) implies that $(\Sigma, \underline{\underline{\sigma}})$ is characteristic. □

2.10. Examples of characteristic submanifolds

Before proceeding further with the theory of the characteristic submanifold, we give some examples.

EXAMPLE 2.10.1. *Admissibly fibered manifolds*

Suppose that $(V, \underline{\underline{v}})$ is admissibly fibered as an I-bundle or Seifert fibered space, and $\underline{\underline{v}}$ is useful. For each free side $(G, \underline{\underline{g}})$ of $(V, \underline{\underline{v}})$, choose a regular neighborhood $(G, \underline{\underline{g}}) \times (\mathrm{I}, \underline{\underline{\emptyset}})$ admissibly imbedded in $(V, \underline{\underline{v}})$. We may select these neighborhoods to be disjoint, and let W be their union. Then $\overline{V - W}$ is an admissibly imbedded fibered submanifold. It is full, since its union with any of its complementary components is not an admissible fibered submanifold of $(V, \underline{\underline{v}})$ (because it would not be admissibly imbedded in $(V, \overline{\underline{\underline{v}}})$). It satisfies the Engulfing and Enclosing Properties, since $(V, \underline{\underline{v}})$ is itself admissibly isotopic into $\overline{V - W}$. Therefore $\overline{V - W}$ is the characteristic submanifold of $(V, \underline{\underline{v}})$. It may seem unnatural that $(V, \underline{\underline{v}})$ is not equal to its characteristic submanifold. This type of example rarely arises in practice, since in most applications (notably, when invoking the Classification Theorem 2.11.1) one uses the characteristic submanifold for the *completed* boundary pattern, and the characteristic submanifold of $(V, \overline{\underline{\underline{v}}})$ is $(V, \overline{\underline{\underline{v}}})$ itself.

EXAMPLE 2.10.2. *Hyperbolizable manifolds*

Recall that a 3-manifold M with incompressible boundary is *hyperbolizable* if its interior admits a complete hyperbolic structure. Our work in section 5.2 will give a complete description of the characteristic submanifold of a hyperbolizable 3-manifold, in a more general context, so here we give only a brief sketch. (See Morgan [**96**] for a discussion of the topological properties of hyperbolic 3-manifolds.)

If M is closed and hyperbolizable, then it contains no essential tori or annuli. Its characteristic submanifold is empty, since it contains no essential I-bundles or Seifert fibered spaces.

Suppose next that M is hyperbolizable, its boundary consists entirely of tori and M is not an I-bundle over the torus or annulus. (In this case, the hyperbolic structure on the interior of M must have finite volume.) Then, M has incompressible boundary and theorem 2.9.3 implies that the characteristic submanifold Σ of $(M, \overline{\underline{\underline{\emptyset}}})$ must contain a regular neighborhood of each component of ∂M. Thus, there can be no I-bundles in Σ unless M is an I-bundle over the torus or Klein bottle. We have explicitly ruled out the I-bundle over the torus and the I-bundle over the Klein bottle is not hyperbolizable. Since there are no essential annuli in M and every essential map of a torus into M is homotopic into ∂M, every Seifert-fibered component of Σ is a regular neighborhood of a component of ∂M, so $(M, \overline{\underline{\underline{\emptyset}}})$ is simple. Conversely, if $(M, \overline{\underline{\underline{\emptyset}}})$ is a simple Haken 3-manifold and each component of ∂M is a torus, then Thurston's Geometrization Theorem (see section 7.1 below) implies that M is hyperbolizable.

Finally, suppose that M is hyperbolizable and has a boundary component which is not a torus. The characteristic submanifold Σ of $(M, \overline{\underline{\underline{\emptyset}}})$ can now contain I-bundles which meet ∂M in their associated ∂I-bundles. Since every essential torus in M is boundary parallel, any Seifert-fibered component of Σ is homeomorphic to either the solid torus or $T^2 \times \mathrm{I}$. Each solid torus component will meet ∂M in a collection of incompressible annuli, and each $T^2 \times \mathrm{I}$ will meet ∂M in one of its boundary tori, together with a possibly empty collection of annuli in its other boundary torus.

EXAMPLE 2.10.3. *Gluing hyperbolizable 3-manifolds to Seifert fibered spaces*

Let N be a hyperbolizable 3-manifold with a single torus boundary component, but not a solid torus. Let V be a Seifert-fibered space whose boundary consists of incompressible tori. We may form a new manifold M by gluing copies of N to some of the boundary components of V. Then $(V, \partial M)$ is a full, essential fibered submanifold of $(M, \underline{\underline{\overline{\emptyset}}})$, and its complement is simple. Therefore, theorem 2.9.3 implies that $(V, \partial M)$ is the characteristic submanifold of $(M, \underline{\underline{\overline{\emptyset}}})$. Notice that examples 1.4.3 and 1.4.4 from chapter 1 are of this form.

EXAMPLE 2.10.4. *Books of I-bundles*

Books of I-bundles were defined and used in [**31**], and provide the basis for the main construction in [**8**]. In example 1.4.5 we gave an explicit construction of one. Following [**31**], a compact connected orientable 3-manifold M is a *book of I-bundles* if $M = E \cup V$ where

  (i) E is an I-bundle over a nonempty compact 2-manifold B, not necessarily connected or orientable,
  (ii) each component of V is homeomorphic to a solid torus,
  (iii) $E \cap V$ is the inverse image of ∂B under the bundle projection $E \to B$, and
  (iv) each component of $E \cap V$ is an annulus in ∂V which is homotopically nontrivial in V.

Assume that no component of B is a disk, 2-sphere, or projective plane, and assume that ∂M is incompressible (that is, no component V_0 of V has $V_0 \cap E$ equal to a single annulus whose core circle generates $\pi_1(V_0)$). Let n be the total number of components of V and E. Suppose some component E_0 of E is the I-bundle over the annulus and has an attaching annulus whose core circle generates the fundamental group of the solid torus V_0 of V that contains it. Let V_1 be the solid torus of V containing the other attaching annulus of E_0, then E_0 may be removed from E and V_1 replaced by $V_1 \cup E_0 \cup V_0$ to give a book of I-bundle structure with smaller n. So we may assume there are no such components of E. If V_0 is a component of V such that $V_0 \cap E$ consists of two annuli whose core circles generate $\pi_1(V_0)$, then V_0 may be added to E to give a book of I-bundles structure with smaller n. So we may assume there are no such V_0. The characteristic submanifold of $(M, \underline{\underline{\overline{\emptyset}}})$ then consists of the union Σ of E with the solid tori in V obtained by removing an open collar of $E \cap V$ in V. For this is a full, essential fibered submanifold containing all tori and annuli of the boundary pattern (of which there are none), and each component of the complement is $A^2 \times I$, so theorem 2.9.3 applies. Proposition 4.3 of [**31**] characterizes books of I-bundles as the atoroidal Haken 3-manifolds with nonempty boundary such that each component of the closure of $\partial M - \partial \Sigma$ is an annulus.

EXAMPLE 2.10.5. *Fiber bundles over S^1*

Let $h \colon F \to F$ be an orientation-preserving homeomorphism of a compact connected orientable 2-manifold. It determines a compact orientable Haken 3-manifold $M(h)$ formed by gluing $M \times \{0\}$ to $M \times \{1\}$ using h, and all 3-manifolds which fiber over S^1 with compact fiber can be described in this way. According to Theorem 4 of Thurston [**122**] or Theorem 0.1 of Handel-Thurston [**46**], F can be written as the union $A \cup B$ of two compact 2-manifolds, one of which may be empty, such that $A \cap B$ consists of noncontractible simple closed curves, and such that h is isotopic to a homeomorphism g such that

  (i) $g(A) = A$ and $g|_A$ has finite order, and
  (ii) $g(B) = B$, and if α is any simple closed curve in B not parallel into ∂B, and $n \neq 0$, then $(g|_B)^n(\alpha)$ is not homotopic to α.

Note that $M(g)$ is homeomorphic to $M(h)$. Take such a decomposition for which the number of components of A and B is minimal. For each boundary component of F that is contained in B, remove an open collar neighborhood of the component from B, and add its closure to A. Put $\Sigma = M(g|_A)$ and $S = M(g|_B)$. Now $M(g)$ decomposes into $\Sigma \cup S$, and if we let $\underline{\underline{\sigma}}$ be the set of boundary components of Σ that are boundary components of $M(g)$, then $(\Sigma, \underline{\underline{\sigma}})$ is an essentially imbedded Seifert fibered manifold in $(M(g), \overline{\underline{\underline{\emptyset}}})$. On the other hand, each component of S is either homeomorphic to $\mathrm{S}^1 \times \mathrm{S}^1 \times \mathrm{I}$ (corresponding to components of B that are annuli) or is hyperbolizable with finite volume (see Thurston [**124**] or Otal [**107**].) When components of S that are homeomorphic to $\mathrm{S}^1 \times \mathrm{S}^1 \times \mathrm{I}$ arise, the Seifert fiberings of the adjacent components cannot be extended over the I-bundle, otherwise the annulus in B could have been added to A. Thus, $(\Sigma, \underline{\underline{\sigma}})$ is full. Since Σ contains all of $\partial M(g)$, theorem 2.9.3 shows that $(\Sigma, \underline{\underline{\sigma}})$ is the characteristic submanifold of $(M(g), \overline{\underline{\underline{\emptyset}}})$.

EXAMPLE 2.10.6. *Drilling out fibers*

Let $(M, \underline{\underline{m}})$ be a Haken 3-manifold with useful boundary pattern and characteristic submanifold $(\Sigma, \underline{\underline{\sigma}})$. Let f_1, $f_2, \ldots,$ f_n be distinct fibers of Σ, not contained in the frontier of Σ, and choose disjoint closed fibered regular neighborhoods $W_1, \ldots,$ W_n of the f_i, also disjoint from the frontier of Σ. Let N be the submanifold which results from removing the topological interior of $\cup W_i$ from M, and let $\underline{\underline{n}}$ be the proper boundary pattern on N. It is immediate to verify that $\Sigma \cap N$, as a submanifold of $(N, \underline{\underline{n}})$, satisfies conditions (i), (ii), and (iii) of theorem 2.9.3, hence $\Sigma \cap N$ is characteristic in $(N, \underline{\underline{n}})$. A special case of this construction is used in [**9**], where the fibers whose neighborhoods are removed are the core circles of fibered solid torus components of the characteristic submanifold of a hyperbolic 3-manifold.

EXAMPLE 2.10.7. *Completions of boundary patterns with disjoint elements*

In many of our applications of the characteristic submanifold theory, the elements of the boundary pattern $(M, \underline{\underline{m}})$ of the 3-manifold will be disjoint. In this case, no element of $\overline{\underline{\underline{m}}}$ can be a square, and any element that is topologically an annulus will have boundary pattern $\overline{\underline{\underline{\emptyset}}}$ and hence will be an annulus as a 2-manifold with boundary pattern. By theorem 2.9.3, the characteristic submanifold of $(M, \overline{\underline{\underline{m}}})$ will contain a regular neighborhood of every element of $\overline{\underline{\underline{m}}}$ that is an annulus or a torus. In particular, these annuli and tori will be elements of the boundary pattern of the characteristic submanifold. For reference, we state this fact as a lemma.

LEMMA 2.10.8. *Let $(M, \underline{\underline{m}})$ be a 3-manifold with boundary pattern whose completion is useful, such that the elements of $\underline{\underline{m}}$ are disjoint. Let $(\Sigma, \widehat{\underline{\underline{\sigma}}})$ be the characteristic submanifold of $(M, \overline{\underline{\underline{m}}})$. Then Σ contains a regular neighborhood of each torus and annulus of $\overline{\underline{\underline{m}}}$, and consequently each such torus or annulus is an element of $\widehat{\underline{\underline{\sigma}}}$.*

Another easy observation is the following:

LEMMA 2.10.9. *Let $(M, \underline{\underline{m}})$ be a 3-manifold with boundary pattern whose completion is useful, such that the elements of $\underline{\underline{m}}$ are disjoint. Let $(\Sigma, \widehat{\underline{\underline{\sigma}}})$ be the characteristic submanifold of $(M, \overline{\underline{\underline{m}}})$, and let $(V, \widehat{\underline{\underline{v}}})$ be a component of $(\Sigma, \widehat{\underline{\underline{\sigma}}})$ which is an I-bundle over $(B, \underline{\underline{b}})$. If a lid of $(V, \widehat{\underline{\underline{v}}})$ is contained in a torus or annulus of $\overline{\underline{\underline{m}}}$, then $M = V$.*

PROOF. By lemma 2.10.8, $(V,\widehat{\underline{\underline{v}}})$ contains a regular neighborhood of the torus or annulus F of $\overline{\underline{\underline{m}}}$ that contains its lid, so the lid must equal F and the sides of $(V,\widehat{\underline{\underline{v}}})$ must lie in ∂M. Since the other lid of V, if any, lies in ∂M, the frontier of V is empty and hence $V=M$. $\square$

EXAMPLE 2.10.10. *A simple 3-manifold whose boundary pattern contains a square*

In example 2.2.2, we constructed a 3-manifold with boundary pattern $(S,\underline{\underline{s}})$ such that one of the elements of $\underline{\underline{s}}$ was a square N, and in example 2.4.7 we verified that when ∂S and the annuli of $\underline{\underline{s}}$ are incompressible, the completion $\overline{\underline{\underline{s}}}$ was useful. Assume now also that S is irreducible, that A_0 and A_1 are not isotopic in ∂S, and that $(S,\underline{\underline{\emptyset}})$ is simple, i. e. $(S,\overline{\underline{\underline{\emptyset}}})$ contains no essential tori or annuli. We will prove that the characteristic submanifold of $(S,\overline{\underline{\underline{s}}})$ consists of a regular neighborhood of N. That is, $(S,\underline{\underline{s}})$ is simple.

We will examine the essential tori, annuli, and squares in $(S,\overline{\underline{\underline{s}}})$. An essential torus would be essential in $(S,\overline{\underline{\underline{\emptyset}}})$, violating simplicity. Also by simplicity, an essential annulus A would have to be parallel to an annulus A' in ∂S. Since ∂A must be disjoint from $J(\underline{\underline{s}})$, $A'\cap J(\underline{\underline{s}})$ must consist of components of $J(\underline{\underline{s}})$. The component G cannot be one of these, since then A_0 and A_1 would be isotopic. Since the two circle components C_0 and C_1 of $J(\underline{\underline{s}})$ are not isotopic, they cannot both lie in A'. But then, A is not essential, since if A' does not meet $J(\underline{\underline{s}})$, then A has a 2-faced compression, and if A' contains one of the C_i, then A has a 3-faced compression.

Consider now an essential square T in $(S,\overline{\underline{\underline{s}}})$. We will show that T is admissibly parallel to N. Since ∂S is incompressible, ∂T bounds a disk T_0 in ∂S, for which $\partial T_0\cap J(\underline{\underline{s}})$ consists of the four corners of T. Suppose for contradiction that some component of $T_0\cap J(\underline{\underline{s}})$ is an arc β. If β connects adjacent corners of T, then a nearby arc in T_0 parallel to β and connecting opposite faces of T is homotopic (rel endpoints) to an essential arc in T, so gives a violation of the essentiality of T. But if β connects nonadjacent corners of T_0, then each of the other corners would have to separate $J(\underline{\underline{s}})$, which is impossible. So no component of $T_0\cap J(\underline{\underline{s}})$ is an arc, and $T_0\cap J(\underline{\underline{s}})$ must be contained in G. If it is not connected, then there is an arc in T_0 connecting opposite sides of T and disjoint from $J(\underline{\underline{s}})$, again violating essentiality. The connected subgraphs of G that are bounded by four points are of two kinds, those containing two vertices of $J(\underline{\underline{s}})$ and those containing four vertices. If $T_0\cap J(\underline{\underline{s}})$ has two vertices, then there is an arc β in T_0 connecting opposite faces of T and meeting $T_0\cap J(\underline{\underline{s}})$ in a single point, so β is admissibly homotopic to a constant map. Since β is homotopic (rel endpoints) to an essential arc in T, it gives a violation of essentiality. So $T_0\cap J(\underline{\underline{s}})$ must contain all four vertices of G. It must appear as four edges bounding a square in the interior of T_0, each of whose corners is joined to a vertex of T by an arc. The only four edges of G that bound a disk in ∂S are the four sides of N, so we conclude that T is admissibly parallel to N.

Now let Σ be the characteristic submanifold of $(S,\overline{\underline{\underline{s}}})$. By theorem 2.9.3(b), Σ contains a regular neighborhood of N. From above, every component of the frontier of Σ is a square admissibly parallel to N, so (using also the fullness property of Σ) it follows that Σ is just an admissible regular neighborhood of N.

EXAMPLE 2.10.11. *I-bundle configurations*

Main Topological Theorem 2, stated in chapter 8, gives an answer to the Finite Index Realization Problem for $(M,\underline{\underline{m}})$ in terms of the characteristic submanifold

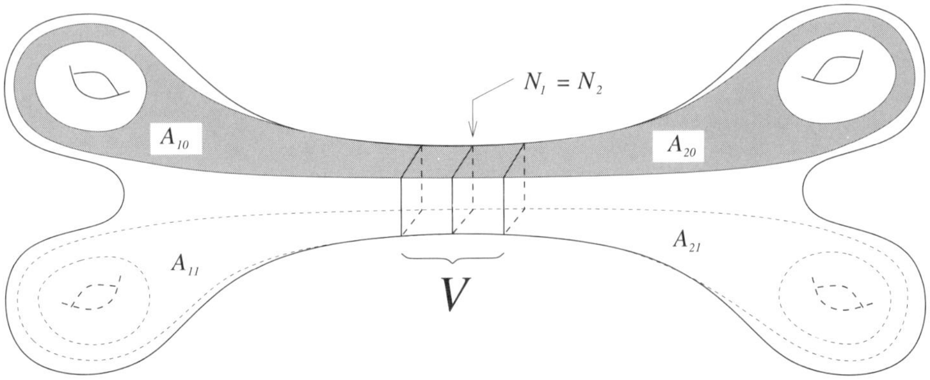

FIGURE 2.6. A 3-ball component of a characteristic submanifold

$(\Sigma,\underline{\underline{\widehat{\sigma}}})$ of $(M,\overline{\underline{\underline{m}}})$. It says that the index is finite if every component V of $(\Sigma,\underline{\underline{\widehat{\sigma}}})$ occurs in a certain list of possible configurations. Here we will give explicit examples illustrating the following three configurations from that list:

(7) V is a 3-ball.
(8) V is I-fibered over a topological annulus or Möbius band and no component of $V\cap\overline{\partial M - |\underline{\underline{m}}|}$ is a square which meets two different components of the frontier of V.
(9) V is I-fibered over the disk with two holes, and $V\cap\overline{\partial M - |\underline{\underline{m}}|}$ is an annulus.

In the theorem, the elements of the boundary pattern $\underline{\underline{m}}$ are assumed to be disjoint submanifolds of ∂M, so our examples will also have this property.

Here is an example where V is a 3-ball, as in item (7). In fact, V will be an I-bundle over the square. Let $(S_1,\underline{\underline{s_1}})$ and $(S_2,\underline{\underline{s_2}})$ be two copies of a simple 3-manifold as in example 2.10.10. Each $\underline{\underline{s_i}}$ has two elements A_{i0} and A_{i1} that are disjoint incompressible topological annuli in ∂S_i. The remaining element of $\underline{\underline{s_i}}$ is a square N_i that meets A_{i0} and A_{i1} in opposite sides. Form M by identifying N_1 and N_2, in such a way that $N_1\cap A_{1j}$ is identified with $N_2\cap A_{2j}$ for $j=0,1$. The boundary pattern $\underline{\underline{m}}$ consists of two elements $A_{1j}\cup A_{2j}$, for $j=0,1$, each of which is a disk with two holes. In example 2.4.7 we checked that each $\overline{\underline{\underline{s_i}}}$ was useful, so lemma 2.4.9 shows that $\overline{\underline{\underline{m}}}$ is useful. Let V be the union of the characteristic submanifolds of $(S_1,\overline{\underline{\underline{s_1}}})$ and $(S_2,\overline{\underline{\underline{s_2}}})$. Then V is a regular neighborhood of $N_1 = N_2$ in M. Let $\underline{\underline{v}}$ be the submanifold boundary pattern of V as a submanifold of $(M,\overline{\underline{\underline{m}}})$. Then $(V,\underline{\underline{v}})$ is admissibly fibered as an I-bundle over a square and is admissibly imbedded in $(M,\overline{\underline{\underline{m}}})$. Theorem 2.9.3 shows that $(V,\underline{\underline{v}})$ is the characteristic submanifold of $(M,\overline{\underline{\underline{m}}})$. Figure 2.6 shows a schematic diagram of M and V.

We next construct a manifold $(M,\underline{\underline{m}})$ such that the characteristic submanifold $(V,\underline{\underline{v}})$ of $(M,\overline{\underline{\underline{m}}})$ is admissibly I-fibered over a topological annulus, and $V\cap\overline{\partial M - |\underline{\underline{m}}|}$ is a square which meets only one component of the frontier of V. This example satisfies the conditions of item (8). We start with a single copy $(S,\underline{\underline{s}})$ of a simple 3-manifold as in example 2.10.10. Let $V=A\times \mathrm{I}$ where A is an annulus, and let β be an arc in ∂A. Obtain M from S and V by identifying N with $\beta\times \mathrm{I}$, in such a way that the arcs $\beta\times\{j\}$ are identified with the arcs $A_j\cap N$ for $j=0,1$. Put $F_j=(A\times\{j\})\cup A_j$, a disk with two holes, and $\underline{\underline{m}}=\{F_0,F_1\}$. Theorem 2.9.3 shows that $(V, \underline{\underline{v}})$ is the characteristic submanifold of $(M, \overline{\underline{\underline{m}}})$ where $\underline{\underline{v}}$ is the submanifold boundary pattern on V as a submanifold of $(M, \overline{\underline{\underline{m}}})$. Notice that $\overline{V \cap \partial M - |\underline{\underline{m}}|}$ is the square $(\partial A \times \mathrm{I}) - N$.

We may similarly construct a manifold $(M, \underline{\underline{m}})$ such that the characteristic submanifold $(V, \underline{\underline{v}})$ of $(M, \overline{\underline{\underline{m}}})$ is admissibly I-fibered over a Möbius band and $\overline{V \cap \partial M - |\underline{\underline{m}}|}$ is a square which meets only one component of the frontier of V. In the construction in the previous paragraph we simply replace V with an I-bundle over the Möbius band and let $\underline{\underline{m}}$ be the single surface F which is a disk with three holes, consisting of the union of A_0 and A_1 with the annulus which is the lid of V, meeting each A_i in an arc in one of its boundary circles.

We can also construct manifolds $(N, \underline{\underline{n}})$ where a component of the characteristic submanifold $(V, \underline{\underline{v}})$ of $(N, \overline{\underline{\underline{n}}})$ is an I-bundle over a topological annulus or Möbius band and $\overline{V \cap \partial N - |\underline{\underline{n}}|}$ is a square which meets two components of the frontier of V, so V fails the conditions of items (8). To do so, we proceed as before but glue two different copies of $(S, \underline{\underline{s}})$ to disjoint squares in a side of V.

Our next examples illustrate the conditions of item (9). One can verify that $(S, \{A_0\})$ is simple. Let V be an I-bundle over the disk with two holes and form M by attaching S to V by identifying a side of V with A_0. If we let $\underline{\underline{m}}$ be the lids of V, then $(V, \underline{\underline{v}})$ is the characteristic submanifold of $(M, \overline{\underline{\underline{m}}})$ where $\underline{\underline{v}}$ is the submanifold boundary pattern. In this case, $\overline{V \cap \partial M - |\underline{\underline{m}}|}$ consists of two annuli, so fails the conditions of item (9). We form N by attaching copies of S to two sides of V and let $\underline{\underline{n}}$ be the lids of V. Then $(V, \underline{\underline{v}})$ is the characteristic submanifold of $(N, \overline{\underline{\underline{n}}})$ and $\overline{V \cap \partial N - |\underline{\underline{n}}|}$ is a single side of V, so the conditions of item (9) hold in this example.

2.11. The Classification Theorem

The central result of Johannson's theory of homotopy equivalences of Haken manifolds is the Classification Theorem, which is given as theorem 24.2 in [**58**]. It will be our main tool for the analysis of homotopy equivalences of 3-manifolds.

THEOREM 2.11.1. (Classification Theorem) *Let $(M_1, \underline{\underline{m_1}})$ and $(M_2, \underline{\underline{m_2}})$ be compact irreducible 3-manifolds with boundary patterns whose completions are useful and nonempty. Let V_1 and V_2 denote the characteristic submanifolds of $(M_1, \overline{\underline{\underline{m_1}}})$ and $(M_2, \overline{\underline{\underline{m_2}}})$ respectively. Let $\underline{\underline{v_1}}$, $\underline{\underline{v_2}}$, $\underline{\underline{w_1}}$ and $\underline{\underline{w_2}}$ denote the proper boundary patterns of V_1, V_2, $\overline{M_1 - V_1}$, and $\overline{M_2 - V_2}$ respectively. Then every admissible homotopy equivalence $f \colon (M_1, \underline{\underline{m_1}}) \to (M_2, \underline{\underline{m_2}})$ is admissibly homotopic to a map $g \colon (M_1, \underline{\underline{m_1}}) \to (M_2, \underline{\underline{m_2}})$ for which $g^{-1}(V_2) = V_1$, $g|_{V_1} : (V_1, \underline{\underline{v_1}}) \to (V_2, \underline{\underline{v_2}})$ is an admissible homotopy equivalence, and $g|_{\overline{M_1 - V_1}} : (\overline{M_1 - V_1}, \underline{\underline{w_1}}) \to (\overline{M_2 - V_2}, \underline{\underline{w_2}})$ is an admissible homeomorphism.*

Note that the V_i are the characteristic submanifolds for the *completed* boundary patterns $\overline{\underline{\underline{m_i}}}$ of the $(M_i, \underline{\underline{m_i}})$. However, their boundary patterns $\underline{\underline{v_i}}$ and $\underline{\underline{w_i}}$ are the *proper* boundary patterns (that is, they include the components of the frontiers of the respective submanifolds), as submanifolds of the $(M_i, \underline{\underline{m_i}})$ with their *original* boundary patterns.

We remark that a stronger statement than the Classification Theorem is true for closed Haken manifolds. For by Waldhausen's Theorem 2.5.6, f is homotopic to a homeomorphism h, and since the characteristic submanifold is unique up to isotopy, we may change h so that it carries V_1 homeomorphically to V_2 and $\overline{M_1 - V_1}$ homeomorphically to $\overline{M_2 - V_2}$.

When $(M, \underline{\underline{m}})$ is simple, the Classifcation Theorem shows that every admissible homotopy equivalence of $(M, \underline{\underline{m}})$ is admissibly homotopic to a homeomorphism. This is augmented by the following result, given as proposition 27.1 in [**58**]. It is the natural generalization of the fact that the mapping class group of an acylindrical 3-manifold is finite.

THEOREM 2.11.2. (Finite Mapping Class Group Theorem) *Let $(M, \underline{\underline{m}})$ be a simple 3-manifold with complete and useful boundary pattern. Then $\mathcal{H}(M, \underline{\underline{m}})$ is finite.*

We will need the following application of the Classification Theorem, in the proof of our characterization of small homotopy types in section 6.2.

LEMMA 2.11.3. *Let $(W, \underline{\underline{w}})$ be a compact orientable irreducible 3-manifold with boundary pattern, such that the completion of $\underline{\underline{w}}$ is useful. Suppose that $f\colon (W, \underline{\underline{w}}) \to (\Sigma, \underline{\underline{\sigma}})$ is an admissible homotopy equivalence, where $(\Sigma, \overline{\overline{\sigma}})$ is an admissibly fibered I-bundle, and no lid of $(\Sigma, \overline{\overline{\sigma}})$ is contained in $\underline{\underline{\sigma}}$. Then there is an admissible fibering of $(W, \overline{\overline{w}})$ as an I-bundle for which f is admissibly homotopic to a fiber-preserving homeomorphism.*

PROOF. We may assume that Σ and W are connected. Since no lid of $(\Sigma, \overline{\overline{\sigma}})$ is contained in $\underline{\underline{\sigma}}$, the elements of $\underline{\underline{\sigma}}$ must be squares and annuli which are all the sides of $(\Sigma, \overline{\overline{\sigma}})$. By the Classification Theorem 2.11.1, $(W, \overline{\overline{w}})$ must equal its characteristic submanifold. We claim that $(W, \overline{\overline{w}})$ may be admissibly fibered as an I-bundle. If it is Seifert-fibered, no homotopy equivalence between W and Σ could be fiber-preserving, since the image of a fiber would be contractible. So the Fiber-preserving Map Theorem 2.8.3 implies that W is either a solid torus or the I-bundle over the torus or Klein bottle. Suppose that W, and hence Σ, is a solid torus. Since W is Seifert-fibered and $\overline{\overline{w}}$ is useful, each component of $\overline{\overline{w}}$ is an annulus. Since Σ is I-fibered and each component of $\overline{\overline{w}}$ is an annulus, $\underline{\underline{\sigma}}$ must consist either of two annuli, A_1 and A_2, such that $\pi_1(A_i) \to \pi_1(\Sigma)$ are isomorphisms, or of a single annulus A such that the image of $\pi_1(A) \to \pi_1(\Sigma)$ has index 2. Therefore $\underline{\underline{w}}$ must be the same, and $(W, \overline{\overline{w}})$ can be admissibly fibered as an I-bundle. If W, and hence Σ, is an I-bundle over the torus or Klein bottle, then $\underline{\underline{\sigma}}$ must be empty (since it contains no lid), hence $\underline{\underline{w}}$ is also empty and again $(W, \overline{\overline{w}})$ can be admissibly fibered as an I-bundle.

Since each element of $\underline{\underline{\sigma}}$ is a square or annulus, no lid of $(W, \overline{\overline{w}})$ can be in $\underline{\underline{w}}$ unless $(W, \overline{\overline{w}})$ is fibered over the square, annulus, or Möbius band, but in these cases the I-fibering may be reselected so that the lids are not in $\underline{\underline{w}}$. So in all cases, we may assume that $\underline{\underline{w}}$ consists of the sides of the I-bundle $(W, \overline{\overline{w}})$, since otherwise the lids would not be free sides of $(W, \underline{\underline{w}})$.

The restriction of f to a free side of $(W, \underline{\underline{w}})$ (i. e. a lid of $(W, \overline{\overline{w}})$) is an admissible map to $(\Sigma, \overline{\overline{\sigma}})$, and carries its boundary into $|\underline{\underline{\sigma}}|$. Also, it is essential, since any essential loop or arc in the lid is also essential in $(W, \underline{\underline{w}})$, and f is an admissible homotopy equivalence. Lemma 5.5 of [**58**] and the remark immediately following it show that every essential map of an orientable surface into an I-bundle, for which the image of the boundary does not meet the lids, is admissibly homotopic to a map into a lid. Applying this, we may assume that f is admissible as a map from $(W, \overline{\overline{w}})$ to $(\Sigma, \overline{\overline{\sigma}})$.

We will show that f is admissibly homotopic to a map which is essential as a map from $(W,\overline{\underline{\underline{w}}})$ to $(\Sigma,\overline{\underline{\underline{\sigma}}})$. Since f is an admissible homotopy equivalence, its restriction to any essential circle in W, or any essential arc in $(W,\overline{\underline{\underline{w}}})$ with both endpoints in elements of $\underline{\underline{w}}$, is essential. There are no essential arcs in $(W,\overline{\underline{\underline{w}}})$ which have one endpoint in a lid and the other in a side, so if f fails to be essential as a map from $(W,\overline{\underline{\underline{w}}})$ to $(\Sigma,\overline{\underline{\underline{\sigma}}})$, there must be an essential arc α with endpoints in a lid or lids of $(W,\overline{\underline{\underline{w}}})$ to which the restriction of f is inessential.

Suppose first that $(W,\overline{\underline{\underline{w}}})$ is a product I-bundle $G \times \mathrm{I}$. Then α must have endpoints in different lids of $(W,\overline{\underline{\underline{w}}})$, and $f(\alpha)$ must have endpoints in the same lid of $(\Sigma,\overline{\underline{\underline{\sigma}}})$, so both lids of $(W,\overline{\underline{\underline{w}}})$ must be carried to a single lid G' of $(\Sigma,\overline{\underline{\underline{\sigma}}})$. Since $G \times \{0\} \to W$ and $f\colon W \to \Sigma$ induce surjections on the fundamental groups, it follows that $G' \to \Sigma$ does as well. Therefore, by the Finite Index Theorem 2.1.1, $(\Sigma,\overline{\underline{\underline{\sigma}}})$ is a product I-bundle $G' \times \mathrm{I}$. So f can be changed by admissible homotopy to take the two lids of $(W,\overline{\underline{\underline{w}}})$ to different lids of $(\Sigma,\overline{\underline{\underline{\sigma}}})$, and f will then be essential on all essential arcs of $(W,\overline{\underline{\underline{w}}})$.

Suppose now that $(W,\overline{\underline{\underline{w}}})$ is twisted with lid L. Let β be an arc in L connecting the endpoints of α. Since the loop $\alpha \cup \beta$ represents the nontrivial coset of $\pi_1(L)$ in $\pi_1(W)$, it follows that f carries $\pi_1(W)$ to $\pi_1(L')$, where L' is a lid of $(\Sigma,\overline{\underline{\underline{\sigma}}})$. Since f is an isomorphism on fundamental groups, $\pi_1(L')$ must equal $\pi_1(\Sigma)$ so Σ is a product I-bundle. Let $(N,\underline{\underline{n}})$ be a cross-section of $(W,\overline{\underline{\underline{w}}})$ and $(F,\underline{\underline{f}})$ a cross-section of $(\Sigma,\overline{\underline{\underline{\sigma}}})$. Note that N is nonorientable and F is orientable. Since $(N,\underline{\underline{n}})$ is admissibly homotopy equivalent to $(W,\underline{\underline{w}})$, and $(F,\underline{\underline{f}})$ to $(N,\underline{\underline{n}})$, it follows that $(N,\underline{\underline{n}})$ is admissibly homotopy equivalent to $(F,\underline{\underline{f}})$. Since these have complete boundary patterns, the Baer-Nielsen Theorem 2.5.5 implies that N and F are homeomorphic, a contradiction.

We have established that after admissible homotopy, f is essential as a map from $(W,\overline{\underline{\underline{w}}})$ to $(\Sigma,\overline{\underline{\underline{\sigma}}})$. By Waldhausen's Theorem 2.5.6, f is admissibly homotopic to a homeomorphism g. Since the elements of $\underline{\underline{w}}$ are the sides of the I-bundle structure on $(W,\overline{\underline{\underline{w}}})$, g takes lids to lids. So by the Unique Fibering Theorem 2.8.1, g is admissibly isotopic to a fiber-preserving homeomorphism. □

Suppose an admissible homotopy is given between two maps of $(M,\underline{\underline{m}})$ which preserve the characteristic submanifold of $(M,\overline{\underline{\underline{m}}})$ and its complement. The next result, corollary 18.2 of [**58**], enables us to deform the homotopy to one which preserves the characteristic submanifold and its complement at each level.

THEOREM 2.11.4. (Homotopy Splitting Theorem) *Let $(M,\underline{\underline{m}})$ be a Haken 3-manifold whose completed boundary pattern $\overline{\underline{\underline{m}}}$ is useful, and let Σ be the characteristic submanifold of $(M,\overline{\underline{\underline{m}}})$. Assume that M is not a torus bundle over S^1. Suppose $H\colon (M \times I, \underline{\underline{m}} \times I) \to (M,\underline{\underline{m}})$ is an admissible homotopy such that $H_0^{-1}(\Sigma)=\Sigma$ and $H_1^{-1}(\Sigma)=\Sigma$. Then H is admissibly homotopic, relative to $M \times \partial I$, to H' such that $(H_t')^{-1}(\Sigma)=\Sigma$ for all $t \in I$.*

Remark: In [**58**], corollary 18.2 is stated only for the case when M has nonempty boundary, but this is used in the proof (of the auxiliary result proposition 18.1) only to conclude that the components Y_1 and Y_2 of $\overline{V}$ in the last full paragraph on p. 156 are distinct, in the case when the manifold called Z is $T^2 \times \mathrm{I}$. But if M is closed, and $Y_1 = Y_2$, then by the argument in the second paragraph of the proof of proposition 18.1, Y_1 is $T^2 \times \mathrm{I}$, so M would be a torus bundle over the circle. Since M may be assumed not to be Seifert-fibered, these are the torus bundles for which the monodromy homeomorphism h of the torus used to construct the bundle $M=(T^2\times \mathrm{I})/(x,0)\sim(h(x),1)$ induces an automorphism of $\pi_1(T^2)\cong\mathbb{Z}\oplus\mathbb{Z}$ whose trace has absolute value at least 3, i. e. the torus bundles which admit a *Sol* geometry (see [**113**], [**121**]). Theorem 2.11.4 fails for these cases; the characteristic submanifold is $T^2\times[0,1/2]$, and the homeomorphism defined by $f([x,t])=[h(x),t]$ is isotopic to the identity, but not homotopic to the identity preserving $T^2\times[0,1/2]$.

2.12. Miscellaneous topological results

In this section we collect some results from low-dimensional topology which will be needed later, but do not fit conveniently in the previous sections.

The proof of the next lemma will use some constructions with homotopies. If $H\colon F\times \mathrm{I}\to F$ is a homotopy, then $\overline{H}$ denotes the reverse homotopy defined by $\overline{H}(x,t)=H(x,1-t)$. The product of two homotopies H and K, such that $H_1=K_0$, is the homotopy $H*K$ that sends (x,t) to $H(x,2t)$ for $0\leq t\leq 1/2$ and to $K(x,2t-1)$ for $1/2\leq t\leq 1$. We will use the concept of the trace of a homotopy at a point. In chapter 10, we will use a more general version of the trace, which will be discussed in detail at that time. For now, we simply define the trace of a homotopy H between two maps that preserve a basepoint x of F to be the element $t(H)$ of $\pi_1(F,x)$ represented by the restriction of H to $x\times \mathrm{I}$. Note that $t(\overline{H})=t(H)^{-1}$ and $t(H*K)=t(H)t(K)$. If both $H|_{F\times\{0\}}$ and $H|_{F\times\{1\}}$ are the identity map of F, then $t(H)$ is a central element of $\pi_1(F)$, since if α is any loop in F based at x, representing an element $a\in\pi_1(F)$, then $\alpha\times\{0\}$ and $(x\times \mathrm{I})*(\alpha\times\{1\})*(\overline{x\times \mathrm{I}})$ are homotopic in $F\times \mathrm{I}$, preserving basepoints, and the composition of H with this homotopy is a homotopy between loops representing a and $t(H)\,a\,t(H)^{-1}$.

Theorem 6.4 of [**35**] asserts, in part, that if h is a homeomorphism of a compact 2-manifold (orientation-preserving, if F is a disk or an annulus), and h is homotopic to the identity map, then h is isotopic to the identity map. We will need a version of this result for surfaces with boundary patterns.

LEMMA 2.12.1. *Let $h\colon(F,\underline{\underline{f}})\to(F,\underline{\underline{f}})$ be a homeomorphism of a compact connected 2-manifold, and let $H\colon F\times I\to F$ be an admissible homotopy from h to the identity. Then there is a deformation of H, relative to $F\times\partial I$ and through admissible homotopies, to an admissible isotopy, unless either*

(i) *$F=D^2$, h is orientation-reversing, and $\underline{\underline{f}}$ is either empty, or a single arc in ∂D^2, or two disjoint arcs in ∂D^2, or*
(ii) *$(F,\underline{\underline{f}})=(S^1\times I,\underline{\underline{\emptyset}})$ and h is orientation-reversing, or*
(iii) *$(F,\underline{\underline{f}})=(\text{Möbius band},\underline{\underline{\emptyset}})$.*

In case (iii), *h is still admissibly isotopic to the identity.*

PROOF. Assuming that h is not as in (i) or (ii), we will show that either (iii) occurs, or H is deformable to an isotopy. Since h is admissibly homotopic to the identity, it preserves each element of $\underline{\underline{f}}$.

Suppose for contradiction that h moves some boundary component of F to a different boundary component. Since h is homotopic to the identity, these components are homotopic in F, so F is an annulus with empty boundary pattern. Since (ii) does not hold, h must be orientation-preserving, but then it induces the nontrivial automorphism on $\pi_1(F)$, so cannot be homotopic to the identity. So we may assume that h preserves each boundary component of F.

Suppose for contradiction that h reverses the orientation on some boundary component of F. Then the component and its reverse represent conjugate elements of $\pi_1(F)$; since no nontrivial element of a free group is conjugate to its inverse, the component must be contractible in F, so F is a 2-disk D^2. If $|\underline{\underline{f}}| = \partial D^2$, then h cannot be admissibly homotopic to the identity, and if $\underline{\underline{f}}$ contains two arcs that meet in one endpoint, then h cannot preserve each of them and also reverse orientation. Therefore $\underline{\underline{f}}$ must consist of disjoint arcs. If there are more than two of them, then since h reverses orientation it cannot preserve each of them. Since (i) does not hold, we may assume that h is orientation-preserving on each element of $\underline{\underline{f}}$.

As a consequence of these initial observations, h must preserve every element of the completed boundary pattern $\overline{\underline{\underline{f}}}$, and every endpoint of every arc of $\overline{\underline{\underline{f}}}$.

Since H is admissible, it too must preserve each intersection point of two arcs of $\underline{\underline{f}}$ at each level, and we may further deform it to preserve any endpoint of an arc of $\overline{\underline{\underline{f}}}$ at each level.

Next, we will deform H to preserve each element of $\overline{\underline{\underline{f}}}$. Suppose k is an arc of $\overline{\underline{\underline{f}}} - \underline{\underline{f}}$. Its endpoints are preserved at each level, so H preserves $\partial(k \times \mathrm{I})$. Since $\overline{F \times I}$ is aspherical, there is a deformation of H relative to $\partial k \times \mathrm{I}$ (and, as with all our deformations, admissibly and relative to $F \times \partial \mathrm{I}$) to make it preserve $k \times I$. Now suppose that k is a circle. Regard the restriction of H to $k \times \mathrm{I}$ as an admissible map from an annulus into $(F, \underline{\underline{f}})$. If this map is essential, then by the Baer-Nielsen Theorem 2.5.5, it is homotopic to a covering map. The only surfaces covered by the annulus are the annulus and the Möbius band. Since $h(k) = k$ and the restriction of H to $k \times \mathrm{I}$ is essential, F must be a Möbius band. Since k was a circle of $\overline{\underline{\underline{f}}} - \underline{\underline{f}}$, $\underline{\underline{f}}$ must be empty, giving case (iii). In that case, every homeomorphism homotopic to the identity is isotopic to the identity, for it is isotopic to be the identity on the boundary and then Theorem 3.4 of [**35**] applies. But H is not deformable to an isotopy, since its trace at a point in ∂F is not homotopic to a loop in ∂F. From now on, we assume that (iii) does not occur. Then the restriction of H to $k \times \mathrm{I}$ is inessential, so we may deform H to preserve each element of $\overline{\underline{\underline{f}}} - \underline{\underline{f}}$. So H is then admissible for $(F, \overline{\underline{\underline{f}}})$.

If two homeomorphisms of a 1-manifold are properly homotopic, then they are isotopic, so using the homotopy extension property we may deform H through admissible homotopies to be an isotopy on k for each $k \in \overline{\underline{\underline{f}}}$; that is, on all of ∂F. Our result then follows from the next lemma:

LEMMA 2.12.2. *Let F be a compact surface. If $G\colon F \times I \to F \times I$ is a homotopy between homeomorphisms, and is an isotopy on $\partial F \times I$, then there is a deformation of G, relative to $\partial(F \times I)$, to an isotopy.*

PROOF. We may assume G is the identity map on $F \times \{0\}$, since if $G(x, 0) = (g(x), 0)$, and there is a deformation D_t from $(g^{-1} \times 1_{\mathrm{I}}) \circ G$ to an isotopy, then $(g \times 1_{\mathrm{I}}) \circ D_t$ is a deformation from G to an isotopy. (Here and in the ensuing argument, subscripts of s or t on maps indicate that they are isotopic deformations starting at $s = 0$ or $t = 0$ and going to $s = 1$ or $t = 1$.)

Since G is admissible for the complete boundary pattern on $(F \times \overline{\underline{\underline{f}}}) \times (\mathrm{I}, \overline{\underline{\underline{\emptyset}}})$, Waldhausen's Theorem 2.5.6 implies that we may assume that G is a homeomorphism (when F is a small-faced disk, the boundary pattern on $F \times \mathrm{I}$ will not be useful, but for these cases one can use the Alexander trick instead). Define maps

$G_t\colon F \times \mathrm{I} \to F \times \mathrm{I}$ as follows. Let $k_s\colon \partial F \to \partial F$ be the restriction of G to $\partial F \times \{s\}$. By the Isotopy Extension Theorem (see for example section 5 of [**108**]), there is an extension of k_s to an isotopy $K_s\colon F \to F$, with K_0 the identity. Define $J_t\colon F \times \mathrm{I} \to F \times \mathrm{I}$ by $J_t(x, s) = (K_{st}^{-1}(x), s)$, and put $G_t = J_t \circ G$. Then $G_0 = G$, and for $x \in \partial F$, $G_1(x, s) = (K_s^{-1} \times 1_{\mathrm{I}})(G(x, s)) = (k_s^{-1}(k_s(x)), s) = (x, s)$.

Lemma 3.5 of [**128**] says that a homeomorphism from $F \times \mathrm{I}$ to $F \times \mathrm{I}$ which is the identity on $F \times \{0\} \cup \partial F \times \mathrm{I}$ is isotopic relative to $\partial(F \times \mathrm{I})$ to a level-preserving homeomorphism. So, there is a deformation L_t from $L_0 = G_1$ to L_1, relative to $\partial(F \times \mathrm{I})$, with L_1 a level-preserving homeomorphism, i. e. an isotopy. Now, define $P_t = J_t^{-1} \circ L_1$. Then P_1 is an isotopy, and agrees with the original G on $\partial(F \times \mathrm{I})$. On $\partial(F \times \mathrm{I})$, the deformation $R_t = (G_t) * (L_t) * (P_t)$ we have constructed from G to P_1 is $J_t \circ G$, followed by the constant deformation which is $J_1 \circ G$ at every time, followed by the reverse of $J_t \circ G$. Letting S be the product $J * (\text{constant } J_1) * \overline{J}$, the deformation defined by putting $T_t = S_t^{-1} \circ R_t$ is a deformation from G to P_1, relative to $\partial(F \times \mathrm{I})$. This completes the proof of Lemma 2.12.2 and hence the proof of Lemma 2.12.1. □

□

Another result concerning isotopies of homeomorphisms of surfaces will find use in chapter 9, specifically in propositions 9.2.1 and 9.2.2. It will allow certain homeomorphisms to be adjusted so that they preserve elements of the boundary pattern that are annuli, thereby making them admissible.

LEMMA 2.12.3. *Let G be a compact 2-manifold, and let G_1 be a 2-dimensional submanifold each of whose components is incompressible and not simply connected, and such that each boundary component of G_1 lies in either the boundary or the interior of G. Suppose that $h\colon G \to G$ is a homeomorphism such that for each component B of G_1, $h_\#(\pi_1(B))$ is conjugate to $\pi_1(B)$. Then h is isotopic to a homeomorphism that preserves G_1.*

PROOF. By replacing each maximal subcollection of parallel annuli by their union with the annuli in G separating them, we may assume that no pair of components of G_1 are homotopic annuli. For each B, both B and $h(B)$ lift to the covering space of G corresponding to the subgroup $\pi_1(B)$, and the lifts are deformation retracts of this covering. Therefore the boundary components of the lifts are pairwise isotopic in the covering, and hence the boundary components of B and $h(B)$ are pairwise homotopic in G. Since homotopic simple closed curves in surfaces are isotopic, $h(\partial B)$ is isotopic to ∂B and hence h may be changed by isotopy to make $h(B) = B$ (if G is a torus and B an annulus, the isotopy taking $\partial(h(B))$ to B might take $h(B)$ to the complementary annulus to B, but in this case a further isotopy takes it to B). We can proceed inductively to make h preserve all the components of G_1; the only complication that could arise would occur when B is an annulus and the isotopy moving $h(B)$ to B pushed $h(B)$ across some annulus in G_1 which is already preserved by h. But this cannot occur, since we have arranged that no two annuli in G_1 are homotopic. □

The next lemma is needed for proposition 2.12.5.

LEMMA 2.12.4. *Let X be a topological space, and let $h\colon X \times [0, 1) \to X \times [0, 1)$ be a homeomorphism which restricts to the identity on $X \times \{0\}$. Then h is isotopic to the identity relative to $X \times \{0\}$.*

PROOF. Use a homeomorphism from $[0,1)$ to $[0,\infty)$ to regard $X \times [0,1)$ as $X \times [0,\infty)$. Write $h(x,s) = (h_1(x,s), h_2(x,s))$. We will define two isotopies of h whose product is the desired isotopy. The first will change h to be the identity on $X \times [0,1]$ by "pushing h onto $X \times [1,\infty)$". Define

$$K_t(x,s) = \begin{cases} (x,s) & 0 \le s \le t \\ (h_1(x,s-t), t + h_2(x,s-t)) & t \le s\,, \end{cases}$$

so that $K_0 = h$ and K_1 restricted to $X \times [0,1]$ equals the identity. The second isotopy conjugates the homeomorphism K_1 by the homeomorphisms of $X \times [0,\infty)$ that multiply by larger and larger numbers $\frac{1}{1-t}$ in the second coordinate. The effect is to make the homeomorphism the identity on $X \times [0, \frac{1}{1-t}]$, so that at time 1 all points will be fixed. Writing $K_1(x,s) = (k_1(x,s), k_2(x,s))$, define

$$L_t(x,s) = \begin{cases} (k_1(x,(1-t)s), k_2(x,(1-t)s)/(1-t)) & 0 \le t < 1 \\ (x,s) & t = 1\,, \end{cases}$$

so that $L_0 = K_1$ and L_1 is the identity. Note that L is continuous at $t = 1$ since $K_1(x,(1-t)s) = (x,(1-t)s)$ when $s < 1/(1-t)$. The product KL is the desired isotopy from h to the identity relative to $X \times \{0\}$. □

The following proposition will be used in chapter 7, in the definition of the function Θ which is a key ingredient in the proof of the Parameterization Theorem.

PROPOSITION 2.12.5. *Let M and M' be compact orientable 3-manifolds, and let C and C' be compact 2-manifolds contained in ∂M and $\partial M'$. Suppose that $j\colon M - C \to M' - C'$ is a homeomorphism. Then j is isotopic to a homeomorphism that extends to a homeomorphism from M to M'.*

PROOF. Assuming that C is connected, we will construct an isotopy from j to a homeomorphism that extends to C. It will be evident from the construction that the isotopy may be chosen to be fixed on any specified compact subset of $M - C$, so the argument can be repeated (taking as the compact subset the union of the components of C to which the homeomorphism has already been extended) until the homeomorphism extends to all of C.

Fix a collar neighborhood $C \times \mathrm{I}$ of C so that $C = C \times \{1\}$. Denote $C \times \{0\}$ by F and $j(F)$ by F'. Now F' separates $M' - C'$ and hence M'. Let W' be the component of M' cut along F' that contains $j(C \times [0,1))$. Then $C' \subseteq W'$ and $W' - C' = j(C \times [0,1))$.

We claim that there are coordinates on W' as $F' \times [0,1]$ so that $F' = F' \times \{0\}$. First, note that if the interior of a 3-manifold is irreducible, then so is the 3-manifold. So if F is not a 2-sphere, W' is irreducible. Now, since $C' \subseteq \partial W'$, we have $\pi_1(W') = \pi_1(W' - C') = \pi_1(j(C \times [0,1)))$, so the inclusion of F' to W' induces an isomorphism on fundamental groups. If F' is not a disk or sphere, then the Finite Index Theorem 2.1.1 applies immediately to establish the claim. If F' is a disk, then since W' is irreducible and simply-connected, it is a 3-ball and has the desired product structure. Suppose that F' is a 2-sphere. Then W' is simply-connected, so $\partial W'$ consists of k 2-spheres for some $k \ge 2$, and $\pi_2(W') \cong H_2(W') \cong \mathbb{Z}^{k-1}$. Since $\pi_2(W') \cong \pi_2(C \times [0,1)) \cong \mathbb{Z}$, $\partial W'$ consists of two 2-spheres. Since $j(C \times [0,1))$ can contain no fake 3-cells, W' is homeomorphic to $\mathrm{S}^2 \times \mathrm{I}$ and has the desired product structure.

Now $j(\partial F \times [0,1))$ is a collar neighborhood of $\partial F' \times \{0\}$ in $\partial F' \times [0,1] \cup F' \times \{1\}$, so by uniqueness of collars we may change the product structure on $F' \times [0,1]$ so that $C' = F' \times \{1\}$, and consequently $j(F \times [0,1)) = F' \times [0,1)$. Identify $F' \times [0,1)$ with $F \times [0,1)$ using the homeomorphism $(j|_F)^{-1} \times 1_{[0,1)}$. Applying lemma 2.12.4, we may change j by isotopy, fixed outside of W, so that j is the identity map of $F \times [0,1)$ with respect to these coordinates. Then, j extends continuously by using $j|_F$ on $C \times \{1\}$. □

CHAPTER 3

Relative Compression Bodies and Cores

In this section we will define and study a second kind of characteristic structure. The submanifolds that determine this structure generalize the characteristic compression body invented by Bonahon [**16**] and subsequently developed by McCullough and Miller [**89**]. A compression body is a 3-manifold which is made by attaching 1-handles to the "tops" of a collection of I-bundles. We will see that each free side of a 3-manifold $(M, \underline{\underline{m}})$ with boundary pattern has a neighborhood in M which is a (relative) compression body. This neighborhood, and moreover the union of a disjoint collection of such neighborhoods for all the free sides, is unique up to admissible isotopy. However, the closure of the complement of such a union has a much stronger characteristic property: it is preserved up to admissible homotopy by admissible homotopy equivalences. We will apply this characteristic property in chapter 4 to obtain finiteness results on homotopy types of 3-manifolds with boundary pattern, and in chapter 6 to classify the "small" pared manifolds needed for our main theorems.

More precisely, an (orientable) compression body is a 3-manifold V which either is a handlebody or can be constructed as follows. Start with a collection $\{F_i \mid 1 \leq i \leq m\}$ of closed orientable connected 2-manifolds, none of which is a 2-sphere. Form a connected irreducible 3-manifold V from $\bigcup_{i=1}^{m} F_i \times \mathrm{I}$ by attaching 1-handles to $\bigcup_{i=1}^{m} F_i \times \{1\}$. The boundary ∂V consists of $\bigcup F_i \times \{0\}$, together with one distinguished boundary component which is the union of the intersection of ∂V with $\bigcup_{i=1}^{m} F_i \times \{1\}$ and the intersection of ∂V with the 1-handles. Bonahon observed that a compact boundary component F of an irreducible 3-manifold M has a neighborhood which is a compression body with distinguished boundary component F, and whose frontier is incompressible and is exactly $\bigcup F_i \times \{0\}$. This compression body is unique up to isotopy in M. One can think of it as the minimal irreducible submanifold of M such that every loop in F which is contractible in M is contractible in the submanifold. The characteristic compression body neighborhood was used by Bonahon to study cobordism of group actions on 2-manifolds, and by McCullough-Miller to study mapping class groups of 3-manifolds.

For compact orientable irreducible 3-manifolds with boundary pattern, we use a relativized version of these ideas. Given a free side F of $(M, \underline{\underline{m}})$, we construct an admissibly imbedded neighborhood $(V, \underline{\underline{v}})$ of F. It has a structure that we call a relative compression body, defined in section 3.1. We actually use two similar but distinct types of relative compression body neighborhood. For each type, the frontier is incompressible, and any loop in F which is contractible in M is contractible in V.

The first type, called a minimally imbedded relative compression body neighborhood, is selected to be as small as possible, in the sense that if W is any other relative compression body neighborhood of F, then V is admissibly isotopic into the topological interior of W. The second, called the normally imbedded relative compression body neighborhood, is as small as possible subject to the condition that no component of its frontier is admissibly homotopic into an element of $\underline{\underline{m}}$. For either of these two types, there exists a disjoint union of such neighborhoods for all the free sides of $(M, \underline{\underline{m}})$. The union of these neighborhoods is unique up to admissible isotopy, so the same is true of the submanifold M' which is the closure of their complement. If the neighborhoods were minimally imbedded, M' is called the maximal incompressible core, and if they were normally imbedded, it is the normal core. Unless all free sides are incompressible, these will not be cores in the sense that their inclusions into M are homotopy equivalences. Each of these two types of incompressible core has the following characteristic property: any admissible homotopy equivalence $f\colon (M, \underline{\underline{m}}) \to (N, \underline{\underline{n}})$ is admissibly homotopic to one which carries the core of $(M, \underline{\underline{m}})$ into the core of $(N, \underline{\underline{n}})$, and restricts to an admissible homotopy equivalence between the cores.

The existence and uniqueness results for these two types of neighborhoods are proven in sections 3.2 and 3.4. Sections 3.3 and 3.5 detail the properties of the maximal and normal incompressible cores.

If the boundary pattern of the normal core has useful completion, the normal core is called the useful core, and the original boundary pattern $\underline{\underline{m}}$ is said to be usable. In particular, we will see in chapter 5 that the boundary pattern associated to a pared structure is usable. In section 4.2, we will use the characteristic property of the normal core to prove that the admissible homotopy type of a manifold $(M, \underline{\underline{m}})$ with usable boundary pattern contains only finitely many elements up to admissible homeomorphism.

In the setting of convex cocompact hyperbolic 3-manifolds, the fundamental groups of normally imbedded compression body neighborhoods of compressible components of the conformal boundary correspond to the function groups in the the Abikoff-Maskit [**3**] decomposition of a convex cocompact Kleinian group into function groups and web groups. In a future paper, we will develop the refined relative compression body neighborhood of the free side of a pared manifold. This decomposition provides the topological analogue of Abikoff and Maskit's decomposition in the setting of geometrically finite hyperbolic 3-manifolds. Our decomposition will be somewhat finer in the setting of geometrically infinite hyperbolic 3-manifolds.

3.1. Relative compression bodies

For $1 \leq i \leq m$ let F_i be a connected (orientable) 2-manifold, not a 2-sphere, with a complete boundary pattern $\underline{\underline{f_i}}$. Form a connected irreducible 3-manifold V from $\bigcup_{i=1}^m F_i \times \mathrm{I}$ by attaching k 1-handles to the manifold interior of $\bigcup_{i=1}^m F_i \times \{1\}$. Denote by F the union of the intersection of ∂V with $\bigcup_{i=1}^m F_i \times \{1\}$ and the intersection of ∂V with the 1-handles. Let $\underline{\underline{v}}$ be a boundary pattern for V such that

(1) F is a free side,
(2) $\bigcup \partial F_i \times \mathrm{I} \subseteq |\underline{\underline{v}}|$, and
(3) if $G \in \underline{\underline{v}}$ and G meets $\partial F_i \times \mathrm{I}$, then each component of $G \cap (\partial F_i \times \mathrm{I})$ is of the form $s \times \mathrm{I}$ for some $s \in \underline{\underline{f_i}}$.

A manifold with boundary pattern $(V, \underline{\underline{v}})$ is called a *relative compression body* if either V is a handlebody and $\underline{\underline{v}}$ is empty, or V is constructed as above and $\underline{\underline{v}}$ satisfies conditions (1), (2), and (3). The free side F is called the *distinguished free side* of $(V, \underline{\underline{v}})$. We denote each $F_i \times \{0\}$ by F_i and call it a *constituent* of V.

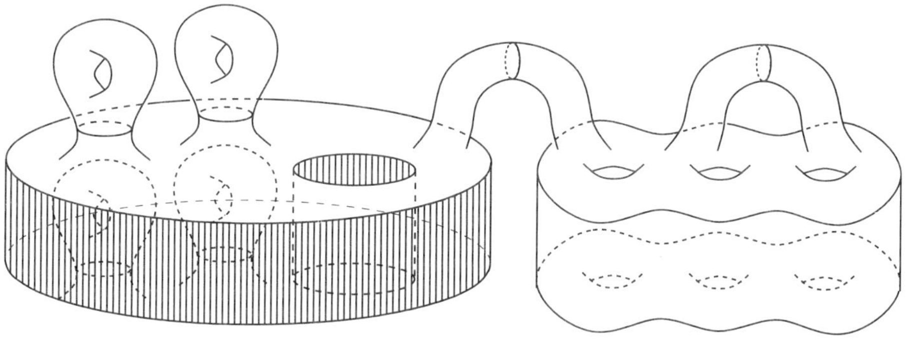

FIGURE 3.1. A compression body with $k\!=\!m\!=\!2$

As indicated above, m denotes the number of constituents of $(V,\underline{\underline{v}})$ and k denotes the number of 1-handles. Figure 3.1 shows a compression body with $m\!=\!2$ and $k=2$: F_1 is a genus-2 surface with two boundary components, and F_2 is a closed surface of genus 3. The distinguished free side F has genus 6 and has two boundary components.

Observe that V is a handlebody if and only if every F_i has nonempty boundary. If V is a handlebody and $F\!=\!\partial V$, then there are no constituents and we define m to be 0 and k to be the genus of V. When $k\!=\!0$, V is either a product $F_1 \times \mathrm{I}$ with $F\!=\!F_1 \times \{1\}$, or V is a 3-ball and $F\!=\!\partial V$; only in these cases is $\pi_1(F) \to \pi_1(V)$ injective.

For a relative compression body $(V,\underline{\underline{v}})$, the homomorphism $\pi_1(F) \to \pi_1(V)$ induced by inclusion is always surjective. If m is positive, then $\pi_1(V) \cong \pi_1(F_1) * \cdots * \pi_1(F_m) * H$, with H a free group of rank $k+1-m$. Lemma 3.1.1 assures us that all admissibly and properly embedded incompressible surfaces are associated to constituent surfaces.

LEMMA 3.1.1. *Let $(V,\underline{\underline{v}})$ be a relative compression body with distinguished free side F, and let $(G,\underline{\underline{g}})$ be a connected surface with complete boundary pattern which is admissibly and properly imbedded in $(V,\underline{\underline{v}})$, with $G \neq S^2$. Assume that $\pi_1(G) \to \pi_1(V)$ is injective. Then there is a unique constituent F_i of $(V,\underline{\underline{v}})$ such that $(G,\underline{\underline{g}})$ is admissibly isotopic into $F_i \times I$. If in addition, there is an admissible isotopy of $(G,\underline{\underline{g}})$ carrying ∂G into ∂F, then $(G,\underline{\underline{g}})$ is admissibly isotopic to $F_i \times \{1/2\}$.*

PROOF. We may assume that V is not a handlebody with empty boundary pattern, since then there are no admissibly imbedded incompressible surfaces with complete boundary pattern. Let E be a union of cocore disks for the 1-handles of V. Since $\underline{\underline{g}}$ is complete, ∂G is disjoint from E. Since G is incompressible, there is an isotopy of G fixed on ∂G that moves G into $V-E$. (This is a standard argument in 3-manifold theory. First move G to be transverse to E, so that the intersection consists of circles. Since $\pi_1(G) \to \pi_1(V)$ is injective, any circle of intersection C must be contractible in G and hence bounds a disk D in G. If C is chosen so that D is innermost among all such disks, then D together with the disk D' that C bounds in E forms an imbedded 2-sphere in V. Since V is irreducible, this bounds a 3-ball in V, and there is an isotopy of G that moves D across the ball and off of E, eliminating C as a circle of intersection, and any other intersection circles that were contained in D'. Repeat this process until G is disjoint from E.) Since $V - E$ admits an admissible deformation retraction to $\bigcup F_i \times \mathrm{I}$, $(G, \underline{\underline{g}})$ is admissibly isotopic into $F_i \times \mathrm{I}$ for some i.

If ∂G and hence $|\underline{\underline{g}}|$ are nonempty, then since the isotopy is admissible, i is uniquely determined. Suppose that G is closed. Since $G \neq \mathrm{S}^2$, $\pi_1(G)$ is nontrivial. Now $\pi_1(V)$ is a free product of the form $\pi_1(F_i) * K$ and $\pi_1(G)$ is conjugate into $\pi_1(F_i)$. Since a subgroup of a free product can be conjugate into at most one free factor (e. g. by considering the normal form of elements), G cannot be homotopic into any other $F_j \times \mathrm{I}$.

To verify the additional assertion, suppose that $(G, \underline{\underline{g}})$ is admissibly isotopic so that $\partial G \subseteq \partial F$, where ∂G may be empty. As before, we may assume that G is contained in some $F_i \times \mathrm{I}$. Using proposition 3.1 of [**128**], any incompressible surface in $F_i \times \mathrm{I}$ with its boundary in $F_i \times \{1\}$ is parallel to $F_i \times \{1\}$. This implies that $(G, \underline{\underline{g}})$ is admissibly isotopic to $F_i \times \{1/2\}$. □

Let F be a free side of $(M, \underline{\underline{m}})$ and suppose that $(V, \underline{\underline{v}})$ is a codimension-zero submanifold of $(M, \underline{\underline{m}})$ with $F \subset V$. We say that $(V, \underline{\underline{v}})$ is a *relative compression body neighborhood of* F if

(a) $(V, \underline{\underline{v}})$ is a relative compression body with distinguished free side F,
(b) the frontier of V is incompressible in M, and
(c) no ∂F_i meets a free side of $(M, \underline{\underline{m}})$.

Condition (c), together with condition (3) in the definition of relative compression body, ensures that ∂F_i lies in the manifold interior of the union of the elements of $\underline{\underline{m}}$ that meet ∂F.

The next theorem is a key property of the complement N of a collection of disjoint relative compression body neighborhoods in M. It guarantees that an admissible homotopy between two maps into M, both of whose images lie in N, has a deformation to a homotopy whose entire image lies in N.

THEOREM 3.1.2. (Homotopy Enclosing Property) *Let $(M, \underline{\underline{m}})$ be a compact irreducible 3-manifold with boundary pattern, and let V be the union of a collection of disjoint relative compression body neighborhoods of some of the free faces of $(M, \underline{\underline{m}})$, for which each constituent either is properly imbedded or is contained in ∂M. Put $N = \overline{M - V}$, with the submanifold boundary pattern $\underline{\underline{n}}$. Let $(X, \underline{\underline{x}})$ be a compact n-manifold with boundary pattern, $1 \leq n \leq 3$. Assume that each component of X either has nonempty boundary pattern or is not simply-connected. Suppose that $f_0, f_1 \colon (X, \underline{\underline{x}}) \to (N, \underline{\underline{n}})$ are essential maps, and $H \colon X \times I \to M$ is an admissible homotopy from f_0 to f_1 as maps into $(M, \underline{\underline{m}})$. Then H is homotopic, relative to $X \times \partial I$ and admissibly with respect to the product boundary pattern on $(X, \underline{\underline{x}}) \times (I, \underline{\underline{\emptyset}})$, to a map $H' \colon X \times I \to N$ which is an admissible homotopy from f_0 to f_1 as maps into $(N, \underline{\underline{n}})$.*

PROOF. We may assume that X is connected, hence that $f_0(X)$ lies in some component $(N_1, \underline{\underline{n_1}})$ of $(N, \underline{\underline{n}})$. Let $\widetilde{M}$ be the covering space of M corresponding to the subgroup $\pi_1(N_1) \subseteq \pi_1(M)$.

Let E_j, $1 \leq j \leq p$ be a collection of cocore 2-disks for the 1-handles of the components of V. Let W be the result of removing from M small open regular neighborhoods $N(E_j)$ (pairwise disjoint, lying in V, with closures disjoint from $|\underline{\underline{m}}|$).

Let W_1 be the component of W that contains N_1, and denote by D_i, $1 \le i \le q$, the components of the frontier of $\bigcup N(E_j)$ that lie in W_1.

Now N_1 meets V in a union $\bigcup C_i$ of constituents that are properly imbedded in M. Since the E_j are a collection of cocore disks for the 1-handles of V, there is a deformation retraction from W_1 to $N_1 \cup (\cup C_i \times \mathrm{I})$, fixed on $|\underline{\underline{w_1}}|$. There is a further admissible deformation retraction from $(N_1 \cup (\cup C_i \times \mathrm{I}), \underline{\underline{w_1}})$ to $(N_1, \underline{\underline{n_1}})$ (this uses the remark after (c) in the definition of relative compression body neighborhood).

The inclusion map of W_1 to M lifts to an imbedding into $\widetilde{M}$ carrying W_1 onto a submanifold $\widetilde{W_1}$. The elements of $\underline{\underline{w_1}}$ lift to $\widetilde{W_1}$, forming a boundary pattern $\underline{\underline{\widetilde{w_1}}}$ for $\widetilde{M}$. Also, the D_i lift to disks $\widetilde{D}_i$ in the free sides of $(\widetilde{W_1}, \underline{\underline{\widetilde{w_1}}})$. Since $\pi_1(\widetilde{W_1}) \to \pi_1(\widetilde{M})$ is an isomorphism, the remainder of $\widetilde{M}$ is obtained by attaching simply-connected 3-manifolds along $\bigcup \widetilde{D}_i$. Therefore $(\widetilde{W_1}, \underline{\underline{\widetilde{w_1}}})$ is a deformation retract of $(\widetilde{M}, \underline{\underline{\widetilde{w_1}}})$.

Let $\widetilde{f_0}\colon X \to \widetilde{W_1} \subset \widetilde{M}$ be the lift of f_0. The admissible homotopy H from f_0 to f_1 lifts to an admissible homotopy $\widetilde{H}$ from $\widetilde{f_0}$ to some lift $\widetilde{f_1}$ of f_1. We claim that $\widetilde{f_1}(X)$ lies in $\widetilde{W_1}$. Since $f_1(X)$ lies in W_1, $\widetilde{f_1}(X)$ is either entirely contained in or entirely disjoint from $\widetilde{W_1}$. If $\underline{\underline{x}}$ is nonempty, then $H(|\underline{\underline{x}}|) \subset |\underline{\underline{w_1}}|$ so $\widetilde{H}(|\underline{\underline{x}}| \times I) \subset |\underline{\underline{\widetilde{w_1}}}|$, and hence $\widetilde{f_1}(X)$ meets $\widetilde{W_1}$. If $\underline{\underline{x}}$ is empty, then $\pi_1(X) \neq 0$ so X contains an essential circle C. But then $\widetilde{f_1}(C)$ cannot lie in the complement of $\widetilde{W_1}$, since the complement is simply-connected. Again, $\widetilde{f_1}(X)$ meets $\widetilde{W_1}$, so $\widetilde{f_1}(X) \subseteq \widetilde{W_1}$, and the claim is established.

Composing $\widetilde{H}$ with the deformation retraction of $\widetilde{M}$ to $\widetilde{W_1}$ and the projection to M defines an admissible deformation from H to a homotopy having image in W_1. Since $(W_1, \underline{\underline{w_1}})$ admits an admissible deformation retraction to $(N_1, \underline{\underline{n_1}})$, there is a further admissible deformation of H yielding a homotopy having image in N_1. □

3.2. Minimally imbedded relative compression bodies

A relative compression body neighborhood $(V, \underline{\underline{v}})$ of F for which each constituent is properly imbedded is said to be *minimally imbedded.* In this case, each constituent is a free face of $(V, \underline{\underline{v}})$.

PROPOSITION 3.2.1. *Let $(M, \underline{\underline{m}})$ be a compact orientable irreducible 3-manifold with boundary pattern. Let F be a free side. Then F has a minimally imbedded relative compression body neighborhood $(V, \underline{\underline{v}})$. Any two minimally imbedded relative compression body neighborhoods of F are isotopic by an admissible ambient isotopy.*

PROOF. By (inductive application of) the Loop Theorem, there exists a sequence $D_1, \ldots, D_\ell$ of disjoint compressing disks with boundary in F so that the frontier of a small regular neighborhood N of $F \cup (\bigcup_{i=1}^{\ell} D_i)$ is incompressible in $\overline{M - N}$. We allow ℓ to be zero, meaning that the collection of disks is empty, when $\pi_1(F) \to \pi_1(M)$ is injective; in this case N is just a product neighborhood of F. We may choose N so that if G is any element of $\underline{\underline{m}}$ that meets F, then $N \cap G$ is a regular neighborhood in G of $G \cap F$, and so that the boundary of the frontier of N lies in the manifold interior of the union of the elements of $\underline{\underline{m}}$ that meet F. If any component of the frontier of N is a 2-sphere, then since M is irreducible it bounds a component of $\overline{M - N}$ which is a 3-ball; adding the union of such balls to N results in a manifold V which is a minimally imbedded relative compression body neighborhood of F.

For later reference, we isolate the next step as a lemma.

LEMMA 3.2.2. *Let F be a free side of $(M, \underline{\underline{m}})$ and let $(V, \underline{\underline{v}})$ be a minimally imbedded relative compression body neighborhood of F in $(M, \underline{\underline{m}})$. Let V' be any irreducible codimension-zero submanifold of M which is a neighborhood of F having incompressible frontier. Then there is an admissible ambient isotopy of $(M, \underline{\underline{m}})$ that moves V into the topological interior of V'.*

PROOF. Suppose first that M is a handlebody and $F = \partial M$. Since M contains no closed incompressible surfaces, the frontiers of V and V' must be empty, so $V = V' = M$. Otherwise, consider a small regular neighborhood $N(V)$ in V of the union of F with the cocore disks for the 1-handles of V. Recall that $\bigcup \partial F_i \times I \subseteq |\underline{\underline{v}}|$, where F_i are the constituents of V, and that if $G \in \underline{\underline{v}}$ and G meets $\partial F_i \times \mathrm{I}$, then each component of $G \cap (\partial F_i \times \mathrm{I})$ is of the form $s \times \mathrm{I}$ for some $s \in \underline{\underline{f_i}}$, a complete boundary pattern for F_i. We may further assume that $N(V)$ meets $\bigcup \partial F_i \times \mathrm{I}$ in $\bigcup \partial F_i \times [\delta, 1]$, where δ is chosen close enough to 1 so that $\bigcup \partial F_i \times [\delta, 1] \subset V'$.

Since the frontier of V' is incompressible, there is an ambient admissible isotopy fixed on ∂M that moves the cocore disks of V into V'. Therefore we may assume that $N(V)$ lies in the topological interior of V'. For each constituent F_i of V, let X_i be the component of $\overline{V - N(V)}$ that contains F_i. It can be given coordinates as a product $F_i \times [0, \delta]$ with $F_i = F_i \times \{0\}$, with $F_i \times \{\delta\}$ a component of the frontier of $N(V)$, and agreeing with the previous coordinates on $\partial F_i \times [0, \delta]$. There is an admissible isotopy of M fixed on $N(V)$ and pulling X_i into any given neighborhood of $N(V)$, so there exists an admissible isotopy moving V into the topological interior of V'. □

To prove the uniqueness of the minimally imbedded relative compression body neighborhood, suppose that V and V' are two minimally imbedded relative compression body neighborhoods of F. By lemma 3.2.2, we may assume that V is contained in the topological interior of V'. By lemma 3.1.1, the first constituent F_1 of V is admissibly isotopic in V' to $G_1 \times \{1/2\}$ for some constituent G_1 of V'. This extends to an admissible isotopy of V in M that is fixed on $\overline{M - V'}$ and hence keeps V in V'. Then, since G_1 is properly imbedded in M, there is an ambient isotopy keeping V in V' that moves F_1 to G_1. Inductively, suppose that the first r constituents $F_1, \ldots, F_r$ of V equal the first r constituents $G_1, \ldots, G_r$ of V'. By lemma 3.1.1, F_{r+1} is admissibly isotopic in V' to some $G_i \times \{1/2\}$. Therefore there is an admissible isotopy of V in V', fixed on $F_1, \ldots, F_r$, that moves F_{r+1} onto $G_i \times \{1/2\}$. If $i \leq r$, then since $G_i \subset V$, V must be contained in $G_i \times [0, 1/2]$, a contradiction since $F \subset V$. So we may assume that $i = r + 1$, and obtain an admissible ambient isotopy of M keeping V in V' and moving F_{r+1} to G_{r+1}. When the induction is completed, each constituent of V is a constituent of V'. Since the union of the constituents of V is the entire frontier of V in M, we have $V = V'$. □

We will often need collections of disjoint minimally imbedded relative compression body neighborhoods of free sides of $(M, \underline{\underline{m}})$.

PROPOSITION 3.2.3. *Let $(M, \underline{\underline{m}})$ be a compact orientable irreducible 3-manifold with a boundary pattern, and let $S_1, \ldots, S_r$ be a collection of free sides of $(M, \underline{\underline{m}})$. Then there exist disjoint minimally imbedded relative compression body neighborhoods for the S_i. Their union is unique up to admissible isotopy of $(M, \underline{\underline{m}})$.*

PROOF. Let V_1 be a minimally imbedded relative compression body neighborhood of S_1 in $(M, \underline{\underline{m}})$. Let M_1 be $\overline{M - V_1}$ with its submanifold boundary pattern $\underline{\underline{m_1}}$. By induction, there exist disjoint minimally imbedded relative compression body neighborhoods $V_2, \ldots, V_r$ of the free sides $S_2, \ldots, S_r$ of $(M_1, \underline{\underline{m_1}})$. Then $V_1, \ldots, V_r$ are disjoint minimally imbedded relative compression body neighborhoods of $S_1, \ldots, S_r$ in $(M, \underline{\underline{m}})$. Given two such collections V_j and V_j', by proposition 3.2.1 we may assume that $V_1 = V_1'$. By induction on r, there is an admissible isotopy of M_1 moving the remaining V_j onto the V_j'. Since this isotopy is admissible, it preserves the free sides of $(M_1, \underline{\underline{m_1}})$ that are the constituents of V_1, so it extends to an admissible isotopy of M preserving V_1. □

We close this section with a characterization of relative compression bodies. We will not use it further in the present work, but it should be included as part of the general theory.

COROLLARY 3.2.4. *Let $(W, \underline{\underline{w}})$ be a compact orientable irreducible 3-manifold with boundary pattern. Then $(W, \underline{\underline{w}})$ is a relative compression body if and only if there exists a free side F of $(W, \underline{\underline{w}})$ such that $\pi_1(F) \to \pi_1(W)$ is surjective.*

PROOF. By construction, if $(W, \underline{\underline{w}})$ is a relative compression body then it has a free side F with $\pi_1(F) \to \pi_1(W)$ surjective. Conversely, given F, let $(V, \underline{\underline{v}})$ be a minimally imbedded relative compression body neighborhood of F. Let V_1 be the component of $\overline{W - V}$ that contains the constituent F_1. Since $\pi_1(F) \to \pi_1(W)$ is surjective, $V_1 \cap V = F_1$. Since F_1 is incompressible, $\pi_1(F_1) \to \pi_1(V_1)$ is injective. Also, $\pi_1(F_1) \to \pi_1(V_1)$ is surjective, otherwise by putting $W' = \overline{W - V_1}$ we would have that $\pi_1(W)$ is a free product with amalgamation $\pi_1(W') *_{\pi_1(F_1)} \pi_1(V_1)$ with the image of $\pi_1(F) \to \pi_1(W)$ contained in $\pi_1(W')$, and $\pi_1(F) \to \pi_1(W)$ could not be surjective. By the Finite Index Theorem 2.1.1, V_1 must be a product $F_1 \times \mathrm{I}$ with $F_1 = F_1 \times \{0\}$. Therefore, if we add this component to V, $(V, \underline{\underline{v}})$ will still be a relative compression body. Repeating with the other constituents of V shows that $(W, \underline{\underline{w}})$ is a relative compression body. □

3.3. The maximal incompressible core

Let $V(M)$ be a union of disjoint minimal relative compression body neighborhoods of the free sides of $(M, \underline{\underline{m}})$, and let $M' = \overline{M - V(M)}$, with the submanifold boundary pattern $\underline{\underline{m}}' = \{\text{components of } F \cap M' \mid F \in \underline{\underline{m}}\}$. Since each component of the frontier of M' is incompressible, and M' cannot be enlarged while retaining this property, we call $(M', \underline{\underline{m}}')$ the *maximal incompressible core* of $(M, \underline{\underline{m}})$. It is not usually a core in the sense that the inclusion $M' \to M$ is a homotopy equivalence. In fact, M' can be empty; this occurs exactly when M is a handlebody and $\underline{\underline{m}}$ is empty. At the other extreme, M' is homotopy equivalent to M exactly when all free faces of M are incompressible; in this case $(M', \underline{\underline{m}}')$ is an admissible deformation retract of $(M, \underline{\underline{m}})$. From proposition 3.2.3, we know that $V(M)$ is unique up to admissible ambient isotopy, and consequently so is M'.

The submanifold $V(M)$ is not characteristic for admissible homotopy equivalences. That is, an admissible homotopy equivalence $(M, \underline{\underline{m}}) \to (N, \underline{\underline{n}})$ need not be admissibly homotopic to one that takes $V(M)$ to $V(N)$. Example 1.4.6 provides a simple example of this. In that example, $(M_1, \underline{\underline{\emptyset}})$ is obtained from $S \times \mathrm{I}$ by attaching both ends of a 1-handle to $S \times \{1\}$, while $(M_2, \underline{\underline{\emptyset}})$ is obtained from $S \times \mathrm{I}$ by attaching opposite ends of a 1-handle to different boundary components of $S \times \mathrm{I}$. For M_1, $V(M_1)$ has two components: $S \times [0, 1/4]$ and the union of the 1-handle with $S \times [3/4, 1]$. On the other hand, $V(M_2)$ is connected: it consists of the union of the 1-handle with $S \times [0, 1/4] \cup S \times [3/4, 1]$. No homotopy equivalence from M_1 to M_2 is homotopic to one taking $V(M_1)$ into $V(M_2)$, since (provided that one selects the basepoint for $V(M_1)$ in the component that contains the 1-handle) $\pi_1(V(M_1)) \to \pi_1(M_1)$ is surjective, but $\pi_1(V(M_2)) \to \pi_1(M_2)$ is not. Example 1.4.1 gives a self-homotopy-equivalence of a 3-manifold $(M, \underline{\underline{\emptyset}})$ which is not homotopic to one preserving $V(M)$. The type of homotopy equivalence used in that example will be studied in detail in section 9.3.

In contrast, the maximal incompressible core M' *is* characteristic for admissible homotopy equivalences. By theorem 3.3.2 below, every admissible homotopy equivalence $f\colon (M, \underline{\underline{m}}) \to (N, \underline{\underline{n}})$ is admissibly homotopic to one that carries M' into the maximal incompressible core N' of N and restricts to an admissible homotopy equivalence from $(M', \underline{\underline{m}}')$ to $(N', \underline{\underline{n}}')$. This characteristic property is a consequence of theorem 3.1.2 and the following property of the maximal incompressible core that applies to all essential maps.

PROPOSITION 3.3.1. *Let $(M, \underline{\underline{m}})$ and $(N, \underline{\underline{n}})$ be compact connected orientable irreducible 3-manifolds with boundary patterns, and let $(M', \underline{\underline{m}}')$ and $(N', \underline{\underline{n}}')$ be their maximal incompressible cores. Let $f\colon (M, \underline{\underline{m}}) \to (N, \underline{\underline{n}})$ be an admissible map which is injective on fundamental groups. Then f is admissibly homotopic to a map which carries $(M', \underline{\underline{m}}')$ into $(N', \underline{\underline{n}}')$.*

PROOF. If M' is empty, there is nothing to prove. When M' is nonempty, N' must also be nonempty, since N' is empty only when N is a handlebody and $\underline{\underline{n}}$ is empty. But either $\pi_1(M')$ is not free so $\pi_1(N)$ is also not free, or $\underline{\underline{m}}$ is nonempty, in which case $\underline{\underline{n}}$ is nonempty.

Let $(V_i, \underline{\underline{v_i}})$ and $(W_j, \underline{\underline{w_j}})$ be the disjoint minimally imbedded relative compression body neighborhoods for the free sides of $(M, \underline{\underline{m}})$ and $(N, \underline{\underline{n}})$ respectively. Let E be the union of the cocores of the 1-handles of the W_j. Since M and N are irreducible, we may change f by admissible homotopy so that each component of $f^{-1}(E)$ is incompressible (as for example in lemma 6.5 of [**51**]). Since $f_\#\colon \pi_1(M) \to \pi_1(N)$ is injective, the components of $f^{-1}(E)$ must be simply-connected, and since M and N are irreducible, components that are 2-spheres may be eliminated by further homotopy of f. So we may assume that each component of the preimage is a properly imbedded 2-disk. Since the boundaries of the disks of E lie in the free sides of $(N, \underline{\underline{n}})$, each disk in $f^{-1}(E)$ has boundary in $\partial M - |\underline{\underline{m}}|$. Since the union of the frontiers of the V_i is incompressible, there is an ambient isotopy, fixed on $|\underline{\underline{m}}|$, which moves $f^{-1}(E)$ into $\bigcup V_i$. Changing f by this isotopy, we may assume that $f^{-1}(E) \subset \bigcup V_i$ and hence that $f(M')$ is disjoint from E. Since N' is nonempty, $N - E$ admits an admissible deformation retraction to N', and we may change f by admissible homotopy so that $f(M') \subset N'$. $\square$

The characteristic property now follows easily.

THEOREM 3.3.2. *Let $(M, \underline{\underline{m}})$ and $(N, \underline{\underline{n}})$ be compact irreducible 3-manifolds with boundary patterns, and let $(M', \underline{\underline{m}}')$ and $(N', \underline{\underline{n}}')$ be their maximal incompressible cores. Let $f\colon (M, \underline{\underline{m}}) \to (N, \underline{\underline{n}})$ be an admissible homotopy equivalence. Then f is admissibly homotopic to a map which carries $(M', \underline{\underline{m}}')$ into $(N', \underline{\underline{n}}')$. Moreover, the restriction of this map to M' is an admissible homotopy equivalence from $(M', \underline{\underline{m}}')$ to $(N', \underline{\underline{n}}')$.*

PROOF. By proposition 3.3.1, we may assume that $f(M') \subset N'$. For an admissible homotopy inverse g of f, we may likewise assume that $g(N') \subseteq M'$. The restriction of gf to M' is admissibly homotopic to the inclusion. By theorem 3.1.2, it is admissibly homotopic to the inclusion by a homotopy with image in M', that is, it is admissibly homotopic to the identity as a map from M' to M'. Similarly, the restriction of fg to N' is admissibly homotopic to the identity on N'. □

3.4. Normally imbedded relative compression bodies

Suppose that $(V, \underline{\underline{v}})$ is a relative compression body neighborhood of a free side F of $(M, \underline{\underline{m}})$. A component R of $\overline{M-V}$ is called *spurious* if $(X, \underline{\underline{x}})$ is of the form $(G \times [-1,0], \{G \times \{-1\} \cup \partial G \times [-1,0]\})$ where $G \times \{0\}$ is a constituent of V. Note that $\underline{\underline{x}}$ consists of a single element, which is homeomorphic to G. The union of V with R is still a relative compression body; the constituent G is replaced by a new constituent $\overline{\partial R - G}$. Conditions (a), (b), and (c) of section 3.1 are still satisfied. We define the *normally imbedded relative compression body neighborhood* of a free side F to be the union of a minimally imbedded relative compression body neighborhood V of F with all spurious components of $\overline{M-V}$.

PROPOSITION 3.4.1. *Let $(M, \underline{\underline{m}})$ be a compact orientable irreducible 3-manifold with a boundary pattern. Let F be a free side. Then F has a normally imbedded relative compression body neighborhood $(W, \underline{\underline{w}})$. Any two normally imbedded relative compression body neighborhoods of F are isotopic by an admissible ambient isotopy.*

PROOF. The existence was explained above. To establish uniqueness, suppose $(W, \underline{\underline{w}})$ is as constructed above, using a minimally imbedded relative compression body neighborhood V, and suppose that $(W', \underline{\underline{w'}})$ is any other normally imbedded relative compression body neighborhood of F. We may form W' by adding the spurious components of $\overline{M-V'}$ to V', for some minimially imbedded relative compression body neighborhood V' of F. By proposition 3.2.1, we may assume that $V = V'$, and hence that $W = W'$. □

We also have the analogue of proposition 3.2.3.

PROPOSITION 3.4.2. *Let $(M, \underline{\underline{m}})$ be a compact orientable irreducible 3-manifold with boundary pattern. Then there exist disjoint normally imbedded relative compression neighborhoods for the free sides of $(M, \underline{\underline{m}})$. Their union is unique up to admissible isotopy of $(M, \underline{\underline{m}})$.*

PROOF. Let $(V_i, \underline{\underline{v_i}})$ be disjoint minimally imbedded relative compression body neighborhoods of the free sides of $(M, \underline{\underline{m}})$. Observe that if X is a spurious component for some V_i, then X is disjoint from all other V_j. For if not, some V_j would be entirely contained in X, but X does not meet any free side of $(M, \underline{\underline{m}})$. Adding in the spurious components, we obtain disjoint normally imbedded compression body neighborhoods of the free sides.

In order to establish uniqueness, consider two collections of disjoint normally imbedded relative compression neighborhoods for the free sides of $(M, \underline{\underline{m}})$. Each may be obtained from a minimally imbedded collection by adding the spurious components. By proposition 3.2.3, we may assume that the minimally imbedded collections are equal, so both normally imbedded collections are formed by adding the spurious complementary components to the same minimally imbedded collection. □

3.5. The normal core and the useful core

Let $V(M)$ be the union of a collection of disjoint normally imbedded relative compression neighborhoods for the free sides of $(M, \underline{\underline{m}})$. We define the *normal core* of $(M, \underline{\underline{m}})$ to be the submanifold $M' = \overline{M - V(M)}$ with submanifold boundary pattern $\underline{\underline{m}}'$. By proposition 3.4.2, the normal core is unique up to ambient isotopy. With respect to admissible homotopy equivalences, we will prove in theorem 3.5.1 that the normal core has the same characteristic property as the maximal incompressible core.

Let $(M', \underline{\underline{m}}')$ be the normal core of $(M, \underline{\underline{m}})$. We say that $\underline{\underline{m}}$ is *usable* if either M' is empty (which occurs when $(M, \underline{\underline{m}})$ is a relative compression body such that $F_i \times \{0\} \cup \partial F_i \times I \in \underline{\underline{m}}$ for every constituent) or $\underline{\underline{m}}'$ has useful completion. When $\underline{\underline{m}}$ is usable, we call $(M', \underline{\underline{m}}')$ the *useful core* of $(M, \underline{\underline{m}})$. After proving the characteristic property of the normal core, we will prove lemma 3.5.2 which will imply that all pared 3-manifolds have usable boundary pattern.

THEOREM 3.5.1. *Let $(M, \underline{\underline{m}})$ and $(N, \underline{\underline{n}})$ be compact connected orientable irreducible 3-manifolds, and let $(M', \underline{\underline{m}}')$ and $(N', \underline{\underline{n}}')$ be their normal cores. Let $f\colon (M, \underline{\underline{m}}) \to (N, \underline{\underline{n}})$ be an admissible homotopy equivalence. Then f is admissibly homotopic to a map which carries $(M', \underline{\underline{m}}')$ into $(N', \underline{\underline{n}}')$. Moreover, the restriction of f is an admissible homotopy equivalence from $(M', \underline{\underline{m}}')$ to $(N', \underline{\underline{n}}')$.*

PROOF. Let $V(M)$ be a disjoint collection of minimally imbedded relative compression body neighborhoods of the free sides of $(M, \underline{\underline{m}})$, and let $M'' = \overline{M - V(M)}$ be the maximal incompressible core. The normal core M' is obtained from M'' by deleting its spurious components. Similarly, we fix $V(N)$, N'', and N'. By theorem 3.3.2, we may assume that f carries M'' to N'' and restricts to an admissible homotopy equivalence $(M'', \underline{\underline{m}}'') \to (N'', \underline{\underline{n}}'')$.

Let $(R, \underline{\underline{r}})$ be a component of $(M'', \underline{\underline{m}}'')$, let $(Y, \underline{\underline{y}})$ be the component of $(N'', \underline{\underline{n}}'')$ that contains $f(R)$, and let $g\colon (R, \underline{\underline{r}}) \to (Y, \underline{\underline{y}})$ be the restriction of f. It suffices to prove that if $(Y, \underline{\underline{y}})$ is spurious, then so is $(R, \underline{\underline{r}})$. Assume that $(Y, \underline{\underline{y}}) = (G \times [-1, 0], \{G \times \{-1\} \cup \partial G \times [-1, 0]\})$ where $G \times \{0\}$ is a constituent of a component of $V(N)$. Since g is an admissible homotopy equivalence, $\underline{\underline{r}}$ contains exactly one element G'. For every essential map of a circle C into R, $g(C)$ is homotopic in Y into $G \times \{0\} \cup \partial G \times \mathrm{I}$, so is inessential in $(Y, \underline{\underline{y}})$. Since g is an admissible homotopy equivalence, this implies that C is inessential in $(R, \underline{\underline{r}})$, so C is homotopic in R into G'. Thus every loop in R is freely homotopic into G', so theorem 3.1 of [**21**] shows that $\pi_1(G') \to \pi_1(R)$ is surjective. Since $G \times \{-1\} \cup \partial G \times [-1, 0]$ is incompressible in Y and g is an admissible homotopy equivalence, it follows that G' is also incompressible, so $\pi_1(G') \to \pi_1(R)$ is injective. By the Finite Index Theorem 2.1.1, R is of the form $G' \times [-1, 0]$ with $G' = G' \times \{-1\}$. Since R cannot meet the free faces of M (since $V(M)$ contains a neighborhood of the free faces), the frontier of R is $\partial G' \times [-1, 0] \cup G' \times \{0\}$. After reselecting the coordinates on $(R, \underline{\underline{r}})$, we see that it is a spurious component of M''. □

We close with the observation that if the boundary pattern consists entirely of closed incompressible surfaces and disjoint incompressible annuli, then the boundary pattern is usable. In particular, the pared 3-manifolds studied in chapter 5 always have usable boundary patterns.

LEMMA 3.5.2. *Let* $(M, \underline{\underline{a}})$ *be an orientable* 3*-manifold with a boundary pattern which consists of closed incompressible surfaces and disjoint incompressible annuli. Then* $\underline{\underline{a}}$ *is usable.*

PROOF. The normal core $(M', \underline{\underline{a}}')$ has incompressible free sides, and the elements of its boundary pattern are disjoint incompressible annuli and closed surfaces. By lemma 2.4.6, each component either has boundary pattern with useful completion, or else with its completed boundary pattern is the product of a 2-faced disk with S^1. But the latter cannot occur, since it would form a spurious component of the normal core. □

CHAPTER 4

Homotopy Types

Homotopy equivalent closed Haken 3-manifolds must be homeomorphic; indeed, Waldhausen's Theorem 2.5.6 shows that admissibly homotopy equivalent Haken 3-manifolds with complete and useful boundary patterns must be admissibly homeomorphic. When the boundary pattern is not complete and useful, the admissible homotopy type may contain distinct homeomorphism types, as seen in examples 1.4.5 and 1.4.6 of chapter 1.

The main results of this chapter, given in section 4.2, form a generalization of a theorem of Johannson [**58, 59**] and Swarup [**118**]. They showed that for a Haken 3-manifold M, there are only finitely many homeomorphism classes of irreducible orientable 3-manifolds homotopy equivalent to M. We extend this in two stages. First, we show in theorem 4.2.1 that the analogous statement holds for the admissible homotopy type of a manifold with a boundary pattern whose completion is useful. The proof follows Swarup's approach closely, and makes essential use of Johannson's Classification Theorem 2.11.1. The second stage is to extend to 3-manifolds whose boundary pattern is usable. This extension follows quickly from the first stage by using some of the results about useful cores developed in section 3.5.

Underlying the proofs of the main results is a technical result on boundary patterns: if two 3-manifolds with boundary pattern are homotopy equivalent (by an admissible homotopy equivalence with admissible homotopy inverse), then the boundary pattern of one is useful if and only if the boundary pattern of the other is useful. Moreover, the same statement is true for the *completions* of the boundary patterns. This invariance is not difficult for the boundary pattern itself, but for the completions the argument is much more involved. We give these arguments in section 4.1. As we explain in that section, our applications to hyperbolic manifolds require this invariance only for the special case of pared manifolds, where a much easier proof is available.

Throughout this chapter, we work only with compact orientable irreducible 3-manifolds.

4.1. Homotopy equivalences preserve usefulness

As discussed above, the main result of this section is the following invariance of usefulness.

THEOREM 4.1.1. *Let $(M, \underline{\underline{m}})$ and $(N, \underline{\underline{n}})$ be compact orientable irreducible 3-manifolds with boundary pattern, which are admissibly homotopy equivalent.*

(i) *If $\underline{\underline{m}}$ is useful, then so is $\underline{\underline{n}}$.*
(ii) *If $\overline{\underline{\underline{m}}}$ is useful, then so is $\overline{\underline{\underline{n}}}$.*

None of our *hyperbolic* applications depends on theorem 4.1.1. In addition to deducing corollary 4.1.2, we will make direct use of it twice. It will be used to deduce proposition 5.2.2, which is essentially a restatement of theorem 4.1.1 in the context of pared 3-manifolds. As we detail in the paragraph before its statement, proposition 5.2.2 can also be deduced from basic topological properties of pared manifolds. The other use of theorem 4.1.1 is in proving the finiteness of the admissible homotopy types for certain classes of 3-manifolds with boundary patterns, in theorems 4.2.1 and 4.2.3. Our hyperbolic applications of this finiteness involve only the case of pared manifolds, for which proposition 5.2.2 can substitute for theorem 4.1.1 in the proofs of theorems 4.2.1 and 4.2.3.

As an immediate consequence of theorem 4.1.1 and theorem 3.5.1, we deduce that the property of having a usable boundary pattern is invariant under admissible homotopy equivalence.

Corollary 4.1.2. *Let $(M, \underline{\underline{m}})$ and $(N, \underline{\underline{n}})$ be compact orientable irreducible 3-manifolds with nonempty boundary, which are admissibly homotopy equivalent. If $\underline{\underline{m}}$ is usable, then $\underline{\underline{n}}$ is usable.*

Proof. By theorem 3.5.1, their normal cores $(M', \underline{\underline{m'}})$ and $(N', \underline{\underline{n'}})$ are admissibly homotopy equivalent. If $\overline{\underline{\underline{m'}}}$ is useful, then part (ii) of theorem 4.1.1 implies that $\overline{\underline{\underline{n'}}}$ is also useful. □

Part (i) of theorem 4.1.1 is an immediate consequence of proposition 2.5.2. Suppose that $\underline{\underline{m}}$ is useful, but that $\underline{\underline{n}}$ is not useful. By proposition 2.5.2, there exists an admissible map $s\colon (D, \underline{\underline{d}}) \to (N, \underline{\underline{n}})$, where $(D, \underline{\underline{d}})$ is a small-faced disk, which is not admissibly homotopic to a constant map. If $g\colon (N, \underline{\underline{n}}) \to (M, \underline{\underline{m}})$ is an admissible homotopy equivalence, then gs is not admissibly homotopic to a constant map of $(D, \underline{\underline{d}})$ to $(M, \underline{\underline{m}})$ (for if it were, then composing the homotopy with an admissible homotopy inverse $f\colon (M, \underline{\underline{m}}) \to (N, \underline{\underline{n}})$ of g would show that fgs is admissibly homotopic to a constant map, but fgs is admissibly homotopic to s). Again by proposition 2.5.2, this shows that $\underline{\underline{m}}$ is not useful. This approach cannot be used in part (ii), since composition with g need not take maps admissible for $(N, \overline{\underline{\underline{n}}})$ to maps admissible for $(M, \overline{\underline{\underline{m}}})$.

The proof of part (ii) of theorem 4.1.1 is an adaptation of the following method for proving that if an irreducible 3-manifold M has incompressible boundary, then any 3-manifold N homotopy equivalent to M also has incompressible boundary (as usual, all 3-manifolds under discussion are assumed to be compact and orientable). Given a homotopy equivalence $f\colon M \to N$, suppose that one has an essential compressing disk E for ∂N. By the usual compression process (as in lemma 6.5 of [**51**], or lemma 2.1 of [**36**]), one may change f by homotopy so that $f^{-1}(E)$ is incompressible. Since f is a homotopy equivalence and E is a disk, this implies that each component of $f^{-1}(E)$ is simply-connected. Since M is irreducible, all 2-sphere components may be eliminated. Since the boundary of M is incompressible, all disk components are parallel into the boundary. By pushing portions of M across these regions of parallelism, one can construct a homotopy starting at the identity of M and ending at a map whose image is disjoint from all the disks in $f^{-1}(E)$. Composing this homotopy with f yields a homotopy from f to a map with image in $N - E$. This implies that f does not induce a surjection on fundamental groups, so could not have been a homotopy equivalence.

The argument above generalizes to give the following result:

LEMMA 4.1.3. *Let* $(M,\underline{\underline{m}})$ *and* $(N,\underline{\underline{n}})$ *be irreducible* 3*-manifolds with boundary patterns, which are admissibly homotopy equivalent. If* $(N,\underline{\underline{n}})$ *has a compressible free side, then so does* $(M,\underline{\underline{m}})$.

PROOF. Let $(D,\partial D)\subseteq(N,\partial N-|\underline{\underline{n}}|)$ be a properly imbedded disk whose boundary is essential in $\partial N-|\underline{\underline{n}}|$. Let $f\colon(M,\underline{\underline{m}})\to(N,\underline{\underline{n}})$ be an admissible homotopy equivalence. Since f is admissible, $f^{-1}(D)\cap|\underline{\underline{m}}|$ is empty. As in the sketch given before the statement of the lemma, we may assume that every component of $f^{-1}(D)$ is a disk whose boundary is essential in $\partial M-|\underline{\underline{m}}|$. To show that $(M,\underline{\underline{m}})$ has a compressible free side, it remains to show that $f^{-1}(D)$ is not empty.

There is an essential circle or arc in $(N,\underline{\underline{n}})$ which is not admissibly homotopic to an arc disjoint from D (if D separates, and a component of the complement is simply-connected, it will be an arc having one or both endpoints in the portion of $|\underline{\underline{n}}|$ that lies in that component). Under an admissible homotopy inverse to f, this circle or arc determines an admissible homotopy class of circles or arcs which are not admissibly homotopic to circles or arcs disjoint from $f^{-1}(D)$. Consequently, $f^{-1}(D)$ cannot be empty. □

PROOF OF THEOREM 4.1.1. We have already explained how part (i) follows from proposition 2.5.2. To establish part (ii), assume that $\overline{\underline{\underline{m}}}$ is useful. Then $\underline{\underline{m}}$ is useful, and by part (i), $\underline{\underline{n}}$ is useful. By lemma 4.1.3, all free sides of $(N,\underline{\underline{n}})$ must be incompressible, so we need only consider an admissibly imbedded i-faced disk D in $(N,\overline{\underline{\underline{n}}})$, with $i=2$ or $i=3$. If all faces of D lie in $|\underline{\underline{n}}|$, then since $\underline{\underline{n}}$ is useful, D cannot violate the usefulness condition for $\overline{\underline{\underline{n}}}$. Therefore we assume that a face of D is contained in a free side of $(N,\underline{\underline{n}})$. Since two free sides cannot meet, D must have exactly one face in a free side of $(N,\underline{\underline{n}})$, and its remaining one or two faces must lie in elements of $\underline{\underline{n}}$.

We may assume that f is transverse to D, first by making it transverse when restricted to elements of $\overline{\underline{\underline{m}}}$, and then on the rest of M. By standard arguments (see for example lemma 6.5 of [**51**] or lemma 2.1 of [**36**]) we may assume that each component of the preimage is simply-connected, and since M is irreducible that each component is a j-faced disk, admissibly imbedded in $(M,\overline{\underline{\underline{m}}})$.

Suppose it happens that each component of $f^{-1}(D)$ is either a 2-faced or a 3-faced disk in $(M,\overline{\underline{\underline{m}}})$, with exactly one of its faces lying in a free side of $(M,\underline{\underline{m}})$. Since $\overline{\underline{\underline{m}}}$ is useful, each such component $(F,\underline{\underline{f}})$ separates M into two components, one of which is a ball B which is the cone on F in such a way that the elements of its boundary pattern are the cones on the elements of $\underline{\underline{f}}$. There is a homotopy of maps from M into M, starting at the identity, that is admissible for $(M,\underline{\underline{m}})$ and moves M into $M-F$. For example, suppose F is a 3-faced disk, so that B is a tetrahedron with base F. One of its triangular faces, say G, is the cone on the element k of $\underline{\underline{f}}$ that lies in a free side of $(M,\underline{\underline{m}})$, so G lies in a free side of $(M,\underline{\underline{m}})$. There is a deformation retraction of B to F that pushes G through B, while preserving the other two triangular faces. Extending this to a map of M using the identity on $\overline{M-B}$, then pushing a little bit farther to move M completely off of B, gives the desired homotopy. The construction is quite similar if F is two-sided. Then, B is a trihedron with base F and two triangular faces, one of which lies in a free side of $(M,\underline{\underline{m}})$, and the deformation retraction pushes that face across B while preserving the other face. Changing the admissible homotopy equivalence f by this homotopy

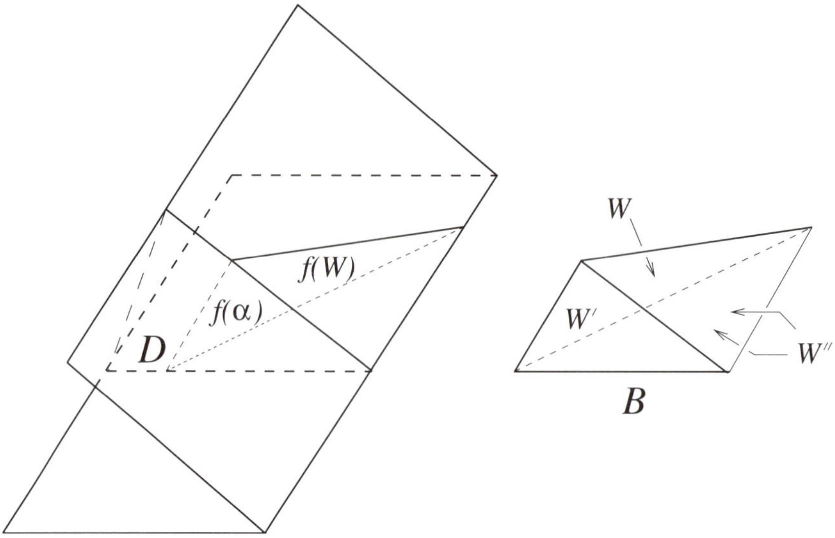

FIGURE 4.1. Construction of a homotopy moving $f(W)$ to D

removes F and possibly other components from the preimage of D. Repeating this operation for the remaining components, we may assume that $f^{-1}(D)$ is empty.

Once we know that $f^{-1}(D)$ is empty, we may derive a contradiction. If D is nonseparating, or separates N into two components neither of which is simply-connected, then f could not induce a surjection on fundamental groups. Suppose, then, that one of the components N_0 of N cut along D is simply-connected, and hence is a 3-ball. We may assume that $f(M) \cap N_0$ is empty. For if not, then $f(M) \subset N_0$, so M and hence N are simply-connected, and the other component of $N - D$ is also simply-connected and may be chosen as N_0. Now ∂D bounds a disk D_0 in the boundary of N_0. Since $f(M) \cap N_0$ is empty and f is an admissible homotopy equivalence, there cannot be any element G of $\underline{\underline{n}}$ completely contained in D_0. Since the elements of $\underline{\underline{n}}$ are incompressible, they must meet D_0 in disks. If $i = 2$, only one component G of $\underline{\underline{n}}$ meets ∂D_0, and $G \cap D_0$ must be a disk, so this shows that $D_0 \cap J(\underline{\underline{n}})$ is the cone on $\partial D_0 \cap J(\underline{\underline{n}})$. If $i = 3$, let G_1 and G_2 be the elements of $\underline{\underline{n}}$ that meet ∂D_0. No component of $G_1 \cap G_2$ can be completely contained in D_0, since the image of f would not meet that component of $G_1 \cap G_2$, and f could not have an admissible homotopy inverse. So the disks $G_1 \cap D_0$ and $G_2 \cap D_0$ meet in a single arc, and again $D_0 \cap J(\underline{\underline{n}})$ is the cone on $\partial D_0 \cap J(\underline{\underline{n}})$. This contradicts the selection of D as a disk that violates the usefulness condition for $\overline{\underline{\underline{n}}}$. Therefore the theorem is reduced to showing that f may be changed by admissible homotopy so that each component of $f^{-1}(D)$ is either a 2-faced or a 3-faced disk in $(M, \overline{\underline{\underline{m}}})$, with exactly one of its faces lying in a free side of $(M, \underline{\underline{m}})$.

The first step is to get rid of preimage disks that have more than 3 faces. To measure the progress in eliminating such disks, define the complexity of f to be the tuple $(\ldots, n_j, n_{j-1}, \ldots, n_4)$, where n_j is the number of components of $f^{-1}(D)$ that are j-faced disks. Since all but finitely many of the n_j are 0, the complexities are well-ordered by the lexicographical ordering, so every decreasing sequence of complexities is finite.

Write $(F, \underline{\underline{f}})$ for $f^{-1}(D)$, where $\underline{\underline{f}}$ is the boundary pattern as a submanifold of $(M, \underline{\underline{m}})$, and suppose that some $n_j > 0$. Then there exists an essential arc μ in $(F, \underline{\underline{f}})$. Since f maps it into $(D, \underline{\underline{d}})$, $f(\mu)$ is inessential. Because f is an admissible homotopy equivalence, this implies that μ is inessential in $(M, \underline{\underline{m}})$. We can now apply the Compression Lemma 2.5.3 to obtain an admissibly imbedded disk $(W, \underline{\underline{w}})$ in $(M, \underline{\underline{m}})$ such that $(W, \overline{\underline{\underline{w}}})$ is a k-faced disk, $2 \leq k \leq 3$, and such that $W \cap F$ is a face α of $(W, \overline{\underline{\underline{w}}})$ which is essential in $(F, \underline{\underline{f}})$.

We claim that $f|_W$ is admissibly homotopic, keeping $f|_\alpha$ fixed, to a map into D. Figure 4.1 illustrates the construction. Let W' be an abstract disk, and write $\partial W'$ as the union $\alpha' \cup \beta'$ of two arcs which meet in their endpoints. Form a disk $W \cup W'$ by identifying α with α'. Let W'' be another abstract disk. Identify its boundary with $\partial(W \cup W')$, and regard $W \cup W' \cup W''$ as the boundary of a 3-ball B.

Using a homotopy of $f|_\alpha$ that moves $f(\alpha)$ through D to an imbedding onto an arc in $D \cap |\underline{\underline{n}}|$, we construct a map $f' : W' \to D$ such that $f'(\alpha') = f(\alpha)$ and $f'(\beta') \subset D \cap |\underline{\underline{n}}|$. The map $f|_W \cup f' \colon W \cup W' \to N$ is admissible as a map into $(N, \underline{\underline{n}})$ for some structure on $W \cup W'$ as a $(k-1)$-faced disk; each face of $W \cup W'$ consists of the union of an element γ of $\underline{\underline{w}}$ with the arc in β' which is the preimage of the element of $\underline{\underline{n}}$ that contains $f(\gamma)$. Since $\underline{\underline{n}}$ is useful, $(f|_W \cup f')|_{\partial(W \cup W')}$ extends to $f'' \colon W'' \to |\underline{\underline{n}}|$ such that $(f'')^{-1}(J(\underline{\underline{n}}))$ is the cone on $(f'')^{-1}(J(\underline{\underline{n}})) \cap \partial W''$ (this cone is an arc, when $k = 3$, and is empty, when $k = 2$). Assuming that f'' is selected to be transverse to D, $(f'')^{-1}(D)$ consists of $W'' \cap W'$ together with some circles in the interior of W''. By cutting and pasting, we may assume that $(f'')^{-1}(D) = W'' \cap W'$.

Since $\pi_2(N) = 0$, $f|_W \cup f' \cup f''$ extends to a map from B into N. The transverse preimage of D consists of W' together with some closed surfaces in the interior of B, and by the usual simplification process we may assume that the preimage of D is just W'. We regard this map of B as a homotopy starting at $f|_W$, through maps from W to N for which the preimage of D consists only of α, and which agree with $f|_\alpha$ on α, and ending at a map from W into D. Using such a homotopy, f may be changed by admissible homotopy first to add W to the preimage of D, and then to perform surgery on F along α. Figure 4.2 illustrates the effect of the homotopy. The component of $(F, \underline{\underline{f}})$ that contained α is a j-faced disk that is replaced by two disks each of which is admissibly homotopic to a disk obtained from a component of $F - \alpha$ by appending a copy of W. Each of the resulting disks has fewer than j sides (since α was essential), so the complexity of f is reduced.

Since all sequences of descending complexities are finite, after finitely many repetitions we arrive at the situation where all $n_j = 0$, so $f^{-1}(D)$ consists of small-faced disks. (Actually, if $i = 2$, then there can be no 3-faced disks in $(F, \underline{\underline{f}})$, since then f would map two adjacent faces of a 3-faced disk to the same face of $(D, \underline{\underline{d}})$, violating admissibility.)

Next, we eliminate 1-faced disks in $f^{-1}(D)$. Such a disk E gives a compression of an element of $\overline{\underline{\underline{m}}}$. Since $\overline{\underline{\underline{m}}}$ is useful, E is admissibly parallel to a disk E_0 in an element of $\overline{\underline{\underline{m}}}$. Since M is irreducible, $E_0 \cup E$ bounds a 3-ball B in M. We will show that there is an admissible homotopy of f that removes E, and possibly some other components, from $f^{-1}(D)$. If ∂E is contained in a free side of $(M, \underline{\underline{m}})$, then there is an admissible homotopy (through imbeddings), starting at the identity map of M and supported in a regular neighborhood of B, that moves M into $M - B$. Composing f with this homotopy gives an admissible homotopy from f to a map from which E has been removed from the preimage of D (along with any

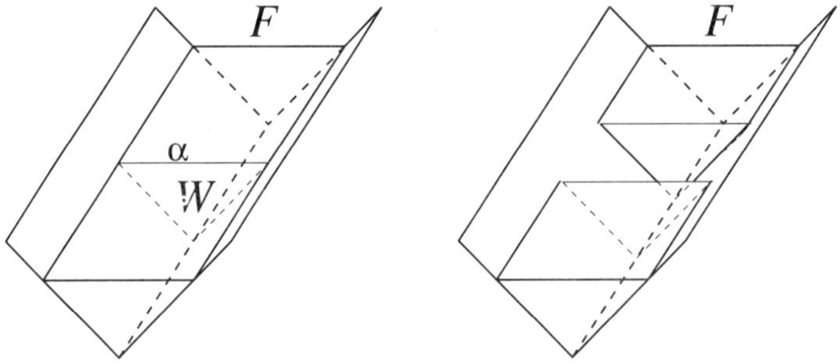

(a) before homotopy (b) after homotopy

FIGURE 4.2. Effect of a homotopy that simplifies $f^{-1}(D)$

other components which may have been contained in B). Suppose now that ∂E is contained in an element F of $\underline{\underline{m}}$. Let γ be the face of D which contains $f(\partial E)$, and let G be the element of $\underline{\underline{n}}$ that contains γ. Since γ is an arc, and G is aspherical (unless $(N,\underline{\underline{n}}) = (D^3,\underline{\underline{\emptyset}})$, in which case $\underline{\underline{m}}$ must consist of one 2-sphere, $(M,\underline{\underline{m}})$ must be $(D^3,\underline{\underline{\emptyset}})$, and the theorem holds), $f|_{E_0}\colon E_0 \to G$ is homotopic relative to ∂E_0 to a map into γ. So we may change f by admissible homotopy to add E_0 to the preimage of D. Since N is aspherical, we may change f by a further admissible homotopy to add the rest of B to the preimage, and by a small further homotopy to remove all of B from the preimage. The net effect is to remove E (and possibly other 1-faced disks contained in B) from the preimage of D. Repeating, we may assume that for each disk $(E,\underline{\underline{e}})$ in $f^{-1}(D)$, $(E,\underline{\underline{e}})$ is j-faced with $2 \leq j \leq 3$.

The last step is to ensure that each of the preimage disks has exactly one of its faces in a free side of $(M,\underline{\underline{m}})$. Suppose E is a 2-faced disk in $f^{-1}(D)$ and both faces lie in $|\underline{\underline{m}}|$. Since f is admissible, we must have $i=3$ and f maps the faces of E to the two faces of D that lie in elements of $\underline{\underline{n}}$. Since $\underline{\underline{m}}$ is useful, ∂E bounds a disk E_0 in $|\underline{\underline{m}}|$ such that $E_0 \cap J(\underline{\underline{m}})$ is the cone on $\partial E_0 \cap J(\underline{\underline{m}})$. Let α be the arc $E_0 \cap J(\underline{\underline{m}})$, let G' and G'' be the two bound sides of $\underline{\underline{n}}$ that contain faces of D, and let p be the point $G' \cap G'' \cap D$. Note that f carries α into $G' \cap G''$ and both endpoints of α are mapped to p, so we may think of $f(\alpha)$ as a loop.

Suppose for contradiction that the loop $f(\alpha)$ based at p is not null-homotopic in $G' \cap G''$. Then the component of $G' \cap G''$ containing p must be a circle, rather than an arc. Since α is homotopic relative to its endpoints into E, $f|_\alpha$ is contractible in N. Since G' and G'' are incompressible, f_α is contractible in each of them, so they are 2-disks whose union is ∂N, which is impossible since then the 3-faced disk D could not exist. We conclude that $f|_\alpha$ is contractible in $G' \cap G''$, so f is admissibly homotopic to a map taking α to p. Since f carries ∂E_0 into the arc $D \cap (G' \cup G'')$, and G' and G'' are aspherical, we may change f by a further admissible homotopy to add E_0 to $f^{-1}(D)$. Since M is irreducible, $E \cup E_0$ bounds a 3-ball in M, and since $\pi_3(N)=0$, f can be changed to add all of this 3-ball to $f^{-1}(D)$. By a small further admissible homotopy, f can be changed to remove E from $f^{-1}(D)$.

Now suppose that E is a 3-faced disk and all faces lie in $|\underline{\underline{m}}|$. Since D has at most two faces in $|\underline{\underline{n}}|$, this would imply that f carries two adjacent elements of $\underline{\underline{m}}$ into a single element of $\underline{\underline{n}}$, which is impossible. So we may assume that every disk in $f^{-1}(D)$ has a face in a free side of $(M, \underline{\underline{m}})$. As we have already observed, this completes the proof. □

4.2. Finiteness of homotopy types

In this section we generalize the Johannson-Swarup finiteness results, as explained in the introduction to this chapter. In the case when the boundary pattern is useful, we follow Swarup's approach closely, making essential use of Johannson's Classification Theorem 2.11.1. We then extend to 3-manifolds whose boundary pattern is usable. This second stage follows quickly from the first by using some of the results about useful cores developed in section 3.5.

THEOREM 4.2.1. *Let $(M, \underline{\underline{m}})$ be a compact, orientable, irreducible 3-manifold with boundary pattern $\underline{\underline{m}}$ whose completion is useful and nonempty. Then the admissible homotopy type of $(M, \underline{\underline{m}})$ contains only finitely many admissible homeomorphism types of compact orientable irreducible 3-manifolds.*

Before beginning the proof, we will discuss Dehn twist homeomorphisms of 3-manifolds, which make their first appearance here, and will be used on various occasions in our later arguments. Suppose that $\mathrm{S}^1 \times \mathrm{S}^1 \times \mathrm{I}$ is a collar neighborhood of a torus $T = S^1 \times S^1 \times \{0\}$ which either lies in the interior of M, or is a torus boundary component of M. Fix a basepoint $x_0 = (1, 1, 0) \in \mathrm{S}^1 \times \mathrm{S}^1 \times \{0\}$, and let γ be the loop that send t to $(e^{2\pi i p t}, e^{2\pi i q t}, 0)$ for some integers p and q. Define a homeomorphism h of M to be the identity off of $\mathrm{S}^1 \times \mathrm{S}^1 \times \mathrm{I}$, while there it will be $h(e^{2\pi i u}, e^{2\pi i v}, w) = (e^{2\pi i(u+pw)}, e^{2\pi i(v+qw)}, w)$. This is called a *Dehn twist* homeomorphism *about* T with *trace* g, where g is the element of $\pi_1(M, x_0)$ represented by γ. The isotopy class of a Dehn twist depends only the torus T, the trace, and (if T is in the interior) the side of T on which the collar neighborhood is selected, but not on the representative path for the trace nor the choice of collar neighborhood.

Similarly, one can define a Dehn twist homeomorphism about any connected 2-manifold F which either is properly and two-sidedly imbedded in M, or lies in ∂M. Fix a collar $F \times \mathrm{I}$, which meets ∂M in $\partial F \times \mathrm{I}$, and a loop j_t in the group of diffeomorphisms $\mathrm{Diff}(F)$, with j_0 and j_1 both equal to the identity id_F, then define $h(w) = w$ for $w \notin F \times \mathrm{I}$ and $h(x, t) = (j_t(x), t)$ for $(x, t) \in F \times \mathrm{I}$. The isotopy class of h depends only on the element of $\pi_1(\mathrm{Diff}(F), \mathrm{id}_F)$ represented by j_t. The cases that we will use will be when F is an annulus, for which $\pi_1(\mathrm{Diff}(F), \mathrm{id}_F) \cong \mathbb{Z}$ with elements classified by the trace, and, in chapter 12, when F is a 2-sphere, for which $\pi_1(\mathrm{Diff}(F), \mathrm{id}_F) \cong \mathbb{Z}/2$ with the nontrivial element represented by letting j_t be an orthogonal rotation through an angle $2\pi t$, fixing the basepoint and its antipode.

PROOF OF THEOREM 4.2.1. We follow the argument in [**118**]. Let $\mathcal{C}$ be the collection of (admissible homeomorphism classes of) admissibly fibered (orientable) I-bundles and Seifert fiber spaces, which are irreducible and have nonempty boundary. Define $\mathcal{C}(G, n)$ to be the collection of elements $(\Sigma, \underline{\underline{\sigma}})$ of $\mathcal{C}$ such that $\pi_1(\Sigma) \cong G$ and $\underline{\underline{\sigma}}$ has cardinality n, and define $\mathcal{C}_0(G, n)$ to be the subset of $\mathcal{C}(G, n)$ consisting of those manifolds for which the boundary pattern has disjoint elements.

Consider the I-bundles in $\mathcal{C}(G, n)$. An I-bundle is completely determined by the admissible homeomorphism type of its quotient surface $(B, \underline{\underline{b}})$, and there are only finitely many admissible homeomorphism types of $(B, \underline{\underline{b}})$ with a given Euler characteristic and cardinality of $\underline{\underline{b}}$. Therefore there are only finitely many I-bundles in each $\mathcal{C}(G, n)$.

Assume that G is not infinite cyclic. By proposition 1.5 of [**118**], there are only finitely many Seifert fibered manifolds $(\Sigma, \underline{\underline{\sigma}})$ in $\mathcal{C}_0(G,n)$. From this we will deduce the same for $\mathcal{C}(G,n)$. If $(\Sigma, \underline{\underline{\sigma}}) \in \mathcal{C}(G,n)$, then an annulus or torus of $|\underline{\underline{\sigma}}|$ may be made up of several annuli of $\underline{\underline{\sigma}}$, but every element of $\mathcal{C}(G,n)$ arises by subdividing some elements of the boundary pattern of one of the finitely many elements $(\Sigma_0, \underline{\underline{\sigma_0}})$ of $\mathcal{C}_0(G,m)$ for some $m \leq n$. If $(\Sigma_0, \underline{\underline{\sigma_0}})$ has a unique admissible Seifert fibering up to isotopy, then it can give rise to only finitely many elements of $C(G,n)$. If its Seifert fibering is not unique, then since Σ is irreducible and has nonempty boundary, the Unique Fibering Theorem 2.8.1 shows that $(\Sigma_0, \underline{\underline{\sigma_0}})$ can be admissibly fibered as the I-bundle over either the Klein bottle or the torus. In the first case, lemma 2.8.5 shows that there are only two isotopy classes of Seifert fiberings. In the second case, there are infinitely many isotopy classes, but all are homeomorphic preserving fibers. In both cases, only finitely many elements of $C(G,n)$ can arise from subdividing elements of $\underline{\underline{\sigma_0}}$. Thus if G is not infinite cyclic, $\mathcal{C}(G,n)$ contains only finitely many Seifert fibered manifolds.

Define $\mathcal{A}_m(n)$ to be the subset of $\mathcal{C}$ consisting of all elements such that $(\Sigma, \underline{\underline{\sigma}})$ is a Seifert fibered solid torus, $\underline{\underline{\sigma}}$ has cardinality n and contains an incompressible annulus F such that the index of $\pi_1(F)$ in $\pi_1(\Sigma)$ has order m, $1 \leq m < \infty$. By proposition 1.5′ of [**118**] and a subdivision argument as in the case of $C(G,n)$, $\mathcal{A}_m(n)$ is finite.

LEMMA 4.2.2. *Let $(\Sigma, \underline{\underline{\sigma}}) \in \mathcal{C}$. Suppose that $\underline{\underline{\sigma}}' \subseteq \underline{\underline{\sigma}}$ and that $\underline{\underline{\sigma}}'$ does not contain a lid of any I-fibered component of $(\Sigma, \underline{\underline{\sigma}})$. Let $\mathcal{E}(\Sigma, \underline{\underline{\sigma}}, \underline{\underline{\sigma}}')$ be the group of path components of the space of admissible self-homotopy equivalences of $(\Sigma, \underline{\underline{\sigma}})$ which take each element of $\underline{\underline{\sigma}}'$ to itself by a homeomorphism. Let c be the number of components of $|\underline{\underline{\sigma}}'|$. There exist a finite group A of order at most 8^c and a homomorphism $\psi\colon \mathcal{E}(\Sigma, \underline{\underline{\sigma}}, \underline{\underline{\sigma}}') \to A$ such that if $\psi(f) = 0$ then the restriction of f to $|\underline{\underline{\sigma}}'|$ extends to an admissible homeomorphism of $(\Sigma, \underline{\underline{\sigma}})$.*

PROOF. We can ignore components of Σ that do not meet $|\underline{\underline{\sigma}}'|$, since on these components the extension can be selected to be the identity map. We may assume that Σ is connected, since if there is a homomorphism as in the lemma for each component, then their direct product satisfies the lemma for their union.

When $(\Sigma, \underline{\underline{\sigma}})$ is an I-bundle, every element of $\underline{\underline{\sigma}}'$ is a square or annulus, so has (admissible) mapping class group of order at most 8. Let A be the direct product of these mapping class groups, and define ψ by restriction. If $\psi(f) = 0$, then the identity map is admissibly isotopic to a homeomorphism whose restriction to $|\underline{\underline{\sigma}}'|$ agrees with the restriction of f, and the lemma is verified. From now on, we assume that $(\Sigma, \underline{\underline{\sigma}})$ is Seifert fibered.

Suppose $|\underline{\underline{\sigma}}'|$ equals $\partial\Sigma$. By Waldhausen's Theorem 2.5.6, f is admissibly homotopic to a homeomorphism h. Thus the lemma holds in this case by taking $A = \{0\}$.

If $\Sigma = S^1 \times S^1 \times I$, and no element of $\underline{\underline{\sigma}}'$ is an annulus, then every homeomorphism of $|\underline{\underline{\sigma}}'|$ that extends to a homotopy equivalence of Σ extends to a homeomorphism, so again we may take $A = \{0\}$.

In all other cases, the Fiber-preserving Self-map Theorem 2.8.6 shows that every element of $\mathcal{E}(\Sigma, \underline{\underline{\sigma}}, \underline{\underline{\sigma}}')$ preserves the fiber up to homotopy. Let A be a direct product of copies of $\mathbb{Z}/2$, two for each component of $|\underline{\underline{\sigma}}'|$, and define $\psi(f)$ to be the nontrivial element in the $(2i-1)^{th}$ coordinate if and only if f reverses the orientation of the ith component of $|\underline{\underline{\sigma}}'|$, and to be the nontrivial element in the $(2i)^{th}$ coordinate if and only if f reverses the orientation in the fiber of the ith component of $|\underline{\underline{\sigma}}'|$. Then if $\psi(f)=0$, f is orientation-preserving on each component of $|\underline{\underline{\sigma}}'|$, and preserves the orientation of the fiber in each component of $|\underline{\underline{\sigma}}'|$. Since these components are fibered tori and annuli, this shows that the restriction of f to $|\underline{\underline{\sigma}}'|$ is isotopic (preserving the elements of $\underline{\underline{\sigma}}'$) to a fiber-preserving homeomorphism. In turn, this is isotopic to a product of Dehn twists about a fiber. Since $|\underline{\underline{\sigma}}'| \neq \partial\Sigma$, we can choose for each component X of $|\underline{\underline{\sigma}}'|$ an annulus imbedded in Σ with one boundary circle a fiber contained in X and the other boundary circle a fiber contained in $\partial\Sigma - |\underline{\underline{\sigma}}'|$. The Dehn twists of the components of $|\underline{\underline{\sigma}}'|$ extend to Dehn twists about these annuli, giving the extension of the restriction of f. □

To complete the proof of theorem 4.2.1, suppose that $f\colon (M,\underline{\underline{m}}) \to (N,\underline{\underline{n}})$ is an admissible homotopy equivalence. Theorem 4.1.1, implies that $\underline{\underline{n}}$ has useful completion. Let $(V,\underline{\underline{v}})$ be the characteristic submanifold of $(M,\overline{\underline{\underline{m}}})$. By the Classification Theorem 2.11.1, we may assume that the restriction of f to $(\overline{M-V},\underline{\underline{m}}')$ is an admissible homeomorphism, and the restriction of f to $(V,\underline{\underline{v}}')$ is an admissible homotopy equivalence, where $\underline{\underline{m}}'$ and $\underline{\underline{v}}'$ are the proper boundary patterns. If V is empty, there is only one admissible homeomorphism type in the admissible homotopy type of $(M,\underline{\underline{m}})$. If $V=M$, then since $\mathcal{C}(G,n)$ is finite if G is not infinite cyclic, and $\mathcal{A}_m(n)$ is finite, the finiteness has already been verified. Therefore we may assume that V is nonempty and $V\neq M$.

Since $\mathcal{C}(G,n)$ and $\mathcal{A}_m(n)$ are finite, the admissible homotopy type of $(V,\underline{\underline{v}})$ consists of some finite number t_1 of admissible homeomorphism types. Let c be the number of elements of $\underline{\underline{v}}'$, and let $t=t_1(c!\,(8^c+1)+1)$. We will show that if $(N_1,\underline{\underline{n_1}}),\ldots,(N_m,\underline{\underline{n_m}})$ is any collection of t compact orientable irreducible 3-manifolds admissibly homotopy equivalent to $(M,\underline{\underline{m}})$, then for some $i\neq j$, $(N_i,\underline{\underline{n_i}})$ is admissibly homeomorphic to $(N_j,\underline{\underline{n_j}})$.

The Classification Theorem 2.11.1 implies that any admissible homotopy equivalence from $(M,\underline{\underline{m}})$ to $(N_i,\underline{\underline{n_i}})$ is admissibly homotopic to a map which restricts to an admissible homotopy equivalence between the characteristic submanifolds of $(M,\overline{\underline{\underline{m}}})$ and $(N_i,\overline{\underline{\underline{n_i}}})$ and an admissible homeomorphism on the complements of the characteristic submanifolds (with their proper boundary patterns). Note that this homeomorphism carries the elements of the proper boundary pattern that lie in the frontier of the characteristic submanifold of $(M,\overline{\underline{\underline{m}}})$ to the elements of the proper boundary pattern that lie in the frontier of the characteristic submanifold of $(N_i,\overline{\underline{\underline{n_i}}})$. Since there are t_1 admissible homeomorphism types in the admissible homotopy type of the characteristic submanifold of $(M,\overline{\underline{\underline{m}}})$, there is a subcollection of at least $c!(8^c+1)+1$ of the N_i such that the characteristic submanifolds are admissibly homeomorphic. So we may select notation so that for $1\leq i\leq c!(8^c+1)+1$, each $(N_i,\underline{\underline{n_i}})$ can be constructed from a fibered manifold $(\Sigma,\underline{\underline{\sigma}})$ and a simple manifold $(S,\underline{\underline{s}})$ by identifying pairs of elements of $\underline{\underline{\sigma}}$ and $\underline{\underline{s}}$ by homeomorphisms (each pair corresponding to a component of the frontier of Σ in N_i).

For each i with $2\leq i\leq c!(8^c+1)+1$, use the Classification Theorem 2.11.1 to select an admissible homotopy equivalence $f_i\colon (N_1,\underline{\underline{n_1}}) \to (N_i,\underline{\underline{n_i}})$ which restricts to a homeomorphism from $(S,\underline{\underline{s}})$ to $(S,\underline{\underline{s}})$ and an admissible homotopy equivalence from $(\Sigma,\underline{\underline{\sigma}})$ to $(\Sigma,\underline{\underline{\sigma}})$. Regard these restrictions as self-homotopy-equivalences of Σ, and for each i let g_i be an admissible homotopy inverse for f_i. Let $\underline{\underline{\sigma}}''$ be the elements of $\underline{\underline{\sigma}}$ that lie in the frontier of Σ in N_1. Since $f_i|_{|\underline{\underline{\sigma''}}|}$ is a homeomorphism, we may assume that $g_i|_{f_i(|\underline{\underline{\sigma''}}|)} = \left(f_i|_{|\underline{\underline{\sigma''}}|}\right)^{-1}$.

There are at least 8^c+1 values of i for which the f_i induce the same permutation on the components of $|\underline{\underline{\sigma''}}|$. Choose one, i_0, and for each of the others define $k_i = f_i g_{i_0}$. Let $\underline{\underline{\sigma'}} = \{f_{i_0}(G) \mid G \in \underline{\underline{\sigma''}}\}$. Notice that σ' does not contain a lid of any I-fibered component V of (Σ, σ), since each element of σ' lies in the frontier of V in N_{i_0} and the lids must lie in $|\overline{\underline{\underline{n_{i_0}}}}|$. Using the notation of lemma 4.2.2, each $k_i|_\Sigma \in \mathcal{E}(\Sigma, \underline{\underline{\sigma}}, \underline{\underline{\sigma'}})$, and that lemma shows that for some i and j with $i \neq j$, the homeomorphism $\left(k_j|_{|\underline{\underline{\sigma'}}|}\right)\left(k_i|_{|\underline{\underline{\sigma'}}|}\right)^{-1}$ extends to a homeomorphism $k\colon (\Sigma, \underline{\underline{\sigma}}) \to (\Sigma, \underline{\underline{\sigma}})$. A homeomorphism $h\colon (N_i, \underline{\underline{n_i}}) \to (N_j, \underline{\underline{n_j}})$ is given by taking k on Σ and $\left(k_j|_S\right)\left(k_i|_S^{-1}\right)$ on S. □

The usable case now follows easily:

THEOREM 4.2.3. *Let $(M, \underline{\underline{m}})$ be a compact orientable irreducible 3-manifold with nonempty boundary, with $\underline{\underline{m}}$ usable. Then the admissible homotopy type of $(M, \underline{\underline{m}})$ contains only finitely many admissible homeomorphism classes of compact orientable irreducible 3-manifolds with boundary pattern.*

PROOF. Let $(M', \underline{\underline{m'}})$ be the useful core of $(M, \underline{\underline{m}})$. By theorem 4.2.1, there are only finitely many admissible homeomorphism classes of orientable 3-manifolds in the admissible homotopy type of $(M', \underline{\underline{m'}})$. By theorem 3.5.1, any compact orientable irreducible manifold admissibly homotopy equivalent to $(M, \underline{\underline{m}})$ is obtained from a manifold admissibly homotopy equivalent to $(M', \underline{\underline{m'}})$ by attaching 1-handles along the free sides. The number of handles added is determined by $\pi_1(M)$ so there are only finitely many resulting manifolds, up to admissible homeomorphism. □

CHAPTER 5

Pared 3-Manifolds

This chapter concerns pared 3-manifolds, which arise naturally in the study of hyperbolic 3-manifolds. Topologically, a pared structure is a special type of boundary pattern, whose elements consist of disjoint incompressible annuli and tori, satisfying some strong additional conditions. Thurston's Geometrization Theorem asserts that these conditions are precisely what is needed to guarantee a hyperbolic structure on the interior of the manifold in which the elements of the boundary pattern correspond to "cusps" of the hyperbolic structure (see chapter 7).

In section 5.1, we will define pared 3-manifolds and give some of their basic properties. Their topological properties are explored more fully in section 5.2. In particular, a pared structure determines a boundary pattern which is usable (as defined in section 3.5), and whose completion is useful if and only if all the free sides are incompressible.

In section 5.3 we investigate the characteristic submanifold of a pared 3-manifold with incompressible free sides. Theorem 5.3.4 shows that its fibering can be chosen so that each component that is an I-bundle has quotient surface of negative Euler characteristic, and each Seifert fibered component is either a solid torus or a thickened torus. If the characteristic submanifold has a $T^2 \times I$ component with at least two frontier annuli, then we say that the pared manifold has double trouble. In theorem 5.3.6, we verify, using the Classification Theorem, that having double trouble is an invariant of pared homotopy type. It will turn out that a pared 3-manifold with no compressible free side has double trouble if and only if the realizable automorphisms have infinite index in the (relative) outer automorphism group. This topological result plays a key role in the proof of the Main Hyperbolic Theorem in the incompressible case.

5.1. Definitions and basic properties

We begin by defining pared manifolds and developing some of their basic properties. Let M be a compact, orientable, irreducible 3-manifold with nonempty boundary which is not a 3-ball, and let $P \subseteq \partial M$. We say that (M, P) is a *pared* 3-manifold (see Morgan [**96**]) if the following three conditions hold.

(P1) Every component of P is an incompressible torus or annulus.

(P2) Every noncyclic abelian subgroup of $\pi_1(M)$ is conjugate into the fundamental group of a component of P.

(P3) Every map $\phi\colon (\mathrm{S}^1 \times \mathrm{I}, \mathrm{S}^1 \times \partial\mathrm{I}) \to (M, P)$ which induces an injection on fundamental groups is homotopic, as a map of pairs, to a map ψ such that $\psi(\mathrm{S}^1 \times \mathrm{I}) \subset P$.

We single out three special cases, called elementary pared manifolds, which are so simple that they are sometimes exceptions to results about pared manifolds, or if not exceptions, may require special arguments. These arguments are never difficult, and will sometimes be ignored in our proofs. Let T^2 and A^2 denote the torus and annulus respectively. If $(M, P) = (T^2 \times \mathrm{I}, T^2 \times \{0\})$, or $(A^2 \times \mathrm{I}, A^2 \times \{0\})$, or $(A^2 \times \mathrm{I}, \emptyset)$, then (M, P) is said to be *elementary,* otherwise it is *nonelementary.* The elementary pared 3-manifolds correspond to the hyperbolic 3-manifolds with abelian fundamental groups.

In the next lemma we collect some basic properties of nonelementary pared 3-manifolds.

LEMMA 5.1.1. *Let (M, P) be a nonelementary pared 3-manifold.*

(i) *Every toroidal component of ∂M is contained in P.*
(ii) *M is not homeomorphic to $T^2 \times I$, to the I-bundle over the Klein bottle, or to the solid torus.*
(iii) *M does not contain an imbedded Klein bottle.*
(iv) *For each component Q of P, the subgroup $\pi_1(Q)$ is a maximal abelian subgroup of $\pi_1(M)$.*

PROOF. Suppose that M has a toroidal boundary component T that is not contained in P. Assume first that T is compressible. Since M is irreducible, it must be a solid torus, and P is nonempty since (M, P) is nonelementary. By (P1) and (P3), P must consist of a single incompressible annulus A. Let A' be the closure of $\partial M - A$. By (P3), A' must be properly homotopic into A, so the pair (M, A) has the form $(A^2 \times \mathrm{I}, A^2 \times \{0\})$ (for the homotopy yields a meridian disk of M that meets A and A' each in a single arc, showing that $\pi_1(A) \to \pi_1(M)$ is surjective). Therefore (M, P) would be elementary. Assume now that T is incompressible. By (P2), $\pi_1(T)$ is conjugate into $\pi_1(T')$ for some torus boundary component T' in P. By lemma 5.1 of [128], this implies that $M = T \times \mathrm{I}$. By (P3), P contains no annulus in T, so $P = T'$ and again (M, P) would be elementary. This proves (i).

If M is a solid torus or I-bundle over the torus or Klein bottle, then each boundary component of M is a torus so would have to be in P. For an I-bundle over the Klein bottle or torus, condition (P3) would be violated. For the solid torus, the incompressibility in condition (P1) would be violated. This verifies (ii).

Suppose that M contains a compressible Klein bottle K. If a separating essential loop in K bounds a compressing disk in M, then surgery on K along this disk yields two disjoint imbedded projective planes. This is impossible since M is irreducible (the projective planes are one-sided since M is orientable, and a regular neighborhood of a one-sided projective plane is an $\mathbb{RP}^3$-summand). If a nonseparating loop bounds a compressing disk, then surgery along this disk yields a 2-sphere which bounds a ball in M, showing that K separates M. This is impossible, since M is orientable.

Suppose now that K is an imbedded incompressible Klein bottle in M. Then $\pi_1(K)$ contains a maximal abelian subgroup isomorphic to $\mathbb{Z} \times \mathbb{Z}$, and by condition (P2) there must be a torus component Q of ∂M such that this subgroup is conjugate into $\pi_1(Q)$. Since M is aspherical, there exists a homotopy $H \colon T^2 \times \mathrm{I} \to M$ mapping $T^2 \times \{0\}$ into Q and $T^2 \times \{1\}$ onto K by the double covering map. This homotopy induces a map from $(N, \partial N)$ into (M, Q), where N is the twisted I-bundle over the Klein bottle, which induces an injection on fundamental groups. By Waldhausen's Theorem 2.5.6, this map is homotopic to a finite covering map, so $\pi_1(Q)$ has finite index in $\pi_1(M)$. The Finite Index Theorem 2.1.1 now shows that M is the I-bundle over the Klein bottle, contradicting part (ii). So (iii) is now verified.

For (iv), consider a component Q of P, such that $\pi_1(Q)$ is not a maximal abelian subgroup of $\pi_1(M)$. Let A be an abelian subgroup of $\pi_1(M)$ containing $\pi_1(Q)$ as a proper subgroup. Suppose first that A is not cyclic. By condition (P2), A is conjugate into $\pi_1(Q')$ for some torus component Q' of P. If $Q'=Q$, then there would be a map from $(Q \times \mathrm{I}, Q \times \partial\mathrm{I})$ to (M,Q) which had degree 1 on $Q \times \{0\}$ and degree not equal to ± 1 on $Q \times \{1\}$. Since the degree is not ± 1 on $Q \times \{1\}$, this map is essential, and we may apply Waldhausen's Theorem and the Finite Index Theorem as in the proof of (iii) to derive a contradiction. If $Q \neq Q'$, then there exists a map of an annulus into M which maps one boundary circle to Q and the other into Q' and which induces an injection on fundamental groups. Such an annulus would contradict (P3), so we have established property (iv) for toroidal components of P.

Suppose now that A is cyclic, so Q is a annulus and there is a loop γ in M whose n^{th} power is homotopic to the core circle of Q, for some $n \geq 2$. There is a homotopy $\mathrm{S}^1 \times \mathrm{I}$ carrying $\mathrm{S}^1 \times \{0\}$ to a core circle of Q and $\mathrm{S}^1 \times \{1\}$ to γ by a degree n map. This induces a map f from the quotient space A_0 obtained from $\mathrm{S}^1 \times \mathrm{I}$ by identifying $(z,1)$ with $(\exp(2\pi i/n)\,z,1)$ for each $z \in \mathrm{S}^1$. Notice that A_0 imbeds in a solid torus T so that the quotient of $\mathrm{S}^1 \times \{1\}$ is the core circle, and $\mathrm{S}^1 \times \{0\}$ lies in ∂T and represents n times the core circle. Then, there is a deformation retraction of T to A_0, so f extends to a map $F\colon T \to M$, in such a way that an annulus neighborhood B of $\mathrm{S}^1 \times \{0\}$ in ∂T is taken to Q. By property (P3) applied to the annulus $\overline{\partial T - B}$, we may change F by a homotopy fixed on B so that it takes all of ∂T to Q. Now, let C be a meridian circle of T, bounding a meridian disk D. The restriction of F to C is null-homotopic in Q, since it is null-homotopic in M and Q is incompressible, so there is a map from D to Q that agrees with F on ∂D. Since $\pi_2(M)=0$, this map and $F|_D$ are homotopic relative to ∂D, so we may change F by homotopy relative to ∂T so that it maps D to Q. Regard T as constructed by attaching a 3-cell E to $D \cup \partial T$ using a map $\phi\colon \partial E \to D \cup \partial T$. Since $\pi_2(Q) = 0$, there is a map from E to Q that agrees with $F \circ \phi$ on ∂E. Since $\pi_3(M)=0$, this map and $F \circ \phi$ are homotopic relative to ∂E, so F may be changed by homotopy to carry all of T to Q. But then, the core circle of Q would be a proper power in $\pi_1(Q)$. This contradiction completes the proof of assertion (iv). □

5.2. The topology of pared manifolds

In this section we will examine the topology of pared 3-manifolds in greater depth. First, note that a pared 3-manifold (M,P) has a boundary pattern $\underline{\underline{p}}$ whose elements are simply the components of P. By lemma 3.5.2, this boundary pattern is usable. We call $\underline{\underline{p}}$ the *boundary pattern associated to* (M,P). We will check that the completion of this boundary pattern is useful if and only if all the free faces are incompressible. This leads to an easy reproof of theorem 4.1.1 for the special case of pared 3-manifolds. This special case is stated as proposition 5.2.2. We also establish a convenient property of pared 3-manifolds with respect to homotopy: homotopic admissible maps are always admissibly homotopic. Next, we will examine some properties related to relative compression body neighborhoods in pared manifolds. Notably, the normal core of a pared 3-manifold is also pared. We then examine the characteristic submanifold of a pared manifold (with respect to its completed boundary pattern).

We say that (M, P) satisfies Bonahon's condition (B) if for every nontrivial free decomposition $H_1 * H_2$ of $\pi_1(M)$, there exists an element g of $\pi_1(M)$ which is conjugate to an element of $\pi_1(P)$ but is not conjugate into either H_1 or H_2. As we see in the next lemma, this is equivalent to usefulness of the completed boundary pattern $\overline{\underline{\underline{p}}}$.

LEMMA 5.2.1. *Let (M, P) be a pared nonelementary 3-manifold, and let $\underline{\underline{p}}$ be its associated boundary pattern. The following are equivalent.*

(i) *(M, P) satisfies Bonahon's condition (B).*
(ii) *The free sides of $(M, \underline{\underline{p}})$ are incompressible.*
(iii) *The completion $\overline{\underline{\underline{p}}}$ is useful.*

PROOF. The equivalence of (i) and (ii) was observed by Bonahon in proposition 1.2 of [**17**]. The equivalence of (ii) and (iii) is immediate from lemma 2.4.6. □

Two pared 3-manifolds (M, P) and (N, Q) are called *pared homotopy equivalent* when they are homotopy equivalent as pairs, and a homotopy equivalence of pairs is called a *pared homotopy equivalence.* These notions coincide with admissibly homotopy equivalent and admissible homotopy equivalence for the associated manifolds $(M, \underline{\underline{p}})$ and $(N, \underline{\underline{q}})$. Theorem 4.1.1 shows that that if two pared manifolds are pared homotopy equivalent, then the completions of their associated boundary patterns are either both useful or both not useful. In the case of pared manifolds, however, this follows just from lemma 5.2.1 by verifying that Bonahon's condition (B) is a pared homotopy invariant, or alternatively it follows by combining lemmas 4.1.3 and 5.2.1. For reference, we state the result here.

PROPOSITION 5.2.2. *Let (M, P) and (N, Q) be pared 3-manifolds which are pared homotopy equivalent. Let $\underline{\underline{p}}$ and $\underline{\underline{q}}$ be their associated boundary patterns. Then $\overline{\underline{\underline{p}}}$ is useful if and only if $\overline{\underline{\underline{q}}}$ is useful.*

The next lemma implies that a homotopy between admissible homotopy equivalences can always be taken to be admissible for the associated boundary patterns.

PROPOSITION 5.2.3. *Let (M, P) and (N, Q) be pared 3-manifolds, and suppose that $f, g\colon (M, P) \to (N, Q)$ are maps which induce injections of $\pi_1(M)$ into $\pi_1(N)$. If f and g are homotopic as maps from M to N, then they are homotopic as maps of pairs.*

PROOF. Let H be a homotopy from f to g. Suppose first that A is an annulus component of P. Let C be the core circle of A. The restriction of H to $C \times \mathrm{I}$ is an incompressible map $(C \times \mathrm{I}, C \times \partial\mathrm{I}) \to (N, Q)$. Since N is pared this is homotopic as a map of pairs to a map into Q. Therefore it is homotopic relative to $C \times \partial\mathrm{I}$ to a map into Q. Using the homotopy extension property, H may be changed near A so that it carries A into Q at all times.

Now consider a torus component T. Choose essential circles C_1 and C_2 in T meeting in one point p. We regard C_1 and C_2 as 1-cells attached to the 0-cell p. As in the annulus case, we may assume that H maps $C_1 \times \mathrm{I}$ to Q. Regard $C_2 \times \mathrm{I}$ as the quotient of a disk D, obtained by identifying two 1-cells in ∂D to form $\{p\} \times \mathrm{I}$. Then, the restriction of H to $C_2 \times \mathrm{I}$ is induced by a map of D into N that carries ∂D

to Q. Since Q is incompressible, there is another map of $C_2 \times \mathrm{I}$ into Q which agrees with H on the boundary of $C_2 \times \mathrm{I}$. Since $\pi_2(N)=0$, this map and the restriction of H are homotopic relative to the boundary of $C_2 \times \mathrm{I}$, so H may be changed so that it maps $C_1 \cup C_2$ to Q at all times. Since T is obtained by attaching a 2-cell E to $C_1 \cup C_2$, the same argument using E and the fact that $\pi_3(N)=0$ allows us to change H so that it maps T to Q at all times.

Repeating these procedures for all components of P results in an admissible homotopy from f to g. □

We now examine some properties related to relative compression body neighborhoods in pared manifolds. First, we note that in pared manifolds, torus and annulus constituents of a minimally imbedded relative compression body can occur only in a very special configuration.

LEMMA 5.2.4. *Let (M, P) be a pared manifold, and let V be a minimal compression body neighborhood of a free side F of (M, P). Suppose that some constituent F_i of V is a torus or annulus. Then the component of $\overline{M-V}$ that contains F_i is of the form $F_i \times [-1, 0]$, where $F_i = F_i \times \{0\}$ and $F_i \times \{-1\} \cup \partial F_i \times [-1, 1]$ is a component of P.*

PROOF. The lemma is obvious if (M, P) is elementary, so assume that (M, P) is nonelementary. Suppose first that F_i is a torus. Then $\pi_1(F_i)$ is conjugate into the fundamental group of a component G of P. This implies that F_i is homotopic into G. By theorem 3.1.2, F_i is homotopic into G in $\overline{M-V}$, so the Parallel Surfaces Theorem 2.5.7 implies that F_i and G cobound a product which is the asserted component of $\overline{M-V}$

Suppose F_i is an annulus. Since $\partial F_i \subset P$, pared condition (P3) implies that F_i is properly homotopic into P and hence into $P \cap \overline{M-V}$ (since P meets V in collar neighborhoods of annuli of P). Again by theorem 3.1.2, F_i is homotopic in $\overline{M-V}$ into P. Let W be the component of $\overline{M-V}$ that contains F_i and let $\underline{\underline{w}}$ be the submanifold boundary pattern. Since $(W, \underline{\underline{w}})$ has no compressible free sides, yet its completion contains a 2-faced disk whose boundary does not bound a disk in ∂W, lemma 2.4.6 implies that W is a product of a 2-faced disk with S^1, so has the asserted form. □

It is immediate from lemma 5.2.4 and the definition of a normally imbedded relative compression body neighborhood that no properly imbedded constituent of a normally imbedded relative compression body neighborhood of a free side is an annulus or torus. Thus we have:

LEMMA 5.2.5. *Let (N, Q) be a pared manifold, and let U be a normally imbedded relative compression body neighborhood of a free side F of (N, Q). If G is a constituent of U that is a torus or annulus, then $G \times \{0\} \cup \partial G \times I$ is a component of Q.*

Lemma 3.5.2 shows that the boundary pattern of the normal core of a pared manifold always has a useful completion. We now check that the normal core is also pared.

PROPOSITION 5.2.6. *Let (M, P) be a pared manifold with normal core $(M', \underline{\underline{p'}})$. Put $P' = |\underline{\underline{p'}}|$. Then each component of (M', P') is a nonelementary pared manifold.*

PROOF. If (M, P) is elementary, then M' is empty, so we may assume that (M, P) is nonelementary. Since M is irreducible and the frontier of M' is incompressible and contains no 2-spheres, M' is irreducible. By construction, each component of P' is either a torus component of P or an annulus which is a deformation retract of an annulus of P, so (M', P') satisfies pared condition (P1). Property (P3) follows from (P3) for (M, P) using the Homotopy Enclosing Theorem 3.1.2.

To establish property (P2), suppose that H is a noncyclic abelian subgroup of $\pi_1(M')$ (where $\pi_1(M')$ is based at a point in some component of M'). By (P2) for M, H must be isomorphic to $\mathbb{Z} \times \mathbb{Z}$ and be conjugate in $\pi_1(M)$ into $\pi_1(T)$, where T is a torus component P. Suppose that T is not in M'. Then there is a normally imbedded relative compression body neighborhood V in M which has T as a constituent and meets M' in some other constituents. Regarding V as the union of $T \times \mathrm{I}$ with some 1-handles, it follows that $\pi_1(M)$ is of the form $\pi_1(T) * G$ where $\pi_1(M')$ is a subgroup of G. But then, the subgroup H cannot be conjugate into $\pi_1(T)$, since a nontrivial subgroup of a free product can be conjugate into at most one free factor. We conclude that $T \subset M'$, and hence in $M' \cap P = P'$. Since M' is aspherical, there is a map of a torus into M' which carries $\pi_1(T)$ isomorphically to H. Since M is aspherical and H is conjugate in $\pi_1(M)$ into $\pi_1(T)$, this map is homotopic in M to a map whose image lies in T. The Homotopy Enclosing Theorem 3.1.2 shows that there is a deformation of this homotopy to a homotopy in M'. Therefore H is conjugate in $\pi_1(M')$ into the fundamental group of a component of P', verifying (P2).

Finally, suppose that a component of (M', P') were elementary. It cannot be $(A^2 \times \mathrm{I}, \underline{\underline{\emptyset}})$, since the free sides of (M', P') are incompressible. If it is $(A^2 \times \mathrm{I}, \{A^2 \times \{0\}\})$ or $(T^2 \times \mathrm{I}, \{T^2 \times \{0\}\})$, it forms a spurious component and cannot be part of the normal core (or alternatively, it violates lemma 5.2.5). □

5.3. The characteristic submanifold of a pared manifold

In this section we investigate the characteristic submanifold of a pared 3-manifold with incompressible free sides. The main result, theorem 5.3.4, shows that both the I-bundle components and the Seifert-fibered components are severely constrained. In fact, the Seifert-fibered components are either solid tori or thickened tori ($\mathrm{S}^1 \times \mathrm{S}^1 \times \mathrm{I}$), and each component of the latter type meets the boundary in a torus boundary component plus some other annuli. If a thickened torus component meets the boundary in at least two annuli, the pared manifold is said to have double trouble, a condition which plays an important role in the main theorems. The final result of the present section is that the property of having double trouble is preserved by admissible homotopy equivalences.

We begin by studying the I-bundle components of the characteristic submanifold.

LEMMA 5.3.1. *Let (M, P) be a pared nonelementary 3-manifold whose associated boundary pattern $\underline{\underline{p}}$ has useful completion $\overline{\underline{\underline{p}}}$. Let $(\Sigma, \widehat{\underline{\underline{\sigma}}})$ denote the characteristic submanifold of $(M, \overline{\underline{\underline{p}}})$. Let $(V, \widehat{\underline{\underline{v}}})$ be a component of $(\Sigma, \widehat{\underline{\underline{\sigma}}})$ which is an I-bundle. Then*

(i) *the lids of $(V, \widehat{\underline{\underline{v}}})$ must be contained in free sides of $(M, \underline{\underline{p}})$, and*
(ii) *every element of $\widehat{\underline{\underline{v}}}$ which is not a lid is an element of $\underline{\underline{p}}$.*

PROOF. Lemma 2.10.9 shows that a lid of $(V, \widehat{\underline{\underline{v}}})$ could lie in $|\underline{\underline{p}}|$ only if M is homeomorphic to an I-bundle over the torus or Klein bottle, or a solid torus, contradicting lemma 5.1.1. So (i) holds. Consequently, any element of $\widehat{\underline{\underline{v}}}$ which is not a lid must be contained in an element of $\underline{\underline{p}}$. By lemma 2.10.8, it must equal the element of $\underline{\underline{p}}$. □

In the pared setting, no element of $\underline{\underline{p}}$ is a square, so part (ii) of lemma 5.3.1 has the following immediate consequence.

COROLLARY 5.3.2. *Let (M, P) be a pared 3-manifold whose associated boundary pattern $\underline{\underline{p}}$ has useful completion $\overline{\underline{\underline{p}}}$. Let $(\Sigma, \widehat{\underline{\underline{\sigma}}})$ denote the characteristic submanifold of $(M, \overline{\underline{\underline{p}}})$. Then no bound side of an I-bundle component of $(\Sigma, \widehat{\underline{\underline{\sigma}}})$ is a square. In particular, no component of $(\Sigma, \widehat{\underline{\underline{\sigma}}})$ is an I-bundle over a disk.*

The following corollary is also an easy consequence of lemma 5.3.1.

COROLLARY 5.3.3. *Let (M, P) be a pared 3-manifold, whose associated boundary pattern $\underline{\underline{p}}$ has useful completion $\overline{\underline{\underline{p}}}$. Let $(\Sigma, \widehat{\underline{\underline{\sigma}}})$ denote the characteristic submanifold of $(M, \overline{\underline{\underline{p}}})$, and let $(V, \underline{\underline{v}})$ be a component of $(\Sigma, \widehat{\underline{\underline{\sigma}}})$ which is an I-bundle over a topological annulus or Möbius band. Then $(V, \underline{\underline{v}})$ can be given an admissible Seifert fibering.*

PROOF. The underlying space of V is a solid torus. By lemma 5.3.1, each bound side of $(V, \underline{\underline{v}})$ is an element of $\underline{\underline{p}}$, and hence must be an annulus (rather than a square). But any solid torus with boundary pattern consisting of incompressible annuli can be given an admissible Seifert fibering. □

Suppose that (M, P) is a pared 3-manifold for which the completion $\overline{\underline{\underline{p}}}$ of the associated boundary pattern $\underline{\underline{p}}$ is useful. Corollary 5.3.3 allows us to adopt the following convention:

Convention: *The fibering of the characteristic submanifold of $(M, \overline{\underline{\underline{p}}})$ is selected so that none of its components is an I-bundle over a topological annulus or Möbius band.*

Equivalently, every solid torus component of the characteristic submanifold is Seifert fibered.

We can now describe the characteristic submanifold of a boundary pattern associated to a pared 3-manifold.

THEOREM 5.3.4. (Pared Characteristic Submanifold Restrictions) *Let (M, P) be a nonelementary pared 3-manifold whose associated boundary pattern $\underline{\underline{p}}$ has useful completion $\overline{\underline{\underline{p}}}$. Let $(\Sigma, \widehat{\underline{\underline{\sigma}}})$ denote the characteristic submanifold of $(M, \overline{\underline{\underline{p}}})$, with fibering selected so that no component of $(\Sigma, \widehat{\underline{\underline{\sigma}}})$ is an I-bundle over an annulus or Möbius band.*

(i) *Suppose $(V, \widehat{\underline{\underline{v}}})$ is an I-bundle component of $(\Sigma, \widehat{\underline{\underline{\sigma}}})$. Then each of its lids lies in a free side of $(M, \underline{\underline{p}})$, its bound sides are elements of $\underline{\underline{p}}$, and its base surface has negative Euler characteristic.*

(ii) *Suppose $(V, \widehat{\underline{\underline{v}}})$ is a Seifert fibered component of $(\Sigma, \widehat{\underline{\underline{\sigma}}})$. Then V is homeomorphic either to $T^2 \times I$ or to a solid torus. If V is homeomorphic to $T^2 \times I$, then one of its boundary components lies in P and the other elements of $\widehat{\underline{\underline{v}}}$ are annuli in free sides of $(M, \underline{\underline{p}})$.*

PROOF. We have already completed most of the work for the consideration of the I-bundle components of $(\Sigma, \widehat{\underline{\underline{\sigma}}})$. Let $(V, \widehat{\underline{\underline{v}}})$ be an I-bundle component of $(\Sigma, \widehat{\underline{\underline{\sigma}}})$. Since M is irreducible, the base surface is not a 2-sphere or projective plane. By corollary 5.3.2, the base surface is not a disk. By hypothesis, the base surface is not a topological annulus or Möbius band, and by lemma 5.1.1 and lemma 5.3.1, the base is not a torus or Klein bottle. The other assertions of statement (i) follow from lemma 5.3.1.

Let $(V, \widehat{\underline{\underline{v}}})$ be a component of $(\Sigma, \widehat{\underline{\underline{\sigma}}})$ which is Seifert fibered over $(B, \underline{\underline{b}})$. By lemma 5.1.1, M cannot contain an imbedded Klein bottle, so B is orientable.

Every essential map of a torus into V must be homotopic into ∂M, and hence into ∂V. In particular, a vertical essential torus, when projected to the base surface B of V, must have image a loop which is homotopic into ∂B. Thus every essential loop in B is homotopic into ∂B, so B is a disk or an annulus.

Suppose that B is a disk. We will prove that V is a solid torus. If not, then V has at least two exceptional fibers. As in section 2.1 we have an exact sequence

$$1 \to \langle t \rangle \to \pi_1(V) \to \langle c_1, \ldots, c_r \mid c_i^{p_i} = 1 \text{ for } 1 \leq i \leq r \rangle \to 1 \ .$$

The boundary circle of B is represented by the loop $c_1 c_2 \cdots c_r$. If $r = 2$ and $p_1 = p_2 = 2$, then V would be homeomorphic to the I-bundle over the Klein bottle, which would contradict assertion (iii) of lemma 5.1.1. Therefore we assume that either $r > 2$ or one of the $p_i > 2$. Then, there is a loop in B_0 which represents an element of $\langle c_1, \ldots, c_r \mid c_i^{p_i} = 1 \text{ for } 1 \leq i \leq r \rangle$ which is not conjugate to a power of one of the c_i or of the boundary circle. The preimage of this loop in V is an immersed incompressible torus which cannot be homotopic into ∂V, contradicting the fact that V is pared. We conclude that there is at most one exceptional fiber, and therefore V is a solid torus.

Suppose now that B is an annulus. Then V has two torus boundary components T_1 and T_2. We claim that $V = T^2 \times \mathrm{I}$. By (P2) each T_i is homotopic into a torus component P_i of P. If $P_1 \neq P_2$, then since a fiber of V in T_1 is homotopic to a fiber in T_2, a violation of (P3) would occur. So $P_1 = P_2$, and by homotopy extension starting at the inclusion map of V we obtain an admissible map from $(V, \underline{\underline{\emptyset}})$ to $(M, \underline{\underline{p}})$ carrying $T_1 \cup T_2$ into P_1. This map cannot be essential, since then Waldhausen's Theorem 2.5.6 would show it is homotopic to a finite covering, and (M, P) would be elementary. So arguing as in the second paragraph of the proof of proposition 5.2.3, the map is homotopic relative to ∂V to a map into P_1 (alternatively, the map lifts to the covering of M corresponding to the subgroup $\pi_1(P_1)$ of $\pi_1(M)$, and P_1 is a deformation retract of this covering). Since $\pi_1(V) \to \pi_1(M)$ is injective, we conclude that $\pi_1(T) \cong \mathbb{Z} \times \mathbb{Z}$, so $V = T^2 \times \mathrm{I}$. Using the Parallel Surfaces Theorem 2.5.7, $T^2 \times \{1/2\}$ is parallel to P_1, and using fullness of the characteristic submanifold, $P_1 \subset V$ and hence P_1 is one of the T_i. Since M is nonelementary, the other boundary component of V cannot lie in ∂M. None of the elements of $\underline{\underline{v}}$ can be annuli of P, since then condition (P3) would be violated. Therefore all other elements of $\underline{\underline{v}}$ are annuli imbedded in free sides of $(M, \underline{\underline{p}})$. □

We will say that (M, P) *has double trouble* if there are two simple closed curves γ_1 and γ_2 in $\partial M - P$ that are not homotopic in ∂M to each other, but are homotopic in M to an essential simple closed curve γ in a torus component of P. The next lemma characterizes double trouble in terms of the characteristic submanifold.

LEMMA 5.3.5. *Let (M, P) be a pared 3-manifold whose associated boundary pattern $(M, \underline{\underline{p}})$ has useful completion. Let $(\Sigma, \widehat{\underline{\underline{\sigma}}})$ denote the characteristic submanifold of $(M, \overline{\underline{\underline{p}}})$. Then (M, P) has double trouble if and only if $(\Sigma, \widehat{\underline{\underline{\sigma}}})$ contains a Seifert fibered component $(V, \widehat{\underline{\underline{v}}})$ which is homeomorphic to $T^2 \times I$ and whose frontier in M has at least two components (which are necessarily annuli).*

PROOF. Suppose $\overline{\underline{\underline{p}}}$ is useful and $(\Sigma, \widehat{\underline{\underline{\sigma}}})$ contains a Seifert fibered component $(V, \widehat{\underline{\underline{v}}})$ which is homeomorphic to $T^2 \times \mathrm{I}$ and whose frontier has at least two components. By the Pared Characteristic Submanifold Restrictions (theorem 5.3.4), one of its boundary components is a torus component of P, and there must be at least two annuli in $\widehat{\underline{\underline{v}}}$ lying in $\partial M - P$.

The core curves of distinct annuli of $\widehat{\underline{\underline{v}}}$ cannot be homotopic to each other in ∂M. For if so, then there are two such curves that cobound an annulus A in ∂M, meeting $\overline{M - V}$ in an annulus A'. The component W of $\overline{M - V}$ that meets A' is bounded by A' and an annulus A'' in the frontier of V. By the Enclosing Property, A' is admissibly homotopic into Σ. The restriction of such a homotopy to an arc in A' connecting its boundary components shows that the proper boundary pattern on W is not useful. Lemma 2.4.6 then shows that W is the product of a 2-faced disk and a circle, violating the fullness of Σ. However, the core curves of the annuli of $\widehat{\underline{\underline{v}}}$ are homotopic to each other in V and hence are homotopic to some curve in the other component of ∂V, which is a toroidal component of P. Thus (M, P) has double trouble.

Conversely, suppose (M, P) has double trouble. Let γ_1 and γ_2 be two simple closed curves in $\partial M - P$ which are not homotopic to each other in ∂M but are both homotopic to a curve γ in a toroidal component T of P. For each i let A_i be a singular annulus spanned by γ_i and γ. Then each A_i is admissibly homotopic to an annulus A_i' in a component $(V_i, \widehat{\underline{\underline{v}}}_i)$ of the characteristic submanifold. But since T is entirely contained in a single component $(V, \widehat{\underline{\underline{v}}})$ of the characteristic submanifold, we must have $V_1 = V_2 = V$. By the Pared Characteristic Submanifold Restrictions 5.3.4, V is homeomorphic to $T^2 \times \mathrm{I}$ and one of its boundary components is a torus component of P. Since γ_1 and γ_2 are homotopically distinct in ∂M, the components of $\partial A_1'$ and $\partial A_2'$ lie in distinct annuli of $\widehat{\underline{\underline{v}}}$. Thus, there must be more than one component of the frontier of V. □

The final result of this section asserts that the property of having double trouble is a pared homotopy type invariant.

THEOREM 5.3.6. *Let (M, P) be a pared 3-manifold whose associated boundary pattern $\underline{\underline{p}}$ has useful completion, and let (N, Q) be a pared 3-manifold which is pared homotopy equivalent to (M, P). Then (M, P) has double trouble if and only if (N, Q) does.*

PROOF. By proposition 5.2.2, $\underline{\underline{q}}$ has useful completion. By the Classification Theorem 2.11.1, the characteristic submanifolds of $(M, \overline{\underline{\underline{p}}})$ and $(N, \overline{\underline{\underline{q}}})$ (with their proper boundary patterns in $(M, \underline{\underline{p}})$ and $(N, \underline{\underline{q}})$) are admissibly homotopy equivalent. Recall that, by the Finite Index Theorem 2.1.1, any component of the characteristic submanifold of $(M, \overline{\underline{\underline{p}}})$ or $(N, \overline{\underline{\underline{q}}})$ which is homotopy equivalent to $T^2 \times \mathrm{I}$ is homeomorphic to $T^2 \times \mathrm{I}$. By the Pared Characteristic Submanifold Restrictions (theorem 5.3.4), each component of their characteristic submanifolds that is homeomorphic to $T^2 \times \mathrm{I}$ has proper boundary pattern consisting of one boundary torus and the components of its frontier. Admissibly homotopy equivalent components have the same number of elements in their boundary patterns, so lemma 5.3.5 implies the assertion of the theorem. □

CHAPTER 6

Small 3-Manifolds

In this section we introduce and study small 3-manifolds. Examples 1.4.1, 1.4.2, and 1.4.3 of chapter 1 suggest that it is quite common for the group $\mathcal{R}(M, \underline{\underline{m}})$ of outer automorphisms which are realizable by admissible homeomorphisms to have infinite index in $\mathrm{Out}(\pi_1(M), \pi_1(\underline{\underline{m}}))$. The Main Topological Theorems, stated in section 8.1, make this precise; they give the relatively short list of manifolds for which the index is finite. Main Topological Theorem 1 shows that the only manifolds on this list which have a compressible free side are the relative compression bodies and a few others which we will call *small.*

We will especially focus on the small pared manifolds, which are characterized in lemma 6.1.1. In chapter 7 we will see that the components of the space $\mathrm{GF}(M, P)$ of geometrically finite uniformizations of a pared 3-manifold (M, P) are enumerated by the cosets of $\mathcal{R}_+(M, P)$ in $\mathrm{Out}(\pi_1(M), \pi_1(P))$ (see corollary 7.3.1.) It will then follow, from Main Topological Theorem 1, that if (M, P) has a compressible free side, then $\mathrm{GF}(M, P)$ has finitely many components if and only if (M, P) is either small or is a relative compression body (see the Main Hyperbolic Theorem in chapter 8).

A pared homotopy type of pared manifolds is said to be *small* if every pared 3-manifold in its homotopy type is either a relative compression body or is small. In theorem 6.2.1, we will identify exactly which pared homotopy types are small. The space $\mathrm{GF}(\pi_1(M), \pi_1(P))$ of all geometrically finite uniformizations of pared manifolds which are pared homotopy equivalent to (M, P) is a natural object of study in the deformation theory of hyperbolic 3-manifolds as it corresponds to the interior of the space of (conjugacy classes of) discrete faithful representations of $\pi_1(M)$ into $\mathrm{PSL}(2, \mathbb{C})$ such that all elements of $\pi_1(P)$ are taken to parabolic elements. It will then follow that if a pared 3-manifold (M, P) has a compressible free side, then $\mathrm{GF}(\pi_1(M), \pi_1(P))$ has finitely many components if and only if the pared homotopy type of (M, P) is small (see the Main Hyperbolic Corollary in chapter 8).

6.1. Small manifolds and small pared manifolds

In this section, we will define small manifolds, and then determine which of them are pared.

Definition: Let $(M, \underline{\underline{a}})$ be an orientable 3-manifold with a boundary pattern whose elements are disjoint and incompressible, and which is not a relative compression body. Assume that $(M, \underline{\underline{a}})$ has a compressible free side F, and that all other free sides of $(M, \underline{\underline{a}})$ are incompressible. Assume further that every element of $\underline{\underline{a}}$ that meets ∂F is an annulus. Let $(V, \underline{\underline{v}})$ be a normally imbedded relative compression body neighborhood of F. We say that $(M, \underline{\underline{a}})$ is *small* if one of the following occurs, where as usual k is the number of 1-handles of V and $F_1, \ldots, F_m$ are the constituents of V.

I. $k=1$, $m=1$, and if $W=\overline{M-V}$, then either
  (a) W is a twisted I-bundle with F_1 as its lid, or
  (b) W is a solid torus, F_1 is an annulus, and $\pi_1(F_1) \to \pi_1(W)$ is not surjective.

II. $k=1$, $m=2$, F_1 separates M, and for each of the one or two components W of $\overline{M-V}$, with frontier G, either
  (a) W is an I-bundle with G as one of its lids,
  (b) W is a solid torus, G is an annulus, and $\pi_1(G) \to \pi_1(W)$ is not surjective, or
  (c) W is either $S^1 \times S^1 \times I$ or the I-bundle over the Klein bottle, and G is an annulus.

III. $k=1$, $m=2$, F_1 does not separate M, and if $W=\overline{M-V}$, then either
  (a) (W, F_1, F_2) is homeomorphic to $(F_1 \times [0,1], F_1 \times \{0\}, F_1 \times \{1\})$,
  (b) W is a solid torus, F_1 and F_2 are essential annuli in ∂W, and $\pi_1(F_1) \to \pi_1(W)$ is not surjective, or
  (c) W is $S^1 \times S^1 \times I$, F_1 and F_2 lie in different boundary components of W, and at least one of F_1 or F_2 is an annulus.

In this definition, the statement that W is an I-bundle does not mean that with its submanifold boundary pattern it is admissibly fibered, only that topologically it is an I-bundle.

We refer to small manifolds by their type, such as type Ia, etc. Type IIa indicates a manifold of type II where $\overline{M-V}$ has only one component (that is, one of the constituents of V lies in ∂M), and this component satisfies condition II(a). Type IIab indicates a manifold of type II where $\overline{M-V}$ has two components, one satisfying condition II(a) and the other condition II(b). Type IIbx indicates a manifold of type II where $\overline{M-V}$ has one or two components, at least one of which satisfies condition II(b).

In a small manifold of type IIa, $\overline{M-V}$ is a twisted I-bundle, since otherwise (M, P) would be a relative compression body. Similarly, in a small manifold of type IIaa, at least one of the components of $\overline{M-V}$ is a twisted I-bundle.

We note that the elements of $\underline{\underline{a}}$ cannot be disks, since the free sides other than F are incompressible. The definition implies that the elements of $\underline{\underline{a}}$ that do not meet F must be of three kinds:

(i) annuli parallel to annuli that meet ∂F,
(ii) incompressible submanifolds of the torus boundary component of W that does not meet F, where M is of type IIcx and $W=S^1 \times S^1 \times I$ as in case II(c), and
(iii) incompressible submanifolds of $F_i \times \{-1\}$, where M is of type IIax and $F_i \times [-1,0]$ is a product I-bundle component of $\overline{M-V}$ with $\partial F_i \times [-1,0]$ contained in P, and $F_i \times \{0\}$ equals $F_i \times [-1,0] \cap V$ and is a constituent of V.

For if a component W of $\overline{M-V}$ is not as in (ii) or (iii), then each component of $\partial W \cap \partial M$ is an annulus, and the only incompressible surfaces in an annulus, other than disks, are annuli. For pared small manifolds, only elements as in (iii) can occur, since annuli as in (i) would violate the definition of pared manifold, while tori as in (ii) cannot occur by lemma 6.1.1 below.

LEMMA 6.1.1. *A small manifold $(M, \underline{\underline{a}})$ is pared if and only if all the following hold:*

(i) *it is of type Ia, IIa, IIaa, or IIIa,*
(ii) *each element of $\underline{\underline{a}}$ is an annulus or torus,*
(iii) *for each constituent F_i of V that is a torus, $F_i \in \underline{\underline{a}}$,*
(iv) *for each constituent F_i of V that is an annulus, $F_i \cup (\partial F_i \times I) \in \underline{\underline{a}}$, and*
(v) *no two annuli in $\underline{\underline{a}}$ are homotopic.*

Conditions (iii) and (iv) are together equivalent to the assertion that any constituent that is a torus or annulus is contained in $|\underline{\underline{a}}|$. Condition (iv) is redundant, since it is implied by (v).

PROOF. Assume that $(M, \underline{\underline{a}})$ is pared. Since $(M, \underline{\underline{a}})$ is small, $M \neq V$ and therefore $(M, \underline{\underline{a}})$ cannot be elementary. We will first rule out the types other than those listed in (i). For a small manifold of type Ib, IIbx, or IIIb, the component of the frontier of $\overline{M - V}$ which is contained in a solid torus is an annulus which violates condition (P3) of the definition of a pared 3-manifold. Consider a small manifold of type IIcx. Suppose first that there is a component W of $\overline{M - V}$ which is homeomorphic to $\mathrm{S}^1 \times \mathrm{S}^1 \times \mathrm{I}$. Its frontier is an annulus G in $\mathrm{S}^1 \times \mathrm{S}^1 \times \{1\}$, and by lemma 5.1.1 its other boundary component $\mathrm{S}^1 \times \mathrm{S}^1 \times \{0\}$ must be an element of $\underline{\underline{a}}$. Then W contains an annulus with one end a boundary component of G and the other end in $\mathrm{S}^1 \times \mathrm{S}^1 \times \{0\}$, violating condition (P3). For the other case of a small manifold of type IIcx, there is a component of $\overline{M - V}$ that is a twisted I-bundle over the Klein bottle, violating lemma 5.1.1(iii). In a manifold of type IIIc, (P2) is violated. This verifies assertion (i).

Condition (ii) is immediate from the definition of pared, conditions (iii) and (iv) follow from lemma 5.2.5, and (v) follows from (P3).

Conversely, assume that the five conditions hold for $(M, \underline{\underline{a}})$. By (ii), every element of $\underline{\underline{a}}$ is an annulus or torus, and these are incompressible by definition since $(M, \underline{\underline{a}})$ is small. Thus (P1) holds. To establish (P2) suppose that A is a noncyclic abelian subgroup of $\pi_1(M)$. Since $(M, \underline{\underline{a}})$ is small and is one of the types allowed in (i), $\pi_1(M)$ is a free product having one of the three forms H, $\pi_1(G_1) * H$, or $\pi_1(G_1) * \pi_1(G_2) * H$ where H is free and each $\pi_1(G_i)$ is the fundamental group of a closed orientable or nonorientable surface S. If S is an orientable surface, then it appears as one of the constituents F_i of V; if a nonorientable surface, then it appears as the 0-section of a twisted I-bundle component of $\overline{M - V}$. Since A is freely indecomposable and not infinite cyclic, the Kurosh Subgroup Theorem shows it is conjugate into one of the $\pi_1(G_i)$. The only closed surfaces whose fundamental group contains a noncyclic abelian group are the Klein bottle and torus. If a Klein bottle occurred, then one of the F_i must be a torus bounding a twisted I-bundle over the Klein bottle in M, which is excluded by condition (iii). Therefore A is conjugate into some $\pi_1(F_i)$ where F_i is a torus, and by condition (iii), $F_i \in \underline{\underline{a}}$. This verifies condition (P2).

Let $f\colon (\mathrm{S}^1 \times \mathrm{I}, \mathrm{S}^1 \times \partial\mathrm{I}) \to (M, |\underline{\underline{a}}|)$ be an incompressible map of pairs. Since f is incompressible and $f(S^1 \times \partial\mathrm{I})$ is disjoint from the 1-handles of V, we may change f by a homotopy of pairs so that its image is disjoint from the cocores of the 1-handles of V. The union of the constituents is a deformation retract of the complement in V of the cocores of its 1-handles, so we may assume by further homotopy of pairs that the image of f lies in $\overline{M - V} \cup (\bigcup F_i)$.

If the image of f lies in some F_i which is contained in ∂M (necessarily a torus or annulus), then since $F_i \subseteq |\underline{\underline{a}}|$, condition (P3) is verified for f. So we may assume that the image of f lies in some component W of $\overline{M - V}$. By condition (i), W is an I-bundle, at least one of whose lids is a constituent of V.

Suppose first that W is a product I-bundle $F_i \times [-1, 0]$ with $F_i \times \{0\}$ a constituent of V and $F_i \times \{-1\} \subset \partial M$. Since M is not a relative compression body, this can occur only when M is of type IIaa. We may assume that the image of f lies in $F_i \times \{-1\}$. By (iii), F_i is not a torus, so $|\underline{\underline{a}}| \cap F_i \times \{-1\}$ is a collection of incompressible annuli. Suppose for contradiction that the loops $f(\mathrm{S}^1 \times \{0\})$ and $f(\mathrm{S}^1 \times \{1\})$ lie in two different annuli in this collection, say A_1 and A_2. In $\pi_1(F_i \times \{-1\})$, their center circles represent elements x and y. By (iii) and (iv), F_i has negative Euler characteristic, so the subgroup H generated by x and y is free of rank at most 2. The map f lifts to the covering of $F_i \times \{-1\}$ determined by this subgroup, so some powers of x and y are conjugate in H and therefore H is free of rank 1. Therefore the covering is an open annulus, and since x and y lift to elements represented by simple closed curves, they are freely homotopic. Therefore A_1 and A_2 are homotopic, violating condition (v). So $f(\mathrm{S}^1 \times \{0\})$ and $f(\mathrm{S}^1 \times \{1\})$ both lie in a single annulus A_1 of $|\underline{\underline{a}}| \cap F_i \times \{-1\}$. An arc γ connecting $\mathrm{S}^1 \times \{0\}$ and $\mathrm{S}^1 \times \{1\}$ determines an element $g \in \pi_1(F_i \times \{-1\})$ with $gx^pg^{-1} = x^q$ for some nonzero p and q. By a similar covering space argument, g lies in the cyclic subgroup generated by x, so the restriction of f to γ is homotopic relative to its endpoints to a map into A_1. This shows that the lift of f to the covering space determined by $\pi_1(A_1)$ carries both boundary components of $\mathrm{S}^1 \times \mathrm{I}$ to the same annulus $\widetilde{A_1}$ of the preimage of A_1. Since there is a deformation retraction from this covering space to $\widetilde{A_1}$, f is homotopic as a map of pairs to a map taking $\mathrm{S}^1 \times \mathrm{I}$ to A_1, and condition (P3) is verified for f.

Suppose that W is a twisted I-bundle over a nonorientable surface N, whose lid is a constituent G of V. This can occur when M is of type Ia, IIa, or IIaa. By (iv), G is not an annulus so N is not a Möbius band. Let $A_1, \ldots, A_r$ be the annuli which are the preimages in the I-bundle W of the boundary components of N. Each boundary component of an A_i meets an element of $\underline{\underline{a}}$ which meets a boundary component of F, and by the definition of small manifold (or by (ii)) this element is an annulus. By (v), A_i must be entirely contained in an annulus of $\underline{\underline{a}}$. So f is homotopic (as a map of pairs) to a map into N, carrying $\mathrm{S}^1 \times \partial\mathrm{I}$ into ∂N. Since N is not an annulus or Möbius band, the Baer-Nielsen Theorem 2.5.5 implies that f is not essential. That is, the restriction of f to an arc connecting the two boundary components of $\mathrm{S}^1 \times \mathrm{I}$ is properly homotopic into ∂N. Again, covering space arguments show that f is homotopic as a map of pairs into $|\underline{\underline{a}}|$.

Finally, suppose that W is a product I-bundle with both lids constituents of V. This occurs when M is of type IIIa. Then $\overline{M - V}$ is of the form $F_1 \times \mathrm{I}$ and by (v), $\partial F_1 \times I \subset |\underline{\underline{a}}|$. Since, by (iv) F_1 is not an annulus, f is (as in the case when W was a twisted I-bundle) homotopic as a map of pairs into $\partial F_1 \times \mathrm{I}$. This completes the verification of pared condition (P3). □

6.2. Small pared homotopy types

A pared homotopy type is called *small* if each of the homeomorphism types that it contains is either a relative compression body or a pared small manifold. To list the small pared homotopy types, it is convenient to introduce a definition. A *remote annulus* of a pared relative compression body or a pared small manifold (M, P) is an annulus of P that does not meet the compressible free side F. Note that pared small manifolds of types Ia, IIa, and IIIa never have remote annuli, since in these cases each component of $\overline{M - V}$ is an I-bundle meeting ∂M only in its sides (a twisted I-bundle for types Ia and IIa and a product I-bundle for type IIIa).

We will denote by $(M', \underline{\underline{p}}')$ and $(M'', \underline{\underline{p}}'')$ respectively the normal core and maximal incompressible core of $(M, \underline{\underline{p}})$. Recall that M' and M'' differ only in that the spurious components of M'' are deleted to obtain M'; when (M, P) is pared, these components must be of the form $F \times [0, 1/2]$ where $F \times \{0\}$ is a torus or annulus constituent of a normally imbedded relative compression body V and $F \times \{0\} \cup \partial F \times [0, 1]$ is a component of P. Observe that for any pared relative compression body or pared small manifold with no remote annuli, every component of $(M', \underline{\underline{p}}')$ is an admissible I-bundle, each of whose lids is a free side of $(M', \underline{\underline{p}}')$.

THEOREM 6.2.1. *The small pared homotopy types of pared 3-manifolds are the following:*

(i) *The homotopy type of a pared relative compression body having each constituent either a torus or annulus. It is the only element in its pared homotopy type.*
(ii) *The homotopy type of a pared relative compression body having exactly two constituents and one 1-handle, and no remote annuli, or a pared small manifold of type Ia, IIa, or IIaa having no remote annuli. It is the only element in its pared homotopy type.*
(iii) *The homotopy type of a pared small manifold of type IIIa. Its pared homotopy type consists of itself together with a relative compression body having exactly one constituent and one 1-handle, no remote annuli, and whose constituent is not an annulus or torus. Each such relative compression body is in one such homotopy type.*

The manifolds in (i) include the case when P is empty and M is a handlebody. An explicit example of a homotopy type as in (iii) was given as example 1.4.6.

PROOF. Fix a pared manifold (M, P) and consider a pared manifold (N, Q) which is pared homotopy equivalent to (M, P). We will first show that if (M, P) is one of the manifolds listed in the theorem, then (N, Q) is a relative compression body or is small. Denote by $(M', \underline{\underline{p}}')$ and $(N', \underline{\underline{q}}')$ the normal cores of $(M, \underline{\underline{p}})$ and $(N, \underline{\underline{q}})$ respectively, and by $(M'', \underline{\underline{p}}'')$ and $(N'', \underline{\underline{q}}'')$ their maximal incompressible cores. By theorem 3.5.1, $(N', \underline{\underline{q}}')$ is admissibly homotopy equivalent to $(M', \underline{\underline{p}}')$, and by theorem 3.3.2, $(N'', \underline{\underline{q}}'')$ is admissibly homotopy equivalent to $(M'', \underline{\underline{p}}'')$.

Case I. (M, P) is a relative compression body with every constituent either an annulus or a torus.

Since M' is empty, so is N', so (N, Q) must be a pared relative compression body and every one of its constituents is a a torus in Q or annulus contained in an annular component of Q. Moreover, every annular component of P and Q must contain exactly one constituent. Since P and Q are homotopy equivalent, (N, Q) has the same constituents as (M, P). Since $\pi_1(M) \cong \pi_1(N)$, they also have the same number of 1-handles, so they are pared homeomorphic. Note that this includes the case when P is empty and M is a handlebody.

For the remaining cases, let V be a normally imbedded relative compression body neighborhood of the compressible free side of (M, P). We may assume that some constituent of V, say F_1, is not a torus or annulus, since otherwise case I applies. As noted above, for any relative compression body or pared small manifold with no remote annuli, $(M', \overline{\underline{\underline{p'}}})$ is an admissible I-bundle, each of whose lids is a free side of $(M', \underline{\underline{p'}})$. So by lemma 2.11.3, $(N', \underline{\underline{q'}})$ and $(M', \underline{\underline{p'}})$ are admissibly homeomorphic. Also, $(N'', \underline{\underline{q''}})$ and $(M'', \underline{\underline{p''}})$ are admissibly homeomorphic. For the spurious components of $(M'', \underline{\underline{p''}})$ are all of the form $(F \times [-1, 0], \{F \times \{-1\} \cup \partial F \times [-1, 0]\})$, with F an annulus or torus. Each spurious component of M'' corresponds under the admissible homotopy equivalence to a spurious component $(G \times [-1, 0], \{G \times \{-1\} \cup \partial G \times [-1, 0]\})$ of N''. Since the only orientable surface homotopy equivalent to an annulus is an annulus, and similarly for the torus, we have F homeomorphic to G. Moreover, any homotopy equivalence from the annulus to itself is homotopic to a homeomorphism, and similarly for the torus. It follows that the admissible homotopy equivalence from $(F \times [-1, 0], \{F \times \{-1\} \cup \partial F \times [-1, 0]\})$ to $(G \times [-1, 0], \{G \times \{-1\} \cup \partial G \times [-1, 0]\})$ is admissibly homotopic to a homeomorphism.

Case II. (M, P) is small of type Ia.

We have $(N', \overline{\underline{\underline{q'}}})$ admissibly homeomorphic to $(M', \overline{\underline{\underline{p'}}})$, so N' is a twisted I-bundle, whose lid is the free side of $(N', \underline{\underline{q'}})$. Since $\pi_1(N) \cong \pi_1(N') * \mathbb{Z}$, N can be formed by attaching a 1-handle to the lid of N', so (N, Q) is homeomorphic to (M, P).

Case III. (M, P) has no remote annuli and is either a relative compression body with exactly one 1-handle and two constituents, or is small of type IIa or IIaa.

In these cases M'' has two components M_1'' and M_2'' which are I-bundles with a constituent of V as one lid. Since $(N'', \underline{\underline{q'}})$ is homeomorphic to $(M'', \underline{\underline{p'}})$, and $\pi_1(N) \cong \pi_1(M) \cong \pi_1(M_1'') * \pi_1(M_2'')$, N can be formed by adding a single 1-handle that connects a lid of M_1'' to a lid of M_2''. Therefore (N, Q) is homeomorphic to (M, P).

Case IV. (M, P) is a pared small manifold of type IIIa.

Note that (M, P) cannot have remote annuli, since such annuli would be homotopic to annuli that meet free sides. Since $(M', \underline{\underline{p'}})$ and $(N', \underline{\underline{q'}})$ are admissibly homeomorphic, (N, Q) can be obtained from $(F_1 \times \mathrm{I}, \partial F_1 \times \mathrm{I})$ by adding 1-handles along $F_1 \times \partial \mathrm{I}$. Since $\pi_1(N) \cong \pi_1(M)$, only one 1-handle is added. If the attaching disks lie in different ends, then (N, Q) is homeomorphic to (M, P), while if they lie in the same end then (N, Q) is a relative compression body with one constituent and one 1-handle, whose pared homeomorphism type is uniquely determined by F_1 and hence by $(M, \underline{\underline{m}})$.

For any relative compression body having exactly one constituent and one 1-handle, no remote annuli, and whose constituent is not an annulus or torus, reattaching one end of its 1-handle yields a homotopy equivalent pared manifold which is a pared small manifold of type IIIa. This gives the last assertion in statement (iii) of theorem 6.2.1.

This completes the proof that the manifolds listed in theorem 6.2.1 have small pared homotopy types. It remains to prove that the homotopy type is not small assuming that (M, P) is either

(1) a pared relative compression body not among those listed in the statement of theorem 6.2.1, that is, some constituent is not an annulus or torus, and either
   (a) $k=1$ and $m=1$, and (M, P) has a remote annulus,
   (b) $k=1$ and $m=2$, and (M, P) has a remote annulus, or
   (c) $k \geq 2$, or
(2) a pared small manifold of type IIaa, and has a remote annulus.

Suppose (M, P) is a pared relative compression body with $k = 1$, $m = 1$, and having a remote annulus. Then its constituent F_1 contains an incompressible free side G of (M, P) that is not homeomorphic to F_1. Form (N, Q) pared homotopy equivalent to (M, P) by reattaching one end of the 1-handle of M to a disk in the interior of G, and letting Q be the copy of P in N. In N there is a normally imbedded relative compression body U having two constituents, homeomorphic to F_1 and G, and one 1-handle; its frontier is the original constituent F_1 and the frontier of a regular neighborhood of G. If S is a free side of (N, Q), then $\pi_1(S) \to \pi_1(N)$ is not surjective, so (N, Q) is not a relative compression body. If (N, Q) were small, then since $\overline{N - U}$ is connected and (N, Q) is pared, it would be of type IIIa. This is impossible because F_1 is not homeomorphic to G. Since (N, Q) is not a relative compression body and is not small, the pared homotopy type of (M, P) is not small.

Assume that (M, P) is a pared relative compression body with $k=1$, $m=2$, and having a remote annulus. Again there is an incompressible free side G contained in a constituent, say F_1, and not homeomorphic to F_1. Form (N, Q) by reattaching the 1-handle of (M, P) that lies in $F_1 \times \{1\}$ to G. As in the previous paragraph, (N, Q) is pared homotopy equivalent to (M, P), and is not a relative compression body. If it were small, then it would have to be of type IIaa, but the component of the normal core that is contained in $F_1 \times \mathrm{I}$ does not satisfy condition II(a), II(b), or II(c) in the definition of small. Again, the pared homotopy type of (M, P) is not small.

Assume now that (M, P) is a pared relative compression body with at least two 1-handles. Let $F_1, \ldots, F_m$ be its constituents, where we may assume that F_1 is not an annulus or a torus. If $m=1$, then attaching one end of one of the 1-handles to $F_1 - P$ yields a manifold which is not a relative compression body and is not small. If $m \geq 2$, regard M as constructed so that for each $1 \leq i < m$, there is a 1-handle with one end attached to a disk in $F_i \times \{1\}$ and the other to a disk in $F_{i+1} \times \{1\}$. If $m=2$, then there is a second 1-handle, and by reattaching one end of it in a component of $F_1 - P$ we obtain a manifold pared homotopy equivalent to (M, P) but not a relative compression body and not small. If $m > 2$, let H be the 1-handle attached to $F_1 \times \{1\}$ and $F_2 \times \{1\}$. Reattach the end of H that is in $F_2 \times \{1\}$ to a component of $F_2 - P$ to obtain a manifold which is pared homotopy eqivalent to (M, P) but not a relative compression body and not small.

Finally, suppose that (M, P) is a pared small manifold of type IIaa which has a remote annulus. We may assume that the component of $\overline{M - V}$ which contains F_1

is of the form $F_1 \times [-1, 0]$, and that there is an annulus component of P lying in the interior of $F_1 \times \{-1\}$. One end of the 1-handle of V lies in $F_1 \times \{1\}$; by reattaching that end to a component of $F_1 \times \{-1\} - P$, we produce a manifold (N, Q) which is pared homotopy equivalent to (M, P). There is no free side S of (N, Q) for which $\pi_1(S) \to \pi_1(N)$ is surjective, so (N, Q) is not a relative compression body. If (N, Q) were small, then it would be of type IIaa, but the component of the normal core that is contained in $F_1 \times \mathrm{I}$ does not satisfy condition II(a), II(b), or II(c) in the definition of small. $\square$

CHAPTER 7

Geometrically Finite Hyperbolic 3-Manifolds

In this section we discuss deformation spaces of geometrically finite hyperbolic 3-manifolds. In section 7.1 we introduce geometrically finite hyperbolic 3-manifolds and define more formally the spaces $\mathrm{GF}(M,P)$ and $\mathrm{GF}(\pi_1(M),\pi_1(P))$ which we will study. In section 7.2 we review the quasiconformal deformation theory of Kleinian groups as developed by Ahlfors, Bers, Kra and Maskit. In section 7.3 we combine this deformation theory with the work of Marden and Thurston to obtain parameterizations of the spaces $\mathrm{GF}(M,P)$ and $\mathrm{GF}(\pi_1(M),\pi_1(P))$.

7.1. Basic definitions

A *Kleinian group* is a discrete faithful representation $\rho\colon G \to \mathrm{PSL}(2,\mathbb{C})$ of a group G into $\mathrm{PSL}(2,\mathbb{C})$. We will always regard $\mathrm{PSL}(2,\mathbb{C})$ as the group of Möbius transformations of the Riemann sphere $\overline{\mathbb{C}} = \mathbb{C} \cup \{\infty\}$. Every Möbius transformation extends continuously to a homeomorphism of $\mathbb{H}^3 \cup \overline{\mathbb{C}}$ which is an orientation-preserving isometry of hyperbolic 3-space $\mathbb{H}^3$. Moreover, every orientation-preserving isometry of $\mathbb{H}^3$ extends continuously to a homeomorphism of $\mathbb{H}^3 \cup \overline{\mathbb{C}}$ whose restriction to $\overline{\mathbb{C}}$ is a Möbius transformation. Hence, we may identify $\mathrm{PSL}(2,\mathbb{C})$ with the group of orientation-preserving isometries of $\mathbb{H}^3$. (See Maskit [**78**] or Kapovich [**61**] for more details on the theory of Kleinian groups.)

If $\rho\colon G \to \mathrm{PSL}(2,\mathbb{C})$ is a Kleinian group, then the action of $\rho(G)$ on $\overline{\mathbb{C}}$ partitions $\overline{\mathbb{C}}$ into two sets. The *domain of discontinuity* $\Omega(\rho)$ is the maximal open subset of $\overline{\mathbb{C}}$ on which $\rho(G)$ acts discontinuously. The *limit set* $\Lambda(\rho)$ is the complement in $\overline{\mathbb{C}}$ of $\Omega(\rho)$. If G does not contain an abelian subgroup of finite index then $\Lambda(\rho)$ is a perfect set, and thus uncountable. In this case $\Omega(\rho)$ inherits a canonical hyperbolic metric, called the Poincaré metric, on which $\rho(G)$ acts as a group of isometries. If G is torsion-free, then $\rho(G)$ acts freely on $\Omega(\rho)$ and $\Omega(\rho)/\rho(G)$ inherits the structure of a hyperbolic Riemann surface.

We define

$$N(\rho)=\mathbb{H}^3/\rho(G)$$

and

$$\widehat{N}(\rho)=(\mathbb{H}^3 \cup \Omega(\rho))/\rho(G).$$

The surface $\partial\widehat{N}(\rho) = \Omega(\rho)/\rho(G)$ is called the *conformal boundary* at infinity for $N(\rho)$. Ahlfors' Finiteness Theorem ([**4**], see also Bers [**15**]) asserts that if G is finitely generated then the conformal boundary is both topologically and analytically finite.

Ahlfors' Finiteness Theorem: *Suppose that G is a finitely generated, non-abelian, torsion-free group and $\rho\colon G \to \mathrm{PSL}(2,\mathbb{C})$ is a discrete faithful representation. Then the conformal boundary $\Omega(\rho)/\rho(G)$ is a finite area hyperbolic surface, possibly disconnected.*

Let (M,P) be an oriented pared 3-manifold. Let $\mathcal{D}(\pi_1(M),\pi_1(P))$ denote the space of discrete faithful representations $\rho\colon \pi_1(M)\to \mathrm{PSL}(2,\mathbb{C})$ such that $\rho(g)$ is parabolic if $g\in\pi_1(P)$. Let

$$\mathrm{AH}(\pi_1(M),\pi_1(P)) = \mathcal{D}(\pi_1(M),\pi_1(P))/\,\mathrm{PSL}(2,\mathbb{C})$$

where $\mathrm{PSL}(2,\mathbb{C})$ acts by conjugation.

A representation $\rho\in \mathrm{AH}(\pi_1(M),\pi_1(P))$ is said to be a *geometrically finite uniformization* of (M,P) if there exists an orientation-preserving homeomorphism from $M-P$ to $\widehat{N}(\rho)$. We will let $\mathrm{GF}(M,P)$ denote the space of geometrically finite uniformizations of (M,P). The representation $\rho\in\mathrm{AH}(\pi_1(M),\pi_1(P))$ lies in $\mathrm{GF}(\pi_1(M),\pi_1(P))$ if and only if it is a geometrically finite uniformization of a pared manifold (M',P') which is pared homotopy equivalent to (M,P). In general we will say that a discrete faithful representation $\rho\colon\pi_1(M)\to\mathrm{PSL}(2,\mathbb{C})$ is *geometrically finite* if it lies in $\mathrm{GF}(M',P')$ for some pared manifold (M',P').

The next result guarantees the existence of hyperbolic 3-manifolds uniformizing pared 3-manifolds with nonempty boundary.

Thurston's Geometrization Theorem: *If (M,P) is an oriented pared 3-manifold with nonempty boundary, then there exists a geometrically finite uniformization of (M,P).*

Remarks: 1) The space $\mathrm{AH}(\pi_1(M),\pi_1(P))$ naturally sits as a closed subset of the character variety $X_T(\pi_1(M),\pi_1(P))$ (which is a quotient of the set of conjugacy classes of representations of $\pi_1(M)$ into $\mathrm{PSL}(2,\mathbb{C})$ such that the image of every element of $\pi_1(P)$ has trace ±2). See Chapter V of Morgan-Shalen [**97**] or Section 4.3 of Kapovich [**61**] for more details. The full version of Marden's Stability Theorem (Proposition 9.1 in [**75**]) implies that $\mathrm{GF}(\pi_1(M),\pi_1(P))$ is an open subset of $\mathrm{AH}(\pi_1(M),\pi_1(P))$ (as a subset of $X_T(\pi_1(M))$). A result of Sullivan [**116**] shows that $\mathrm{GF}(\pi_1(M),\pi_1(P))$ is the interior of $\mathrm{AH}(\pi_1(M),\pi_1(P))$, again as a subset of $X_T(\pi_1(M),\pi_1(P))$. Conjecturally (see Bers [**12**], Sullivan [**116**], and Thurston [**121**]) $\mathrm{GF}(\pi_1(M),\pi_1(P))$ is dense in $\mathrm{AH}(\pi_1(M),\pi_1(P))$. In the epilogue, we will discuss further the global topology of $\mathrm{AH}(\pi_1(M),\pi_1(P))$.

2) Our definition of geometric finiteness differs somewhat from the standard definitions. The most classical definition is that a Kleinian group Γ is geometrically finite if there is a finite-sided convex fundamental polyhedron for its action on $\mathbb{H}^3$, see Bowditch [**20**] for a complete discussion of various other equivalent definitions.

We will give a brief outline of a proof that our definition is equivalent to the standard definitions. If $N(\rho)$ is homeomorphic to $M-P$ where (M,P) is a pared 3-manifold, then it follows immediately from Theorem 1 in Abikoff [**2**] that $\rho(\pi_1(M))$ is geometrically finite in the classical definition. (In Abikoff's language our definition implies immediately that $\widehat{N}(\rho)$ has a well-positioned ample submanifold R such that $\partial R-\partial C$ has characteristic zero. The submanifold R may be constructed by simply taking the portion of $\widehat{N}$ which is identified with the complement in M of a regular neighborhood of P.) Corollary 6.10 in Morgan [**96**] asserts that if ρ is geometrically finite, in the standard definition, then $\widehat{N}(\rho)$ is homeomorphic to $M-P$ where P is a pared manifold. (In [**96**], a hyperbolic 3-manifold $N=\mathbb{H}^3/\Gamma$ is said to be geometrically finite if any ϵ-neighborhood of its convex core $C(N)$ has finite volume; this is shown to be equivalent to the classical definition in [**20**]. Corollary 6.10 in [**96**] asserts that, if Γ does not have a finite index Fuchsian subgroup, then $C(N)$ is homeomorphic to $M-P$ where (M,P) is a pared 3-manifold.

However, Morgan's statement that N is homeomorphic to $C(N)$ is incorrect if $\partial\widehat{N}$ is nonempty. It should assert that $\widehat{N}$ is homeomorphic to $C(N)$, which follows from the fact that $\widehat{N} - C(N)$ is homeomorphic to $\partial\widehat{N} \times (0,1]$. If Γ contains a Fuchsian subgroup of finite index, then it is easily checked that $\widehat{N}$ is homeomorphic to $M-P$ where M is an I-bundle and $M - P$ is the associated ∂I-bundle.)

7.2. Quasiconformal deformation theory: a review

In this section we briefly review the quasiconformal deformation theory of Kleinian groups. In particular, we will outline the proof of the Quasiconformal Parameterization Theorem which gives a complete description of the space of Kleinian groups which are quasiconformally conjugate to a given geometrically finite Kleinian group.

Good references for the theory of quasiconformal maps are the books of Lehto-Virtanen [**71**] and Lehto [**70**]. Good references for Teichmüller theory are the books of Abikoff [**1**], Gardiner [**44**] and Lehto [**70**]. An excellent, analytically oriented, survey of the quasiconformal deformation theory of Kleinian groups is given in a paper of Bers [**14**]. We will take a more topological viewpoint.

7.2.1. Quasiconformal maps and Beltrami differentials. Given a function $f\colon D \to \overline{\mathbb{C}}$ defined on a domain D in $\overline{\mathbb{C}}$, we may write it as $f(x,y) = u(x,y) + iv(x,y)$. We say f is *ACL* (absolutely continuous on lines) if given any rectangle $R=[a,b] \times [c,d]$ in D both u and v are absolutely continuous restricted to almost every vertical and almost every horizontal line segment in R. If f is ACL then the partial derivatives of u and v exist almost everywhere and we define $f_x = u_x + iv_x$ and $f_y = u_y + iv_y$. Then, we let $f_z = \frac{1}{2}(f_x - if_y)$ and $f_{\bar z} = \frac{1}{2}(f_x + if_y)$. (Recall that the Cauchy-Riemann equations assert that if f is analytic then $f_{\bar z} = 0$ for all $z \in D$.) We define the *Beltrami differential* of f to be $\mu_f = \dfrac{f_{\bar z}}{f_z}$. Notice that if f is differentiable at a point z and $Jf(z)$ is its Jacobian, then the image of the unit circle (in the tangent space $T_z(D)$) under $Jf(z)$ is an ellipse, the ratio of the lengths of the axes is given by $K(z) = \dfrac{1+|\mu_f(z)|}{1-|\mu_f(z)|}$, and the angle that the preimage of the (longer) axis makes with the x-axis is $\frac{1}{2}\arg(\mu_f(z))$.

One says that an orientation-preserving homeomorphism $f\colon D \to D'$ is K-*quasiconformal* if f is ACL and $|\mu_f| \le \frac{K-1}{K+1}$ almost everywhere. This says that, typically, very small circles are taken to curves very much like ellipses with eccentricity at most K. One way of formalizing this is by defining

$$H(z) = \limsup_{r\to 0} \frac{\max_\theta |f(z+re^{i\theta}) - f(z)|}{\min_\theta |f(z+re^{i\theta}) - f(z)|}.$$

An orientation-preserving homeomorphism $f\colon D \to \mathbb{C} \cup \{\infty\}$ is K-quasiconformal if and only if H is bounded on $D - \{\infty, f^{-1}(\infty)\}$ and $H(z) \le K$ almost everywhere in D (see pages 177 and 178 in Lehto [**70**]). If one uses the spherical metric on $\overline{\mathbb{C}}$, then one need not exclude ∞ and $f^{-1}(\infty)$ from consideration.

One may check that the composition of a K_1-quasiconformal map and a K_2-quasiconformal map is a K_1K_2-quasiconformal map. Another useful fact is:

PROPOSITION 7.2.1. (Theorem 1.5.1 in Lehto-Virtanen [**71**]) *A quasiconformal map is conformal if and only if it is 1-quasiconformal.*

The most fundamental result concerning quasiconformal maps is the Measurable Riemann Mapping Theorem (see Ahlfors-Bers [**6**] or Lehto [**70**]) which asserts that the Beltrami differential determines the quasiconformal map (up to normalization) and that every Beltrami differential (of norm less than 1) determines a quasiconformal map.

Measurable Riemann Mapping Theorem: *Suppose that $\mu \in L_\infty(\overline{\mathbb{C}}, \mathbb{C})$ and $||\mu||_\infty < 1$. Then there exists a unique quasiconformal map $\phi_\mu\colon \overline{\mathbb{C}} \to \overline{\mathbb{C}}$ whose Beltrami differential is μ and such that ϕ_μ fixes 0, 1, and ∞. Moreover, ϕ_μ depends analytically on μ.*

Notice that one may combine the Measurable Riemann Mapping Theorem and the traditional Riemann Mapping Theorem to observe that the same result holds for the upper half-plane $\mathbb{H}^2$. This version of the result is used in traditional Teichmüller theory and also plays a role in our proof of the Quasiconformal Parameterization Theorem.

Measurable Riemann Mapping Theorem (Disk version): *Suppose that $\mu \in L_\infty(\mathbb{H}^2, \mathbb{C})$ and $||\mu||_\infty < 1$. Then there exists a unique quasiconformal map $\phi_\mu\colon \mathbb{H}^2 \to \mathbb{H}^2$ whose Beltrami differential is μ and such that ϕ_μ fixes i, $2i$, and $3i$. Moreover, ϕ_μ depends analytically on μ.*

An alternative characterization of quasiconformal mappings of $\overline{\mathbb{C}}$ is obtained by considering biLipschitz homeomorphisms of $\mathbb{H}^3$. Any orientation-preserving biLipschitz homeomorphism of $\mathbb{H}^3$ extends continuously to a homeomorphism of $\mathbb{H}^3 \cup \overline{\mathbb{C}}$ whose restriction to $\overline{\mathbb{C}}$ is quasiconformal (see for example Theorem 3.22 in Matsuzaki-Taniguchi [**82**]). On the other hand, any quasiconformal map $\phi\colon \overline{\mathbb{C}} \to \overline{\mathbb{C}}$ extends to an orientation-preserving homeomorphism $\Phi\colon \mathbb{H}^3 \cup \overline{\mathbb{C}} \to \mathbb{H}^3 \cup \overline{\mathbb{C}}$ such that the restriction of Φ to $\mathbb{H}^3$ is a biLipschitz homeomorphism (see, for example, Theorem 5.31 in Matsuzaki-Taniguchi [**82**]). Combining these results gives:

PROPOSITION 7.2.2. *Let $\phi\colon \overline{\mathbb{C}} \to \overline{\mathbb{C}}$ be an orientation-preserving homeomorphism. Then ϕ is quasiconformal if and only if it extends to a homeomorphism $\Phi\colon \mathbb{H}^3 \cup \overline{\mathbb{C}} \to \mathbb{H}^3 \cup \overline{\mathbb{C}}$ whose restriction to $\mathbb{H}^3$ is biLipschitz (with respect to the hyperbolic metric).*

7.2.2. Teichmüller spaces and modular groups. Let S_0 and S be hyperbolic Riemann surfaces. An orientation-preserving homeomorphism $\phi\colon S_0 \to S$ is K-quasiconformal if whenever $U_0 \subset S_0$ and $U \subset S$ are local coordinates so that $\phi(U_0) \subset U$, then $\phi|_{U_0}$ is K-quasiconformal. (Equivalently, we could have required that the lift $\widetilde{\phi}\colon \mathbb{H}^2 \to \mathbb{H}^2$ of ϕ be K-quasiconformal.) If ϕ is quasiconformal, we will call (S, ϕ) a *quasiconformal deformation* of S_0. Two quasiconformal deformations (S_1, ϕ_1) and (S_2, ϕ_2) of S_0 are said to be equivalent if there exists a conformal map $g\colon S_1 \to S_2$ which is homotopic to $\phi_2 \circ \phi_1^{-1}$. The *Teichmüller space* $\mathcal{T}(S_0)$ of S_0 is the space of equivalence classes of quasiconformal deformations of S_0. (Actually if S_0 does not have finite area this definition gives rise to what is usually called the *reduced* Teichmüller space.)

Let $\psi\colon S \to S'$ and $\psi'\colon S \to S'$ be two quasiconformal maps between Riemann surfaces. We say that $\{\psi_t\}_{t\in[0,1]}$ is a *strong quasiconformal isotopy* between ψ and ψ' if it is an isotopy such that each ψ_t is a quasiconformal map and the Beltrami differential μ_t of ψ_t depends continuously on t. The following proposition is the quasiconformal analogue of the Baer-Nielsen theorem (Theorem 2.5.5).

PROPOSITION 7.2.3. (Earle-McMullen [**33**]) *Let S and S' be two finite area Riemann surfaces. If $\psi\colon S \to S'$ and $\psi'\colon S \to S'$ are homotopic quasiconformal maps, then there exists a strong quasiconformal isotopy between ψ and ψ'.*

As an immediate corollary we obtain the following stronger characterization of the equivalence of two quasiconformal deformations.

COROLLARY 7.2.4. *Let S_0 be a finite area hyperbolic Riemann surface. Two quasiconformal deformations (S_1, ϕ_1) and (S_2, ϕ_2) of S_0 are equivalent if and only if there exists a conformal map $g\colon S_1 \to S_2$ which is strongly quasiconformally isotopic to $\phi_2 \circ \phi_1^{-1}$.*

Teichmüller proved, among other things, that the Teichmüller space of a finite area Riemann surface is a cell. (See any of the above references on Teichmüller theory for a proof of Teichmüller's theorem, or see Bers [**11**].)

Teichmüller's Theorem: *If S_0 is a finite area, hyperbolic Riemann surface which is homeomorphic to a closed surface of genus g with p punctures, then $\mathcal{T}(S_0)$ is homeomorphic to $\mathbb{R}^{6g+2p-6}$.*

Notice that if $h\colon S_1 \to S_2$ is a homeomorphism between two finite area Riemann surfaces, then one can find an isotopic quasiconformal homeomorphism $j\colon S_1 \to S_2$. If S_1 is compact any isotopic diffeomorphism will do. If S_1 is noncompact, we recall that each end of S_1 and S_2 has a neighborhood which is conformally equivalent to a punctured disk. Hence, one may choose j to be an isotopic diffeomorphism which is conformal on a neighborhood of each end of S_1. Notice that j gives rise to a homeomorphism between $\mathcal{T}(S_1)$ and $\mathcal{T}(S_2)$ simply by identifying $(S, \phi\colon S_1 \to S)$ with $(S, \phi \circ j^{-1})$.

If F is an oriented topological surface, then the Teichmüller space $\mathcal{T}(F)$ of (equivalence classes of) marked hyperbolic structures on F consists of pairs (S, ϕ) where S is a finite area hyperbolic surface and $\phi\colon F \to S$ is an orientation-preserving homeomorphism. Two pairs (S_1, ϕ_1) and (S_2, ϕ_2) are called equivalent if there exists a conformal map $g\colon S_1 \to S_2$ which is homotopic to $\phi_2 \circ \phi^{-1}$. If S is a hyperbolic surface and $f\colon F \to S$ is an orientation-preserving homeomorphism, then f induces an identification of $\mathcal{T}(F)$ and $\mathcal{T}(S)$.

The *modular group* $\mathrm{Mod}(S_0)$ is the group of (isotopy classes of) orientation-preserving diffeomorphisms of S_0. If $[h]$ is an isotopy class in $\mathrm{Mod}(S_0)$, then we may choose a representative h which is quasiconformal. Then $[h]$ acts on $\mathcal{T}(S_0)$ by taking (the equivalence class of) (S, ϕ) to (the equivalence class of) $(S, \phi \circ h)$. One may readily check that the action of $[h]$ is independent of our choice of representative and that $\mathrm{Mod}(S_0)$ acts as a group of homeomorphisms of $\mathcal{T}(S_0)$. Harvey [**47**] gives a more complete discussion of the modular group. Notice that if $f\colon F \to S$ is an orientation-preserving homeomorphism of F, then f induces an identification of $\mathrm{Mod}(F)$ with $\mathrm{Mod}(S)$.

The following theorem combines results of various authors. We will say that a group G of automorphisms of a space acts *almost effectively* if there is a finite normal subgroup G_0 such that G/G_0 acts effectively.

THEOREM 7.2.5. *Let S be a finite area hyperbolic Riemann surface which is homeomorphic to the interior of a compact surface F. Then* $\mathrm{Mod}(S)$ *is isomorphic to an index two subgroup of* $\mathrm{Out}(\pi_1(F), \pi_1(\partial F))$, *and* $\mathrm{Mod}(S)$ *acts almost effectively and properly discontinuously on $\mathcal{T}(S)$.*

We note that, in fact, $\text{Mod}(S)$ usually acts effectively on $\mathcal{T}(S)$. In particular, it always acts effectively if g has genus at least 3. For a complete enumeration of situations when $\text{Mod}(S)$ does not act effectively see MacBeath-Singerman [**74**].

7.2.3. Quasiconformal deformation theory of Kleinian groups. For the remainder of the section, we will assume that (M, P) is an oriented pared manifold and $\rho_0 \in \text{AH}(\pi_1(M), \pi_1(P))$. A *quasiconformal deformation* of ρ_0 is a pair $(\rho, \widetilde{\phi})$ where $\rho\colon \pi_1(M) \to \text{PSL}(2,\mathbb{C})$ is a representation of $\pi_1(M)$ and $\widetilde{\phi}\colon \overline{\mathbb{C}} \to \overline{\mathbb{C}}$ is a quasiconformal map such that $\rho(g) = \widetilde{\phi} \circ \rho_0(g) \circ \widetilde{\phi}^{-1}$ for all $g \in \pi_1(M)$. Note that if $(\rho, \widetilde{\phi})$ is a quasiconformal deformation of ρ_0, then $\rho \in \text{AH}(\pi_1(M), \pi_1(P))$.

We now define the quasiconformal deformation space $\widehat{\text{QC}}(\rho_0)$ to be the space of all quasiconformal deformations of ρ_0, where $(\rho, \widetilde{\phi})$ and $(\rho', \widetilde{\phi}')$ are said to be equivalent if

(1) there exists $\gamma \in \text{PSL}(2,\mathbb{C})$ such that $\rho' = \gamma \circ \rho \circ \gamma^{-1}$, and
(2) there exists an equivariant strong quasiconformal isotopy $\{\widetilde{\psi}_t\}$ between $\widetilde{\phi}$ and $\gamma^{-1} \circ \widetilde{\phi}'$, where equivariant means that $\rho(g) \circ \widetilde{\psi}_t = \widetilde{\psi}_t \circ \rho(g)$ for all $g \in \pi_1(M)$ and all t.

Notice that condition (1) is equivalent to saying that there exists $\gamma \in \text{PSL}(2,\mathbb{C})$ such that $(\rho, \widetilde{\phi}'')$ is a quasiconformal deformation of ρ_0, where $\widetilde{\phi}'' = \gamma^{-1} \circ \widetilde{\phi}'$. Condition (2) then assures us that there is a quasiconformal isotopy between the maps $\phi\colon \partial\widehat{N}(\rho_0) \to \partial\widehat{N}(\rho)$ and $\phi''\colon \partial\widehat{N}(\rho_0) \to \partial\widehat{N}(\rho)$. A 3-dimensional characterization of this equivalence is given by corollary 7.2.8.

If $(\rho, \widetilde{\phi})$ is a quasiconformal deformation of ρ_0, then $\widetilde{\phi}$ descends to a homeomorphism $\phi\colon \partial\widehat{N}(\rho_0) \to \partial\widehat{N}(\rho)$. If ρ_0 is geometrically finite, then Marden [**75**] observed that ϕ extends to a "quasiconformal" homeomorphism $\Phi\colon \widehat{N}(\rho_0) \to \widehat{N}(\rho)$. Later, Douady-Earle [**32**], Reimann [**111**], Thurston [**120**] and Tukia [**125**] proved that ϕ extends to a biLipschitz homeomorphism of $N(\rho_0)$ to $N(\rho)$, with no hypotheses on ρ_0.

Proposition 7.2.6. (Douady-Earle, Reimann, Thurston, Tukia) *Let $(\rho, \widetilde{\phi})$ be a quasiconformal deformation of ρ_0. Then there exists a continuous extension of $\phi\colon \partial\widehat{N}(\rho_0) \to \partial\widehat{N}(\rho)$ to a homeomorphism $\Phi\colon \widehat{N}(\rho_0) \to \widehat{N}(\rho)$, whose restriction to $N(\rho_0)$ is biLipschitz. Moreover, Φ depends continuously on the Beltrami differential $\mu_{\widetilde{\phi}}$ of $\widetilde{\phi}$, and Φ is an isometry if $\widetilde{\phi}$ is conformal.*

An immediate corollary of the above result is the fact that any quasiconformal deformation of a geometrically finite Kleinian group is itself geometrically finite.

Corollary 7.2.7. *If $\rho_0 \in \text{GF}(M, P)$ and $(\rho, \widetilde{\phi})$ is a quasiconformal deformation of ρ_0, then $\rho \in \text{GF}(M, P)$.*

As another corollary of proposition 7.2.6 we obtain a 3-dimensional characterization of the equivalence of two quasiconformal deformations of ρ (compare with corollary 7.2.4).

Corollary 7.2.8. *Let ρ_0 be a finitely generated Kleinian group. Two quasiconformal deformations $(\rho, \widetilde{\phi})$ and $(\rho', \widetilde{\phi}')$ are equivalent in $\widehat{\text{QC}}(\rho_0)$ if and only if there exists a homeomorphism $G\colon \widehat{N}(\rho) \to \widehat{N}(\rho')$ which is an isometry with respect to the hyperbolic metric on $N(\rho)$ and is isotopic to $\Phi' \circ \Phi^{-1}$.*

PROOF. First suppose that $(\rho, \widetilde{\phi})$ and $(\rho', \widetilde{\phi}')$ are equivalent. Let γ be the element of $\mathrm{PSL}(2,\mathbb{C})$ such that $\rho' = \gamma \circ \rho \circ \gamma^{-1}$ and there exists a strong equivariant quasiconformal isotopy $\{\widetilde{\psi}_t\}$ between $\widetilde{\phi}$ and $\gamma^{-1} \circ \widetilde{\phi}'$. Then γ descends to a homeomorphism $G\colon \widehat{N}(\rho) \to \widehat{N}(\rho')$ which is an isometry on $N(\rho)$ and $\partial \widehat{N}(\rho)$ and $\{\widetilde{\psi}_t\}$ descends to a strong equivariant quasiconformal isotopy $\{\psi_t\}$ between ϕ and $G^{-1} \circ \phi'$. Proposition 7.2.6 can then be used to extend $\{\psi_t\}$ to an isotopy between Φ and $G^{-1} \circ \Phi'$. Hence, there exists a homeomorphism $G\colon \widehat{N}(\rho) \to \widehat{N}(\rho')$ which is an isometry with respect to the hyperbolic metrics on both $N(\rho)$ and $\partial \widehat{N}(\rho)$ and is isotopic to $\Phi' \circ \Phi^{-1}$.

Now suppose that there exists a homeomorphism $G\colon \widehat{N}(\rho) \to \widehat{N}(\rho')$ such that G is an isometry with respect to the hyperbolic metrics on both $\partial \widehat{N}(\rho)$ and $N(\rho)$ and G is isotopic to $\Phi' \circ \Phi^{-1}$. Let S be a component of $\partial \widehat{N}(\rho)$, then $(\phi' \circ \phi^{-1})|_S$ is isotopic to the isometry $G|_S$. Hence, by proposition 7.2.3, there exists a strong quasiconformal isotopy between $(\phi' \circ \phi^{-1})|_S$ and $G|_S$.

Let γ be the lift of $G|_{\partial \widehat{N}(\rho)}$ to a conformal map from $\Omega(\rho)$ to $\Omega(\rho')$. We note that γ extends to a conformal map defined on all of $\overline{\mathbb{C}}$, since it is the extension of the isometry $\widetilde{G}$ which is the lift of $G|_{N(\rho)}$ to $\mathbb{H}^3$. In particular, $\rho' = \gamma \circ \rho \circ \gamma^{-1}$.

We may lift the strong quasiconformal isotopies between $\phi' \circ \phi^{-1}$ and G on each component of $\partial \widehat{N}_\rho$ to obtain a strong quasiconformal isotopy $\{\widetilde{\psi}_t\}$, defined only on $\Omega(\rho)$, between $\widetilde{\phi}' \circ \widetilde{\phi}^{-1}$ and γ. We then extend each $\widetilde{\psi}_t$ to a function defined on all of $\overline{\mathbb{C}}$ by setting $\widetilde{\psi}_t(\xi) = \widetilde{\phi}' \circ \widetilde{\phi}^{-1}(\xi) = \gamma(\xi)$ for all $\xi \in \Lambda(\rho)$. Maskit's Extension Theorem (see [**76**]), which we state below, insures that $\{\widetilde{\psi}_t^{-1} \circ \widetilde{\phi}'\}$ is a strong equivariant quasiconformal isotopy between $\widetilde{\phi}$ and $\gamma^{-1} \circ \widetilde{\phi}'$.

Maskit's Extension Theorem: *Let $\rho\colon \pi_1(M) \to \mathrm{PSL}(2,\mathbb{C})$ be a finitely generated Kleinian group. Suppose that a function $f\colon \overline{\mathbb{C}} \to \overline{\mathbb{C}}$ is quasiconformal on $\Omega(\rho)$, $f(\Omega(\rho)) = \Omega(\rho)$, f is equal to the identity map on $\Lambda(\rho)$, and*

$$\rho(g) \circ f = f \circ \rho(g)$$

for all $g \in \pi_1(M)$. Then f is a quasiconformal homeomorphism.

We have completed the proof that $(\rho, \widetilde{\phi})$ and $(\rho', \widetilde{\phi}')$ are equivalent, and hence the proof of corollary 7.2.8. □

Remark: Instead of invoking Maskit's Extension Theorem to produce the strong quasiconformal isotopy above, one may use the structure of geometrically finite hyperbolic 3-manifolds to produce an isotopy of biLipschitz homeomorphisms between G and $\Phi' \circ \Phi^{-1}$. This isotopy then lifts to an isotopy of biLipschitz homeomorphisms of $\mathbb{H}^3$ between $\widetilde{G}$ and $\widetilde{\Phi}' \circ \widetilde{\Phi}^{-1}$. The biLipschitz homeomorphisms in the lifted isotopy then extend to quasiconformal homeomorphisms of $\overline{\mathbb{C}}$ which give a quasiconformal isotopy between γ and $\widetilde{\phi}' \circ \widetilde{\phi}^{-1}$. With a little more care, one can make sure that the quasiconformal isotopy produced agrees with $\{\widetilde{\psi}_t\}$ on $\Omega(\rho_0)$ and hence is strong. This argument is more topological than the argument given above, but also more complicated.

Let $S_1, \ldots, S_n$ denote the components of $\partial \widehat{N}_{\rho_0}$. We can then define

$$\mathcal{T}(\rho_0) = \mathcal{T}(S_1) \times \cdots \times \mathcal{T}(S_n)\ .$$

If $(\rho, \widetilde{\phi})$ is a quasiconformal deformation of ρ_0, then $\widetilde{\phi}$ induces a quasiconformal homeomorphism $\phi\colon \partial\widehat{N}(\rho_0) \to \partial\widehat{N}(\rho)$. Hence, if S_i is any component of $\partial\widehat{N}(\rho_0)$, then $(\phi(S_i), \phi|_{S_i})$ is an element of $\mathcal{T}(S_i)$. We define a map $Z\colon \widehat{\mathrm{QC}}(\rho_0) \to \mathcal{T}(\rho_0)$ by setting

$$Z(\rho, \widetilde{\phi}) = ((\phi(S_1), \phi|_{S_1}), \ldots, (\phi(S_n), \phi|_{S_n}))\,.$$

The first fundamental theorem of the quasiconformal deformation theory of Kleinian groups is due to Bers [**13**]. It asserts that if ρ_0 is geometrically finite, then the map Z is a homeomorphism.

THEOREM 7.2.9. (Bers [**13**]) *Let (M, P) be an oriented pared 3-manifold with $\partial M - P$ nonempty. If $\rho_0 \in \mathrm{GF}(M, P)$, then $Z\colon \widehat{\mathrm{QC}}(\rho_0) \to \mathcal{T}(\partial\rho_0)$ is a homeomorphism.*

SKETCH OF THE PROOF OF THEOREM 7.2.9. Assume that $Z(\rho, \widetilde{\phi}) = Z(\rho', \widetilde{\phi}')$. Then for every component S_i of $\partial\widehat{N}(\rho_0)$, there exists a conformal map from $\phi(S_i)$ to $\phi'(S_i)$ which is strongly quasiconformally isotopic to $\phi' \circ \phi^{-1}$. We may then proceed exactly as in the proof of corollary 7.2.8, to construct a strong quasiconformal isotopy $\{\psi_t\}$, defined only on $\Omega(\rho)$, between $\widetilde{\phi}' \circ \widetilde{\phi}^{-1}$ and a conformal map $\widetilde{\psi}_1$, such that $\widetilde{\psi}_t \circ \rho(g) = \rho'(g) \circ \widetilde{\psi}_t$ for all $g \in \pi_1(M)$. As in the proof of corollary 7.2.8, we may extend $\widetilde{\psi}_t$ to a strong equivariant quasiconformal isotopy, defined on all of $\overline{\mathbb{C}}$, such that $\widetilde{\psi}_1$ is conformal on $\Omega(\rho)$.

We now make our only use of the fact that ρ_0 (and hence ρ) is geometrically finite. Ahlfors [**5**] proved that if ρ is geometrically finite, then $\Lambda(\rho)$ has measure zero. Proposition 7.2.1 then guarantees that $\widetilde{\psi}_1$ is conformal and thus gives an element of $\mathrm{PSL}(2, \mathbb{C})$. Hence, again exactly as in the proof of corollary 7.2.8, we see that $(\rho, \widetilde{\phi})$ and $(\rho', \widetilde{\phi}')$ are equivalent. Therefore, Z is injective.

We now show that Z is surjective. Let

$$\sigma = ((S_1', \phi_1'), \ldots, (S_n', \phi_n')) \in \mathcal{T}(\rho_0).$$

If U is a component of $\Omega(\rho)$ which covers S_i, then we may lift ϕ_i' to a quasiconformal map $\widetilde{\phi}_U : U \to U'$ where U' is the appropriate cover of S_i'. Let μ_U be the Beltrami differential of $\widetilde{\phi}_i'$. (Although the image of $\widetilde{\phi}_U$ does not, a priori, lie in $\overline{\mathbb{C}}$, we may still use local coordinates to construct a Beltrami differential.) By examining each component separately we obtain a Beltrami differential μ defined on all of $\Omega(\rho_0)$. We then set $\mu = 0$ on $\Lambda(\rho_0)$ to obtain a Beltrami differential defined on $\overline{\mathbb{C}}$. Choose $\widetilde{\phi}$ to be a quasiconformal map with Beltrami differential μ. By construction, for any $g \in \pi_1(M)$, $\widetilde{\phi}$ and $\widetilde{\phi} \circ \rho_0(g)$ have the same Beltrami differential. Hence, by the uniqueness portion of the Measurable Riemann Mapping Theorem, $\widetilde{\phi} \circ \rho_0(g) \circ \widetilde{\phi}^{-1}$ is a Möbius transformation. Thus, we obtain a quasiconformal deformation $(\rho, \widetilde{\phi})$ of ρ_0 by setting $\rho(g) = \widetilde{\phi} \circ \rho_0(g) \circ \widetilde{\phi}^{-1}$ for all $g \in \pi_1(M)$. One may use the uniqueness portion of the disk version of the Measurable Riemann Mapping Theorem to check that $Z(\rho, \widetilde{\phi}) = \sigma$. Therefore, Z is surjective and we have completed our sketch of proof. □

Let $\mathrm{QC}(\rho_0)$ be the space consisting of all (conjugacy classes of) representations $\rho\colon \pi_1(M) \to \mathrm{PSL}(2, \mathbb{C})$ such that there exists a quasiconformal map $\widetilde{\phi}\colon \overline{\mathbb{C}} \to \overline{\mathbb{C}}$ such that $(\rho, \widetilde{\phi}) \in \widehat{\mathrm{QC}}(\rho_0)$. It is this space which we will be directly interested in, since it arises as a subset of $\mathrm{AH}(\pi_1(M), \pi_1(P))$.

We let $\mathrm{Mod}_0(M,P)$ denote the group of (isotopy classes of) orientation-preserving homeomorphisms of $\partial M - P$ which extend to homeomorphisms of M which are homotopic to the identity. If $f\colon M - P \to \widehat{N}(\rho_0)$ is a homeomorphism, then $\mathrm{Mod}_0(M,P)$ is naturally identified with a subgroup $\mathrm{Mod}_0(\rho_0)$ of $\mathrm{Mod}(S_1) \times \cdots \times \mathrm{Mod}(S_n)$. Explicitly, if $[h] \in \mathrm{Mod}_0(M,P)$, then we let $\bar{h}\colon \partial\widehat{N}(\rho_0) \to \partial\widehat{N}(\rho_0)$ be a quasiconformal map which is isotopic to fhf^{-1} and we identify $[h]$ with

$$([\bar{h}|_{S_1}], \ldots, [\bar{h}|_{S_n}]) \in \mathrm{Mod}(S_1) \times \cdots \times \mathrm{Mod}(S_n).$$

Maskit [**76**] showed that $\mathrm{Mod}_0(\rho_0)$ acts freely on $\mathcal{T}(\rho_0)$.

If $\rho_0 \in \mathrm{GF}(M,P)$, then $\mathcal{T}(\rho_0)$ can be identified with the Teichmüller space $\mathcal{T}(\partial M, P)$ of all marked hyperbolic structures (of finite area) on $\partial M - P$. The group $\mathrm{Mod}_0(M,P)$ acts on $\mathcal{T}(\partial M, P)$ and the homeomorphism $f\colon M - P \to N(\rho_0)$ gives rise to an identification of $\mathcal{T}(\rho_0)/\mathrm{Mod}_0(\rho_0)$ with $\mathcal{T}(\partial M,P)/\mathrm{Mod}_0(M,P)$.

It is the second fundamental theorem of the quasiconformal deformation theory of Kleinian groups that if ρ_0 is a geometrically finite uniformization of (M,P) then $\mathrm{QC}(\rho_0)$ is homeomorphic to $\mathcal{T}(\rho_0)/\mathrm{Mod}_0(\rho_0)$. This result combines Bers' work [**13**] with work of Kra [**66**] and Maskit [**76**]. Since $\mathrm{Mod}_0(\rho_0)$ acts freely and properly discontinuously on $\mathcal{T}(\rho_0)$, $QC(\rho_0)$ is a manifold. Corollary 7.2.7 implies that $\mathrm{QC}(\rho_0) \subset \mathrm{GF}(M,P)$.

Quasiconformal Parameterization Theorem: *Let (M,P) be an oriented pared 3-manifold with $\partial M - P$ is nonempty. If $\rho_0 \in \mathrm{GF}(M,P)$, then*

$$\mathrm{QC}(\rho_0) \cong \mathcal{T}(\rho_0)/\mathrm{Mod}_0(\rho_0).$$

Moreover, $\mathcal{T}(\rho_0)/\mathrm{Mod}_0(\rho_0)$ is identified with $\mathcal{T}(\partial M,P)/\mathrm{Mod}_0(M,P)$.

Sketch of proof. We define a forgetful map $F\colon \widehat{\mathrm{QC}}(\rho_0) \to \mathrm{QC}(\rho_0)$. Clearly F is surjective, so the result will follow once we check that $F(\rho,\widetilde{\phi}) = F(\rho',\widetilde{\phi}')$ if and only if there exists $[h] \in \mathrm{Mod}_0(\rho_0)$ such that $[h](Z(\rho,\widetilde{\phi})) = Z(\rho',\widetilde{\phi}')$.

First take $[h] \in \mathrm{Mod}_0(\rho_0)$ and $(\rho,\widetilde{\phi}) \in \widehat{\mathrm{QC}}(\rho_0)$. Let $h'\colon \partial N(\rho_0) \to \partial N(\rho_0)$ be a quasiconformal map in the homotopy class of $[h]$. Theorem 8.1 in Marden [**75**] implies that there exists a quasiconformal map $\widetilde{\psi}\colon \overline{\mathbb{C}} \to \overline{\mathbb{C}}$ whose restriction to $\Omega(\rho)$ is a lift of h' and such that $\widetilde{\psi}|_{\Lambda(\rho)}$ is the identity map. Then $(\rho, \widetilde{\phi}\circ\widetilde{\psi})$ is a quasiconformal deformation of ρ_0, $[h](Z(\rho,\widetilde{\phi})) = Z(\rho,\widetilde{\phi}\circ\widetilde{\psi})$, and $F(\rho,\widetilde{\phi}) = F(\rho,\widetilde{\phi}\circ\widetilde{\psi}) = \rho$.

Now suppose that $F(\rho,\widetilde{\phi}) = F(\rho',\widetilde{\phi}')$. By proposition 7.2.6, there are extensions $\Phi\colon \widehat{N}(\rho_0) \to \widehat{N}(\rho)$ and $\Phi'\colon \widehat{N}(\rho_0) \to \widehat{N}(\rho')$ of ϕ and ϕ'. Since ρ' is conjugate to ρ, we may assume, perhaps after altering $\widetilde{\phi}$ by post-composing by a Möbius transformation, that $\rho' = \rho$. Thus $\widehat{N}(\rho)$ and $\widehat{N}(\rho')$ may be canonically identified and Φ' is homotopic to Φ. This implies that $\phi^{-1}\circ\phi'$ has an extension to a homeomorphism $\Phi^{-1}\circ\Phi'$ which is homotopic to the identity and hence gives rise to an element $[h] \in \mathrm{Mod}_0(\rho_0)$ such that $[h](Z(\rho,\widetilde{\phi})) = Z(\rho',\widetilde{\phi}')$, where $h = \phi^{-1}\circ\phi'$. $\square$

Remark: Sullivan [**115**] extended the Quasiconformal Parameterization theorem to all finitely generated Kleinian groups. In particular, if $\rho_0\colon \pi_1(M) \to \mathrm{PSL}(2,\mathbb{C})$ is any finitely generated, torsion-free Kleinian group, then $\widehat{\mathrm{QC}}(\rho_0)$ is homeomorphic to $\mathcal{T}(\rho_0)$. The key added ingredient in Sullivan's approach is a theorem stating that if $(\rho,\widetilde{\phi})$ is a quasiconformal deformation of a finitely generated Kleinian group

ρ_0 such that $\widetilde{\phi}$ is conformal on $\Omega(\rho_0)$, then $\widetilde{\phi}$ is a Möbius transformation. This theorem replaces the use of the fact that $\Lambda(\rho_0)$ has measure zero in our approach.

If we let $\mathrm{Mod}_0(\rho_0)$ denote the group of isotopy classes of homeomorphisms of $\partial\widehat{N}(\rho_0)$ which extend to homeomorphisms of $\widehat{N}(\rho_0)$ which are homotopic to the identity, then one may extend the proof of the Quasiconformal Parameterization Theorem to show that $\mathrm{QC}(\rho_0)$ is homeomorphic to $\mathcal{T}(\partial\widehat{N}(\rho_0))/\mathrm{Mod}_0(\rho_0)$. In order to do so, one must show that every element $[h]$ of $\mathrm{Mod}_0(\rho_0)$ has a representative h whose lift $\widetilde{h}$ to $\Omega(\rho_0)$ admits an extension to a quasiconformal map of $\overline{\mathbb{C}}$. (To accomplish this, let $j : \widehat{N}(\rho_0) \to \widehat{N}(\rho_0)$ be a diffeomorphism which is homotopic to the identity such that $[j|_{\partial\widehat{N}(\rho_0)}] = [h]$. If $\partial\widehat{N}(\rho_0)$ is compact, then $h = j|_{\partial\widehat{N}(\rho_0)}$ will be quasiconformal. If not, we may deform j so that h is conformal on a neighborhood of each end of $\partial\widehat{N}(\rho_0)$, which suffices to guarantee that h is quasiconformal on $\partial\widehat{N}(\rho_0)$. Since j is homotopic to the identity map, we can choose a lift $\widetilde{h}\colon \mathbb{H}^3 \cup \Omega(\rho_0) \to \mathbb{H}^3 \cup \Omega(\rho_0)$ of h such that $\widetilde{h} \circ \rho_0(g) = \rho_0(g) \circ \widetilde{h}$ for all $g \in \pi_1(M)$. If we let $\widetilde{h}(z) = z$ for all $z \in \Lambda(\rho_0)$, then Maskit's Extension Theorem guarantees that the restriction of $\widetilde{h}$ to $\overline{\mathbb{C}}$ is the desired quasiconformal homeomorphism of $\overline{\mathbb{C}}$.)

7.3. The Parameterization Theorem

In this section we make use of the Quasiconformal Parameterization Theorem, Marden's Isomorphism Theorem and Marden's Stability Theorem to give parameterizations of the spaces $\mathrm{GF}(M,P)$ and $\mathrm{GF}(\pi_1(M),\pi_1(P))$.

Let (M,P) be an oriented pared manifold. We first consider the space $\mathrm{A}(M,P)$ of oriented pared manifolds homotopy equivalent to (M,P). We will consider two oriented pared manifolds (M_1,P_1) and (M_2,P_2) to be equivalent if there exists an orientation-preserving pared homeomorphism $h\colon (M_1,P_1) \to (M_2,P_2)$.

We will also consider the space $\mathcal{A}(M,P)$ of marked, oriented pared manifolds homotopy equivalent to (M,P). Its basic objects are pairs $((M',P'),h')$ where $(M',P') \in \mathrm{A}(M,P)$ and $h'\colon (M,P) \to (M',P')$ is a pared homotopy equivalence. We will consider two pairs $((M_1,P_1),h_1)$ and $((M_2,P_2),h_2)$ to be equivalent if there exists an orientation-preserving pared homeomorphism $j\colon (M_1,P_1) \to (M_2,P_2)$ such that $j \circ h_1$ is admissibly homotopic to h_2. $\mathcal{A}(M,P)$ is the space of equivalence classes of such pairs.

We next define a function $\Theta\colon \mathrm{GF}(\pi_1(M),\pi_1(P)) \to \mathcal{A}(M,P)$. Given an element $\rho \in \mathrm{GF}(\pi_1(M),\pi_1(P))$, there exists an orientation-preserving homeomorphism $j\colon \widehat{N}(\rho) \to M' - P'$ where $(M',P') \in \mathrm{A}(M,P)$. On the other hand, ρ gives rise to an identification of $\pi_1(M)$ with $\pi_1(\widehat{N}(\rho))$ and hence to a homotopy equivalence $r_\rho\colon M \to \widehat{N}(\rho)$, well-defined up to homotopy. In order to construct Θ we must show that $j \circ r_\rho$ is homotopic to a pared homotopy equivalence $h_\rho\colon (M,P) \to (M',P')$. Let A be a component of P. Our assumptions imply that $j \circ r_\rho(A)$ is homotopic to an unique component A' of P'. Hence, we may change $j \circ r_\rho$ by homotopy to a homotopy equivalence such that $j \circ r_\rho(A) = A'$ (without changing the image of $j \circ r_\rho$ on any component of $P - A$). We may therefore, by applying the above procedure to each component, change $j \circ r_\rho$ to a pared homotopy equivalence $h_\rho\colon (M,P) \to (M',P')$. We then set $\Theta(\rho) = ((M',P'),h_\rho)$. It remains to check that this gives a well-defined element of $\mathcal{A}(M,P)$. Suppose h_ρ and h'_ρ are two pared homotopy equivalences constructed as above. Then suppose that h_ρ is homotopic to $j \circ r_\rho$ and h'_ρ is homotopic to $j' \circ r_\rho$ for some, perhaps different, orientation-preserving homeomorphism $j' \colon \widehat{N}(\rho) \to M' - P'$. Proposition 2.12.5 implies that $j' \circ j^{-1}$ is homotopic to a map which extends to a orientation-preserving pared homeomorphism $\phi \colon (M', P') \to (M', P')$. We then note that $\phi \circ h_\rho$ is homotopic to h'_ρ. By proposition 5.2.3 they are admissibly homotopic. It follows that $((M', P'), h_\rho)$ is equivalent to $((M', P'), h'_\rho)$ and hence that our map Θ is well-defined.

The two key tools in our proof of the Parameterization Theorem both come from Marden's seminal paper [**75**]. Marden's Isomorphism Theorem (Theorem 8.1 in [**75**]) will allow us to conclude that if $\Theta(\rho) = \Theta(\rho')$, then ρ and ρ' are quasiconformally conjugate.

Marden's Isomorphism Theorem: *Suppose that ρ_0 and ρ are elements of $\mathrm{GF}(\pi_1(M), \pi_1(P))$. Then $\rho \in \mathrm{QC}(\rho_0)$ if and only if there exists an orientation-preserving homeomorphism $s \colon \widehat{N}(\rho_0) \to \widehat{N}(\rho)$ such that $s \circ r_{\rho_0}$ is homotopic to r_ρ.*

Marden's Stability Theorem (Proposition 9.1 in [**75**]) will allow us to conclude that Θ is a continuous (i. e. locally constant) map.

Marden's Stability Theorem: *If $\rho_0 \in \mathrm{GF}(\pi_1(M), \pi_1(P))$, then there exists a neighborhood U of ρ in $\mathrm{AH}(\pi_1(M), \pi_1(P))$ such that if $\rho \in U$, then ρ is quasiconformally conjugate to ρ_0.*

We are now ready to give our parameterization:

Parameterization Theorem: *Let (M, P) be an oriented pared 3-manifold such that $\partial M - P$ is nonempty. Then $\mathrm{GF}(\pi_1(M), \pi_1(P))$ is homeomorphic to*

$$\bigsqcup_{((M',P'),h') \in \mathcal{A}(M,P)} \mathcal{T}(M', P') / \mathrm{Mod}_0(M', P').$$

PROOF. We first show that Θ is continuous, i. e. locally constant. Marden's Stability Theorem implies that if $\rho_0 \in \mathrm{GF}(\pi_1(M), \pi_1(P))$, then there exists a neighborhood U of ρ_0 such that if $\rho \in U$, then ρ is quasiconformally conjugate to ρ_0. Marden's Isomorphism Theorem (or proposition 7.2.6) then implies that if $\rho \in U$ then there exists an orientation-preserving homeomorphism $s \colon \widehat{N}(\rho_0) \to \widehat{N}(\rho)$ such that $s \circ r_{\rho_0}$ is homotopic to r_ρ. Let $j \colon \widehat{N}(\rho) \to M' - P'$ be a homeomorphism. Then, h_{ρ_0} can be chosen to be homotopic to $j \circ s \circ r_{\rho_0}$ and h_ρ can be chosen to be homotopic to $j \circ r_\rho$. It follows that h_ρ and h_{ρ_0} can be chosen to be homotopic and, by proposition 5.2.3, to be admissibly homotopic. Thus, $\Theta(\rho) = \Theta(\rho_0)$ if $\rho \in U$. Therefore, Θ is locally constant.

Next we will use Marden's Isomorphism Theorem to show that if $\Theta(\rho_1) = \Theta(\rho_2)$, then $\rho_2 \in \mathrm{QC}(\rho_1)$. Suppose that $\Theta(\rho_1) = \Theta(\rho_2)$ and let $j_i \colon \widehat{N}(\rho_i) \to M_i - P_i$ be homeomorphisms and let h_{ρ_i} be pared homotopy equivalences homotopic to $j_i \circ r_{\rho_i}$. Since $\Theta(\rho_1) = \Theta(\rho_2)$ there exists an orientation-preserving pared homeomorphism $s \colon (M_1, P_1) \to (M_2, P_2)$ such that $s \circ j_1 \circ r_{\rho_1}$ is homotopic to $j_2 \circ r_{\rho_2}$. Thus, $s' = j_2^{-1} \circ s \circ j_1 \colon \widehat{N}(\rho_1) \to \widehat{N}(\rho_2)$ is an orientation-preserving homeomorphism such that $s' \circ r_{\rho_1}$ is homotopic to r_{ρ_2}. Hence, Marden's Isomorphism Theorem guarantees that $\rho_2 \in \mathrm{QC}(\rho_1)$.

We will use Thurston's Geometrization Theorem to show that Θ is surjective. When $((M', P'), h') \in \mathcal{A}(M, P)$, it implies that there exist a geometrically finite hyperbolic 3-manifold N and an orientation-preserving homeomorphism $j \colon M' - P' \to \widehat{N}$ from $M' - P'$ to the conformal extension $\widehat{N}$ of N. Since we may identify $\pi_1(\widehat{N})$ with a subgroup of $\mathrm{PSL}(2, \mathbb{C})$, we may think of $j_\# \colon \pi_1(M') \to \pi_1(\widehat{N})$ as an element of $\mathrm{GF}(M', P')$. We then let $\rho = (j \circ h')_{\#}$. It is easy to check that $\rho \in \mathrm{GF}(M, P)$ and $\Theta(\rho) = ((M', P'), h')$.

Thus, if $((M', P'), h')$ is any element of $\mathcal{A}(M, P)$, then $\Theta^{-1}((M', P'), h')$ is nonempty and is of the form $\mathrm{QC}(\rho)$ for any $\rho \in \Theta^{-1}((M', P'), h')$. The Quasiconformal Parameterization Theorem implies that, in this circumstance, $\mathrm{QC}(\rho)$ is homeomorphic to $\mathcal{T}(M', P')/\mathrm{Mod}_0(M', P')$. The result follows. □

We will actually make most use of the following nearly immediate corollary of the Parameterization Theorem.

COROLLARY 7.3.1. *Let (M, P) be an oriented pared 3-manifold. The components of* $\mathrm{GF}(\pi_1(M), \pi(P))$ *are enumerated by elements of* $\mathcal{A}(M, P)$. *The components of* $\mathrm{GF}(M, P)$ *are in a one-to-one correspondence with the cosets of* $\mathcal{R}_+(M, P)$ *in* $\mathrm{Out}(\pi_1(M), \pi_1(P))$.

PROOF. The first assertion of the corollary follows immediately from the Parameterization Theorem and the fact that $\mathcal{T}(M', P')/\mathrm{Mod}_0(M', P')$ is connected for all oriented pared manifolds (M', P').

To prove the second assertion we first note that $\rho \in \mathrm{GF}(M, P)$ if and only if $\Theta(\rho) = ((M, P), h')$ for some pared homotopy equivalence h'. Hence the components of $\mathrm{GF}(M, P)$ can be identified with the set $\mathcal{A}_0(M, P)$ of elements of the form $((M, P), h')$ in $\mathcal{A}(M, P)$. We first note that every element $\alpha \in \mathrm{Out}(\pi_1(M), \pi_1(P))$ is realized by a pared homotopy equivalence $h_\alpha\colon (M, P) \to (M, P)$. Moreover, by proposition 5.2.3, any two pared homotopy equivalences realizing α are admissibly homotopic. Hence, there is a well-defined surjection $J\colon \mathrm{Out}(\pi_1(M), \pi_1(P)) \to \mathcal{A}_0(M, P)$ given by taking α to $((M, P), h_\alpha)$. Moreover, $J(\alpha) = J(\alpha')$ if and only if there is an orientation-preserving pared homeomorphism $s\colon (M, P) \to (M, P)$ with $h_{\alpha'}$ homotopic, hence admissibly homotopic, to $s \circ h_\alpha$, i. e. if and only if $\alpha' \circ \alpha^{-1} \in \mathcal{R}_+(M, P)$. Therefore, $J(\alpha) = J(\alpha')$ if and only if α and α' lie in the same right coset of $\mathcal{R}_+(M, P)$ in $\mathrm{Out}(\pi_1(M), \pi_1(P))$ and the result follows. □

Remarks: 1) One may extend Θ to a map $\overline{\Theta}\colon \mathrm{AH}(\pi_1(M)) \to \mathcal{A}(M, P)$. If $\rho \in AH(\pi_1(M))$, then $\Theta(\rho)$ records the marked homeomorphism type of a (relative) compact core for $\widehat{N}(\rho)$. Anderson and Canary [**8**] showed that $\overline{\Theta}$ is not always continuous on $\mathrm{AH}(\pi_1(M), \pi_1(P))$, even if we assume that P is empty. This phenomenon shows that distinct components of $\mathrm{GF}(M, P)$ may "bump," i. e. have intersecting closures. This phenomenon has been completely analyzed when M has incompressible boundary and P is empty by Anderson, Canary and McCullough [**9**]. See the epilogue for more details.

2) Marden's original isomorphism theorem incorporates Waldhausen's Theorem. It asserts that if $\rho_1, \rho_2 \in GF(\pi_1(M), \pi_1(P))$ and there exists an orientation-preserving homeomorphism $f : \Omega(\rho_1) \to \Omega(\rho_2)$ which induces an isomorphism $\phi : \rho_1(\pi_1(M)) \to \rho_2(\pi_1(M))$, then there exists a quasiconformal homeomorphism of $\mathbb{H}^3 \cup \overline{\mathbb{C}}$ which induces ϕ. Our version is an immediate corollary of the original statement and Proposition 7.2.6.

CHAPTER 8

Statements of Main Theorems

8.1. Statements of Main Topological Theorems

We are now ready to state the complete versions of our main topological theorems which characterize when the group $\mathcal{R}(M, \underline{\underline{a}})$ of outer automorphisms which are realizable by homeomorphisms has finite index in the group $\mathrm{Out}(\pi_1(M), \pi_1(\underline{\underline{a}}))$ of all outer automorphisms.

Main Topological Theorem 1 concerns the case where $(M, \underline{\underline{a}})$ has a compressible free side. It requires strong restrictions on the boundary pattern, but nonetheless applies to all pared 3-manifolds which have a compressible free side.

Main Topological Theorem 1: *Let $(M, \underline{\underline{a}})$ be a compact orientable irreducible 3-manifold with boundary pattern consisting of a (possibly empty) collection of disjoint incompressible submanifolds. Suppose there exists a free side F which is compressible in M, such that each element of $\underline{\underline{a}}$ that meets F is an annulus. Then $\mathcal{R}(M, \underline{\underline{a}})$ has finite index in $\mathrm{Out}(\pi_1(M), \pi_1(\underline{\underline{a}}))$ if and only if either $(M, \underline{\underline{a}})$ is a relative compression body or $(M, \underline{\underline{a}})$ is small.*

Main Topological Theorem 2 applies when the elements of the boundary pattern of M are disjoint, and the completion of the boundary pattern is useful. In particular, by lemma 5.2.1, it will apply to all pared 3-manifolds which do not have a compressible free side.

Main Topological Theorem 2: *Let M be a compact orientable irreducible 3-manifold with nonempty boundary and a (possibly empty) boundary pattern $\underline{\underline{m}}$ whose completion is useful. Assume that the elements of $\underline{\underline{m}}$ are disjoint. Let $\underline{\underline{m}}'$ be the set of elements of $\underline{\underline{m}}$ that are not annuli. Then $\mathcal{R}(M, \underline{\underline{m}})$ has finite index in $\mathrm{Out}(\pi_1(M), \pi_1(\underline{\underline{m}}))$ if and only if every Seifert-fibered component V of the characteristic submanifold of $(M, \overline{\underline{\underline{m}}})$ that meets $\partial M - |\underline{\underline{m}}|$ satisfies one of the following:*

1. *V is a solid torus, or*
2. *V is either $S^1 \times S^1 \times I$ or the I-bundle over the Klein bottle, and no boundary component of V contains more than one component of $\overline{V \cap \partial M - |\underline{\underline{m}}'|}$, or*
3. *V is fibered over the annulus with one exceptional fiber, and no component of $\overline{V \cap \partial M - |\underline{\underline{m}}'|}$ is an annulus, or*
4. *V is fibered over the disk with two holes with no exceptional fibers, and $\overline{V \cap \partial M - |\underline{\underline{m}}'|}$ is one of the boundary tori of V, or*
5. *$V = M$ and V is fibered either over the disk with two exceptional fibers, or over the Möbius band with one exceptional fiber, or over the torus with one hole with no exceptional fiber, or*
6. *$V = M$ and V is fibered over the disk with three exceptional fibers, each of type $(2, 1)$,*

and every I-bundle component V of the characteristic submanifold of $(M, \overline{\underline{\underline{m}}})$ which has all of its lids contained in $|\underline{\underline{m}}|$ and meets $\partial M - |\underline{\underline{m}}|$ satisfies one of the following:

(7) *V is a 3-ball, or*

(8) *V is I-fibered over a topological annulus or Möbius band and no component of $V \cap \overline{\partial M - |\underline{\underline{m}}|}$ is a square which meets two different components of the frontier of V, or*

(9) *V is I-fibered over the disk with two holes, and $V \cap \overline{\partial M - |\underline{\underline{m}}|}$ is an annulus, or*

(10) *$V = M$ and V is I-fibered over the torus with one hole.*

We note that since the elements of $\underline{\underline{m}}$ are disjoint, if an I-bundle component of the characteristic submanifold of $(M, \overline{\underline{\underline{m}}})$ has a lid contained in $|\underline{\underline{m}}|$ and meets $\partial M - |\underline{\underline{m}}|$, then its other lid (if it has one) must also be contained in $|\underline{\underline{m}}|$. Examples 1.4.3 and 1.4.4 illustrate item (4) of the theorem, and example 2.10.11 illustrates items (7), (8), and (9).

Although Main Topological Theorem 2 is stated for Haken manifolds with nonempty boundary, it holds true for closed Haken manifolds as well. In the closed case, of course, no component of the characteristic submanifold can meet the boundary, so the topological conditions in Main Topological Theorem 2 are trivially satisfied. On the other hand, since Haken manifolds are aspherical, every outer automorphism of the fundamental group is induced by a homotopy equivalence, and in the closed case Waldhausen's Theorem 2.5.6 shows that every homotopy equivalence is homotopic to a homeomorphism. That is, $\mathcal{R}(M) = \mathrm{Out}(\pi_1(M))$.

The proofs of Main Topological Theorems 1 and 2 comprise chapters 9 and 10 respectively.

8.2. Statements of Main Hyperbolic Theorem and Corollary

We are now ready to state our main hyperbolic theorem which characterizes exactly when there are finitely many components of the space $\mathrm{GF}(M, P)$ of geometrically finite uniformizations of a pared manifold. Recall that small pared manifolds were completely described in lemma 6.1.1.

Main Hyperbolic Theorem: *Let (M, P) be a pared 3-manifold and let $\underline{\underline{p}}$ be the associated boundary pattern on M.*

(1) *If $(M, \underline{\underline{p}})$ has a compressible free side then $\mathrm{GF}(M, P)$ has finitely many components if and only if $(M, \underline{\underline{p}})$ is either small or a relative compression body.*

(2) *If $(M, \underline{\underline{p}})$ has no compressible free side then $\mathrm{GF}(M, P)$ has finitely many components if and only if (M, P) does not have double trouble.*

We may use our Main Hyperbolic Theorem to completely characterize when $\mathrm{GF}(\pi_1(M), \pi_1(P))$ has finitely many components. Recall that the homotopy type of (M, P) is said to be small if and only if every element of $\mathrm{A}(M, P)$ is either small or a relative compression body. The small homotopy types of pared 3-manifolds were completely described in theorem 6.2.1.

Main Hyperbolic Corollary: *Let (M, P) be a pared 3-manifold and let $\underline{\underline{p}}$ be the associated boundary pattern on M.*

(1) *If $(M, \underline{\underline{p}})$ has a compressible free side, then $\mathrm{GF}(\pi_1(M), \pi_1(P))$ has finitely many components if and only if the pared homotopy type of (M, P) is small.*
(2) *If $(M, \underline{\underline{p}})$ has no compressible free side, then $\mathrm{GF}(\pi_1(M), \pi_1(P))$ has finitely many path components if and only if (M, P) does not have double trouble.*

8.3. Derivation of hyperbolic results

In this section we derive the Main Hyperbolic Theorem and the Main Hyperbolic Corollary from the Parameterization Theorem and the Main Topological Theorems.

PROOF OF MAIN HYPERBOLIC THEOREM. If $\partial M - P$ is empty, then the Mostow Rigidity Theorem [**98**], as extended by Prasad [**110**], asserts that $\mathrm{GF}(M, P)$ is either empty or a single point. Since (M, P) cannot have double trouble if $\partial M - P$ is empty, the theorem holds in this case. So we may assume that $\partial M - P$ is nonempty.

Corollary 7.3.1 shows that the components of $\mathrm{GF}(M, P)$ are in one-to-one correspondence with the cosets of $\mathcal{R}_+(M, P)$ in $\mathrm{Out}(\pi_1(M), \pi_1(P))$. Thus $\mathrm{GF}(M, P)$ has finitely many components exactly when $\mathcal{R}(M, P)$ has finite index in $\mathrm{Out}(\pi_1(M), \pi_1(P))$.

Suppose $(M, \underline{\underline{p}})$ has a compressible free side. By Main Topological Theorem 1, $\mathcal{R}_+(M, P)$ has finite index in $\mathrm{Out}(\pi_1(M), \pi_1(P))$ if and only if $(M, \underline{\underline{p}})$ is either small or a relative compression body.

Suppose $(M, \underline{\underline{p}})$ does not have a compressible free side. By lemma 5.2.1, the completion of $(\overline{M}, \underline{\underline{p}})$ is useful. Therefore Main Topological Theorem 2 applies to show that $\mathcal{R}(M, P)$ has finite index in $\mathrm{Out}(\pi_1(M), \pi_1(P))$ if and only if every Seifert-fibered component of the characteristic submanifold $(\Sigma, \underline{\underline{\sigma}})$ of $(M, \underline{\underline{\overline{p}}})$ which meets $\partial M - P$ satisfies one of conditions (1)-(6) and every I-bundle component of $(\Sigma, \underline{\underline{\sigma}})$ either has all lids contained in $\partial M - P$ or satisfies one of conditions (7)-(10). For the I-bundle components, the Pared Characteristic Submanifold Restrictions (theorem 5.3.4) guarantee that every lid is contained in $\partial M - P$. They also guarantee that each Seifert-fibered component is either a solid torus or is a $T^2 \times \mathrm{I}$ that meets P in one if its boundary tori, and meets $\partial M - P$ in a possibly empty collection of annuli in its other boundary torus. The solid torus components satisfy condition (1) of Main Topological Theorem 2. The $T^2 \times \mathrm{I}$ components satisfy condition (2) precisely when each meets $\partial M - P$ in at most one annulus. By lemma 5.3.5, this is equivalent to saying that (M, P) does not have double trouble. □

We now turn our attention to the Main Hyperbolic Corollary.

PROOF OF MAIN HYPERBOLIC COROLLARY. We first note that by definition,

$$\mathrm{GF}(\pi_1(M), \pi_1(P)) = \bigsqcup_{(M', P') \in \mathrm{A}(M, P)} \mathrm{GF}(M', P').$$

Lemma 3.5.2 implies that the boundary pattern associated to (M', P') is usable whenever $(M', P') \in \mathrm{A}(M, P)$, so theorem 4.2.3 implies that $\mathrm{A}(M, P)$ is finite. Thus, $\mathrm{GF}(\pi_1(M), \pi_1(P))$ has finitely many components if and only if $\mathrm{GF}(M', P')$ has finitely many components for all $(M', P') \in \mathrm{A}(M, P)$.

We recall from lemma 4.1.3 that (M, P) has a compressible free side if and only if (M', P') has a compressible free side for all $(M', P') \in \mathrm{A}(M, P)$. Thus, the Main Hyperbolic Theorem implies that if (M, P) has a compressible free side, then $\mathrm{GF}(\pi_1(M), \pi_1(P))$ has finitely many components if and only if every $(M', P') \in \mathrm{A}(M, P)$ is either small or a relative compression body, i.e. if and only if the pared homotopy type of (M, P) is small. Statement (1) follows.

If (M, P) does not have a compressible free side, then the Main Hyperbolic Theorem guarantees that $\mathrm{GF}(\pi_1(M), \pi_1(P))$ has finitely many components if and only if every element of $\mathrm{A}(M, P)$ does not have double trouble. However, by theorem 5.3.6, (M, P) has double trouble if and only if every element of $\mathrm{A}(M, P)$ has double trouble. Statement (2) follows. $\square$

CHAPTER 9

The Case When There Is a Compressible Free Side

In this section, we give the proof of our first main topological result, which characterizes when the group $\mathcal{R}(M, \underline{\underline{a}})$ of realizable automorphisms has finite index in the group $\mathrm{Out}(\pi_1(M), \pi_1(\underline{\underline{a}}))$ of all outer automorphisms in the case that $(M, \underline{\underline{a}})$ has a compressible free side. Recall that small manifolds were defined in section 6.1.

Main Topological Theorem 1: *Let $(M, \underline{\underline{a}})$ be a compact orientable irreducible 3-manifold with boundary pattern consisting of a (possibly empty) collection of disjoint incompressible submanifolds. Suppose there exists a free side F which is compressible in M, such that each element of $\underline{\underline{a}}$ that meets F is an annulus. Then $\mathcal{R}(M, \underline{\underline{a}})$ has finite index in $\mathrm{Out}(\pi_1(M), \pi_1(\underline{\underline{a}}))$ if and only if either $(M, \underline{\underline{a}})$ is a relative compression body or $(M, \underline{\underline{a}})$ is small.*

Maskit [**77**] proved that $\mathcal{R}(M, \underline{\underline{a}})$ has finite index in $\mathrm{Out}(\pi_1(M), \pi_1(\underline{\underline{a}}))$ for relative compression bodies which are pared manifolds, and McCullough and Miller [**89**] showed this for compression bodies (i. e. relative compression bodies with empty boundary pattern).

Even when $(M, \underline{\underline{a}})$ is a compression body, it rarely happens that $\mathcal{R}(M, \underline{\underline{a}})$ is all of $\mathrm{Out}(\pi_1(M), \pi_1(\underline{\underline{a}}))$.

EXAMPLE 9.0.1. *A compression body $(V, \underline{\underline{\emptyset}})$ for which $\mathcal{R}(V, \underline{\underline{\emptyset}})$ is a proper subgroup of $\mathrm{Out}(\pi_1(V))$.*

Let V be a compression body with two constituents F_1 and F_2 which are closed surfaces, so that V is obtained from $F_1 \times \mathrm{I}$ and $F_2 \times \mathrm{I}$ by attaching a 1-handle connecting a disk in $F_1 \times \{1\}$ to a disk in $F_2 \times \{1\}$. Let h be an orientation-reversing homeomorphism of F_1 which leaves the disk in $F_1 \times \{1\}$ invariant. The homeomorphisms $h \times \mathrm{id}_{\mathrm{I}}$ on $F_1 \times \mathrm{I}$ and $\mathrm{id}_{F_2} \times \mathrm{id}_{\mathrm{I}}$ on $F_2 \times \mathrm{I}$ extend over the 1-handle to a homotopy equivalence of V. This homotopy equivalence is not homotopic to a homeomorphism, for such a homeomorphism would have to be orientation-reversing on the boundary component $F_1 \times \{0\}$ and orientation-preserving on the boundary component $F_2 \times \{0\}$. One can construct similar examples where the boundary pattern is nonempty.

Our next example shows that Main Topological Theorem 1 can fail without the assumption that every element of $\underline{\underline{a}}$ that meets F is an annulus.

EXAMPLE 9.0.2. *A relative compression body $(V, \underline{\underline{v}})$ such that $\mathcal{R}(V, \underline{\underline{v}})$ has infinite index in $\mathrm{Out}(\pi_1(V), \pi_1(\underline{\underline{v}}))$.*

Let V be a compression body with one constituent F_1, which is a compact surface of genus 2 with one boundary component. Give V the boundary pattern $\underline{\underline{v}} = \{F_1 \times \{0\} \cup \partial F_1 \times \mathrm{I}\}$, so that $(V, \underline{\underline{v}})$ is a relative compression body. Now $\pi_1(F_1)$

is a free group of rank 4 and we may choose generators a, b, c, and d so that the boundary circle of F_1 is $[a,b][c,d]$. For each n, there is an automorphism α_n which fixes b, c, and d, and sends a to ac^n. Since F_1 is aspherical, α_n can be realized by a homotopy equivalence of $F_1 \times \{0\} \cup \partial F_1 \times \mathrm{I}$. This can be extended to $F_1 \times \mathrm{I}$, preserving each attaching disk of a 1-handle of V, and then extended to the 1-handles. The result is an admissible homotopy equivalence of $(V, \underline{\underline{v}})$. If $n \neq 0$, the automorphism cannot be realized by a homeomorphism of $F_1 \times \{0\} \cup \partial F_1 \times \mathrm{I}$, since α_n does not preserve the boundary circle of F_1 up to conjugacy. Consequently, α_n cannot be realized by an admissible homeomorphism of $(V, \underline{\underline{v}})$. Thus, $\alpha_m^{-1}\alpha_n = \alpha_{n-m}$ can only be realized by an admissible homeomorphism when $m = n$. Therefore, the α_n represent infinitely many distinct cosets of $\mathcal{R}(V, \underline{\underline{v}})$ in $\mathrm{Out}(\pi_1(V), \pi_1(\underline{\underline{v}}))$.

The proof of Main Topological Theorem 1 will occupy the remainder of this section. In section 9.1, we collect some algebraic facts that will be needed in the later arguments. Section 9.2 contains the proof of Main Topological Theorem 1 in the cases when the index is finite, while in section 9.3, the infinite index cases are treated.

To prove that the index is infinite in the cases when $(M, \underline{\underline{a}})$ is not a relative compression body and not small, we need only a much weaker condition on the boundary pattern than the one given in our Main Topological Theorem 1. It is sufficient to assume that no boundary circle of F is contractible in M. In section 9.3, we use only this weaker assumption, emphasizing it by using $\underline{\underline{m}}$ to denote the boundary pattern. In fact, as we explain in that section, one can dispense even with this assumption, but this would require the definition of some additional types of small manifolds.

9.1. Algebraic lemmas

We collect here a few algebraic lemmas that will be used in the proof of proposition 9.2.2, which treats the case where $(M, \underline{\underline{a}})$ is a small 3-manifold. Some of these lemmas will also be useful in chapter 10 in the proofs of theorem 10.1.1 and lemma 10.3.6.

LEMMA 9.1.1. *Let $G = G_1 * G_2$ be a free product of groups. Let H be any subgroup of G that contains G_1. Then G_1 is a free factor of H.*

PROOF. Form a $K(G,1)$-complex K by taking the 1-point union of a $K(G_1,1)$-complex (K_1, k_1) and a $K(G_2,1)$-complex (K_2, k_2). Take the basepoint to be the join point $k_0 = k_1 = k_2$, and let $(\widetilde{K}, \widetilde{k}_0)$ be the covering space corresponding to H. Then the inclusion map from (K_1, k_1) to (K, k_0) lifts to an imbedding $\widetilde{i}\colon (K_1, k_1) \to (\widetilde{K}, \widetilde{k}_0)$, and $(\widetilde{K}, \widetilde{k}_0)$ is the 1-point union of $\widetilde{i}(K_1)$ and another complex. Therefore G_1 is a free factor of H. □

In the next lemma, we find particularly nice representative automorphisms for certain elements of the outer automorphism groups of a free product $G_1 * G_2$. Recall that two automorphisms ϕ_1 and ϕ_2 represent the same outer automorphism if there is an inner automorphism μ such that $\phi_1 = \mu\phi_2$. Note that precomposition and postcomposition by inner automorphisms have the same possible effects, since if $\mu(x)$ is conjugation by x, then $\phi\mu(x) = \mu(\phi(x))\phi$. Also, we recall the normal form for elements in the free product $G_1 * G_2$. It says that each nontrivial element can be written uniquely as a product $g_1 g_2 \cdots g_n$ where each g_i lies in one of the factors G_1 or G_2, but for no i do g_i and g_{i+1} lie in the same factor (in particular, no g_i is equal to 1).

LEMMA 9.1.2. *Let $G_1 * G_2$ be a free product of two groups, and let ϕ be an automorphism of $G_1 * G_2$.*

(i) *Suppose that each $\phi(G_i)$ is conjugate to G_i. Then one may alter ϕ, by composing by an inner automorphism, so that $\phi(G_i) = G_i$ for each i. Moreover, the restriction of ϕ to each G_i is uniquely determined by the element of* $\mathrm{Out}(G_1 * G_2)$ *represented by ϕ.*
(ii) *Suppose that $\phi(G_1)$ is conjugate to G_1, and G_2 is infinite cyclic. Then one may alter ϕ, by composing by an inner automorphism, so that $\phi(G_1)=G_1$ and, if ω generates G_2, $\phi(\omega) = \gamma\omega^{\pm 1}$ for some $\gamma \in G_1$. Moreover, the element γ, the exponent of ω, and the restriction of ϕ to G_1 are uniquely determined by the element of* $\mathrm{Out}(G_1 * G_2)$ *represented by ϕ.*

PROOF. In case (i), for $i = 1, 2$ there exist elements $\alpha_i \in G_1 * G_2$ so that $\phi(G_i) = \alpha_i\, G_i\, \alpha_i^{-1}$ for $i = 1, 2$. Changing ϕ by postcomposing by conjugation by α_1^{-1}, we have $\phi(G_1) = G_1$ and $\phi(G_2) = \alpha_1^{-1}\alpha_2\, G_2\, (\alpha_1^{-1}\alpha_2)^{-1}$. Since ϕ is surjective, elements of the forms g_1 and $\alpha_1^{-1}\alpha_2\, g_2\, (\alpha_1^{-1}\alpha_2)^{-1}$ must still generate $G_1 * G_2$. It follows, by normal form considerations, that $\alpha_1^{-1}\alpha_2 = h_1 h_2$ with $h_i \in G_i$ (where either of the h_i might equal 1). Changing ϕ by postcomposing by conjugation by h_1^{-1}, we have $\phi(G_i) = G_i$ for $i = 1, 2$. That is, there are automorphisms φ_i of G_i so that for all $g_i \in G_i$, $\phi(g_i) = \varphi_i(g_i)$. Further conjugating by any nontrivial element results in an automorphism that no longer preserves at least one of G_1 and G_2, so the φ_i are uniquely determined.

In case (ii), we have, after composing ϕ by an inner automorphism, that $\phi(g_1) = \varphi_1(g_1)$ for all $g_1 \in G_1$ and some $\varphi_1 \in \mathrm{Aut}(G_1)$. Since $\phi(G_1)$ and $\phi(\omega)$ generate $G_1 * G_2$, normal form considerations show that $\phi(\omega) = \gamma_1\omega^{\pm 1}\gamma_2$ for $\gamma_1, \gamma_2 \in G_1$. Postcomposing with the inner automorphism that conjugates by γ_2, we have $\phi(\omega) = (\gamma_2\gamma_1)\omega^{\pm 1}$. As in case (i), $\gamma = \gamma_2\gamma_1$, the exponent of ω, and φ_1 are uniquely determined. □

The next lemma will be used in the proof of proposition 9.2.2 to realize automorphisms of the fundamental groups of small manifolds of type II and III by homeomorphisms, and also at the very end of the proof of theorem 10.1.1. For a collection $\{H_1, \ldots H_k\}$ of subgroups of a group G, denote by $\mathrm{Aut}(G; H_1, \ldots, H_k)$ the subgroup of $\mathrm{Aut}(G)$ consisting of those automorphisms that preserve each H_j.

LEMMA 9.1.3. *Let T be a torus.*

(i) *Let C be an essential simple loop in T. Then* $\mathrm{Aut}(\pi_1(T \times I); \pi_1(C))$ *is a semidirect product $\mathbb{Z} \circ (\mathbb{Z}/2 \times \mathbb{Z}/2)$, where the infinite cyclic subgroup is generated by the induced automorphism of a Dehn twist in the annulus $C \times I$ in $T \times I$.*
(ii) *Let C_1 and C_2 be essential simple loops in T with $C_1 \neq \pm C_2$ in $\pi_1(T)$. Then* $\mathrm{Aut}(\pi_1(T \times I); \pi_1(C_1), \pi_1(C_2))$ *is finite.*

PROOF. In case (i), choose a basis of $\pi_1(T) \cong \mathbb{Z} \times \mathbb{Z}$ so that $\pi_1(C)$ equals $\mathbb{Z} \times \{0\}$. Now $\mathrm{Aut}(\mathbb{Z} \times \mathbb{Z}) \cong \mathrm{GL}(2, \mathbb{Z})$, and the automorphisms that preserve $\pi_1(C)$ are of the form $\begin{pmatrix} a & b \\ 0 & d \end{pmatrix}$. Sending this matrix to (a, d) defines a homomorphism to

$\mathbb{Z}/2 \times \mathbb{Z}/2$ which splits and has kernel generated by $\begin{pmatrix} 1 & 1 \\ 0 & 1 \end{pmatrix}$, which is induced by the Dehn twist about C.

Part (ii) follows since C_1 and C_2 must represent linearly independent primitive elements in $\pi_1(T)$. □

9.2. The finite-index cases

In this section we prove that the index of $\mathcal{R}(M, \underline{\underline{m}})$ in $\mathrm{Out}(\pi_1(M), \pi_1(\underline{\underline{m}}))$ is finite when $(M, \underline{\underline{m}})$ is a relative compression body (proposition 9.2.1) or when $(M, \underline{\underline{m}})$ is small (proposition 9.2.2).

Homeomorphisms of compression bodies (i. e. the cases when $\underline{\underline{m}}$ is empty, and consequently all constituents are closed) were extensively studied in [**89**]. It seems likely that the entire theory there could be adapted to the relative case, but for brevity we will treat here only the portion needed for our present purposes.

Recall that $\pi_1(V)$ is a free product of the form

$$G_1 * \cdots * G_m * G_{m+1} * \cdots * G_{m+\ell},$$

where $G_i = \pi_1(F_i)$ for $1 \leq i \leq m$ and G_{m+j} is infinite cyclic for $1 \leq j \leq \ell$. The latter correspond to 1-handles of V. We first recall some algebraic work of Fouxe-Rabinovitch [**40, 41**] which gives generators for the subgroup $\mathrm{Aut}_p(\pi_1(V))$ of $\mathrm{Aut}(\pi_1(V))$ which consists of the automorphisms which take each G_i to a conjugate of itself, $1 \leq i \leq m$. We then show that most of these generators can be realized by "slide homeomorphisms" of V which are the identity on $\partial V - F$. These "slide homeomorphisms" will be used in the proofs of both propositions 9.2.1 and 9.2.2.

For $1 \leq j \leq \ell$, fix a generator a_j of G_{m+j}. Extend this collection to a set of generators for $\pi_1(V)$ by adding generators for each G_i with $i \leq m$. Fouxe-Rabinovitch [**40, 41**], showed that the collection of all automorphisms of the following type generate $\mathrm{Aut}_p(\pi_1(V))$:

(1) right slide automorphisms which slide 1-handle factors: If $1 \leq j \leq \ell$, $1 \leq i \leq m+\ell$, $i \neq m+j$ and $x \in G_i$, then we can define $\rho_{i,m+j}(x)\colon \pi_1(V) \to \pi_1(V)$ by setting $\rho_{i,m+j}(x)(a_j) = a_j x$ and letting $\rho_{i,m+j}(x)$ fix all other generators of $\pi_1(V)$.
(2) left slide automorphisms which slide 1-handle factors: If $1 \leq j \leq \ell$, $1 \leq i \leq m+\ell$, $i \neq m+j$ and $x \in G_i$, then we can define $\lambda_{i,m+j}(x)\colon \pi_1(V) \to \pi_1(V)$ by letting $\lambda_{i,m+j}(x)(a_j) = x^{-1} a_j$ and letting $\lambda_{i,m+j}(x)$ fix all other generators of $\pi_1(V)$.
(3) slide automorphisms which slide constituent factors: If $1 \leq j \leq m$, $1 \leq i \leq m+\ell$, $i \neq j$ and $x \in G_i$, we define $\mu_{i,j}(x)\colon \pi_1(V) \to \pi_1(V)$ by letting $\mu_{i,j}(x)(g) = x^{-1} g x$ for all $g \in G_j$ and letting $\mu_{i,j}(x)$ fix all other generators of $\pi_1(V)$.
(4) interchange automorphisms which interchange infinite cyclic factors corresponding to 1-handles: If $1 \leq i, j \leq \ell$ and $i \neq j$, then we define $\omega_{m+i,m+j}\colon \pi_1(V) \to \pi_1(V)$ by letting $\omega_{m+i,m+j}(a_i) = a_j$, $\omega_{m+i,m+j}(a_j) = a_i$, and letting $\omega_{m+i,m+j}$ fix all other generators of $\pi_1(V)$.
(5) automorphisms which flip an infinite cyclic factor corresponding to a 1-handle: If $1 \leq j \leq \ell$, we define $\sigma_{m+j}\colon \pi_1(V) \to \pi_1(V)$ by setting $\sigma_{m+j}(a_j) = a_j^{-1}$ and letting σ_{m+j} fix all other generators of $\pi_1(V)$.

(6) factor automorphisms: If $1 \leq i \leq m$ and $\phi \in \mathrm{Aut}(G_i)$, then we can define $\phi_i\colon \pi_1(V) \to \pi_1(V)$ by setting $\phi_i(g) = \phi(g)$ if $g \in G_i$ and letting ϕ_i fix all other generators of $\pi_1(V)$.

Remark: The papers of Fouxe-Rabinovitch actually concern free products $G_1 * \cdots * G_m * G_{m+1} * \cdots * G_{m+\ell}$ in which each G_i for $i > m$ is infinite cyclic and each G_i for $i \leq m$ is indecomposable and not infinite cyclic, but the latter hypothesis is used only to know that every automorphism preserves the conjugacy classes of these first m factors, possibly permuting them. In that case, the subgroup $\mathrm{Aut}_p(G_1 * \cdots * G_{m+\ell})$ has finite index in $\mathrm{Aut}(G_1 * \cdots * G_{m+\ell})$. The Fouxe-Rabinovitch generators used in the original papers include interchanges of any two isomorphic free factors, and they generate all of $\mathrm{Aut}(G_1 * \cdots * G_{m+\ell})$. We will define and use these additional interchange automorphisms in chapter 12.

Every automorphism φ in $\mathrm{Aut}_p(G_1 * \cdots * G_{m+\ell})$ can be written as $\varphi_1\varphi_2$, where φ_1 is a product of slide automorphisms, flip automorphisms and interchange automorphisms, and $\varphi_2 = \prod_{i=1}^{m} \phi_i$ where each ϕ_i is a factor automorphism associated to G_i. This is an immediate consequence of the following easily checked relations among our generators for $\mathrm{Aut}_p(\pi_1(V))$. In this list, ϕ stands for an element of $\mathrm{Aut}(G_i)$, and i, j, and k represent indices with $i \neq j$. In this list and throughout the rest of this chapter, we assume that the subscripts of these generating automorphisms lie in the appropriate ranges for which the generating automorphisms are defined. That is, for $\rho_{j,k}(x)$ and $\lambda_{j,k}(x)$, we have $1 \leq j \leq m + \ell$ and $m+1 \leq k \leq m+\ell$, while for $\mu_{j,k}(x)$, we have $1 \leq j \leq m+\ell$ and $1 \leq k \leq m$. Also, for ϕ_i, $1 \leq i \leq m$, while for $\omega_{i,j}$ and σ_j, $m+1 \leq i, j \leq m+\ell$.

(1) If $\rho_{j,k}(x)$ is a right slide automorphism, then $\phi_i\rho_{j,k}(x) = \rho_{j,k}(x)\phi_i$.
(2) If $\rho_{i,k}(x)$ is a right slide automorphism, then $\phi_i\rho_{i,k}(x) = \rho_{i,k}(\phi(x))\phi_i$.
(3) If $\lambda_{j,k}(x)$ is a left slide automorphism, then $\phi_i\lambda_{j,k}(x) = \lambda_{j,k}(x)\phi_i$.
(4) If $\lambda_{i,k}(x)$ is a left slide automorphism, then $\phi_i\lambda_{i,k}(x) = \lambda_{i,k}(\phi(x))\phi_i$.
(5) If $\mu_{j,k}(x)$ is a slide automorphism, then $\phi_i\mu_{j,k}(x) = \mu_{j,k}(x)\phi_i$.
(6) If $\mu_{i,k}(x)$ is a slide automorphism, then $\phi_i\mu_{i,k}(x) = \mu_{i,k}(\phi(x))\phi_i$.
(7) If $\omega_{j,k}$ is an interchange automorphism, then $\phi_i\omega_{j,k} = \omega_{j,k}\phi_i$.
(8) If σ_j is a flip automorphism, then $\phi_i\sigma_j = \sigma_j\phi_i$.
(9) If $\psi \in \mathrm{Aut}(G_j)$, then $\phi_i\psi_j = \psi_j\phi_i$.
(10) If $\psi \in \mathrm{Aut}(G_i)$, then $\phi_i\psi_i = (\phi\psi)_i$.

Remark: Fouxe-Rabinovitch gave a complete list of defining relations among his generators for $\mathrm{Aut}(G_1 * \cdots * G_{m+\ell})$, for the case of no infinite cyclic free factors in [**40**] and for the general case in [**41**]. These relations are listed in a convenient form in [**89**]. The proofs of the completeness of the sets of relations in [**40**] and [**41**] are quite complicated, but they were confirmed by Gilbert [**45**] using methods developed by McCool [**83, 84**]. We will not rely on any of these results, since we will use only certain easily checked relations (those given above and some additional ones used in chapter 12), and will never need the fact that the Fouxe-Rabinovitch relations are a complete set of defining relations.

We will now explain how to realize each slide automorphism, interchange automorphism and flip automorphism of the type described above, by a homeomorphism of V that fixes $\partial V - F$, and hence is admissible.

First we give a general description of a slide homeomorphism of V. Let $D \times \mathrm{I}$ be a 1-handle of V. Let $\{J_t\}$ be an isotopy of $\overline{V - (D \times \mathrm{I})}$ such that J_0 is the identity, each J_t fixes a basepoint v and is the identity on $\partial V - F$, and J_1 is the identity on $D \times \partial I$. Moreover, each J_t fixes one of the attaching disks for $D \times \mathrm{I}$. If $D \times \{k\}$ ($k \in \{0,1\}$) is the other attaching disk, and α is the loop in ∂V traced out by the image under the homotopy of the basepoint of $D \times \{k\}$, (where $k = 0$ or $k = 1$) then we say that J_t "slides" the attaching disk $D \times \{k\}$ around the loop α. A slide homeomorphism of V which slides $D \times \{k\}$ around α is obtained by extending J_1 over $D \times \mathrm{I}$ by using the identity map.

Now we describe how most of the specific generating automorphisms for $\mathrm{Aut}(\pi_1(V))$ can be realized. When $m \geq 1$, it will be convenient to regard $m-1$ of the 1-handles $h'_2, \ldots, h'_m$ as connecting $F_1 \times \{1\}$ to $F_i \times \{1\}$, and the remaining ℓ of the 1-handles $h_1, \ldots, h_\ell$ as having both ends attached in $F_1 \times \{1\}$.

Choose the basepoint v of V to lie in the interior of V, and in $F_1 \times \{1/2\}$ if $m \geq 1$.

Each generator a_j is represented by a loop which is the path product of an arc α_j^0 that runs from v to the "left" attaching disk for h_j, the core arc α_j^c of h_j, and an arc α_j^1 from the "right" attaching disk back to v. For each j with $2 \leq j \leq m$, let β_j be obtained as the product of an arc β_j^0 in $F_1 \times I$ joining the v to the left attaching disk of h'_j, the core arc β_j^c of h'_j, and an arc β_j^1 in $F_j \times I$ from the endpoint of β_j^c to a basepoint $v_j \in F_j \times \{1/2\}$. We regard $\pi_1(F_j, v_j)$ as a subgroup of $\pi_1(V, v)$ by identifying a loop γ based at v_j with the loop $\beta_j * \gamma * \overline{\beta_j}$ based at v.

If $\rho_{i,m+j}(x)$ is a right slide automorphism, we choose the sliding isotopy J_t on $\overline{V - h_j}$ so that $\overline{\alpha_j^1} * J_1(\alpha_j^1)$ represents x and each J_t fixes α_j^0. The right attaching disk of h_j slides around x^{-1}.

If $\lambda_{i,m+j}(x)$ is a left slide automorphism, we choose J_t so that $J_1(\alpha_j^0) * \overline{\alpha_j^0}$ represents x^{-1} and each J_t fixes α_j^1. The left attaching disk for h_j slides around x^{-1}.

We next consider a slide automorphism $\mu_{i,j}(x)$. If $2 \leq j \leq m$, we use an isotopy J_t so that $J_1(\beta_j^0) * \overline{\beta_j^0}$ represents x^{-1}. Suppose now that $j = 1$. Let f be a homeomorphism of V which fixes ∂V and realizes the inner automorphism of $\pi_1(V, v)$ given by conjugation by the element x^{-1}, i. e. $f_\#(g) = x^{-1} g x$ for all $g \in \pi_1(V, v)$. Compose f with the slide homeomorphisms inducing $\mu_{i,k}(x^{-1})$ for each $2 \leq k \leq m$ such that $k \neq i$ and $\rho_{i,k}(x^{-1})$ and $\lambda_{i,k}(x^{-1})$ for each $m+1 \leq k \leq m+\ell$ such that $k \neq i$. (Since the automorphisms induced by these homeomorphisms commute, the order of composition is irrelevant.) If $i > m$, the resulting composition conjugates the elements of G_1 by x^{-1} and fixes all other generators, hence induces $\mu_{i,1}(x)$. Moreover, the composition is a homeomorphism which is the identity on $\partial V - F$. If $2 \leq i \leq m$, however, it still conjugates each element of G_i by x^{-1}. In this case, identify x with $x' \in \pi_1(F_i, v_i)$ and let r_0 be a homeomorphism of $F_i \times \mathrm{I}$ which is the identity on $\partial(F_i \times \mathrm{I})$, induces the identity automorphism on $\pi_1(F_i \times \mathrm{I}, v_i)$, and such that $\overline{\beta_i^1} * r_0(\beta_i^1)$ represents x'. Extend r_0 to a homeomorphism r of V by letting it be the identity off of $F_i \times \mathrm{I}$. The composition of r with the homeomorphism already constructed is the identity on $\partial V - F$ and induces $\mu_{i,1}(x)$.

To realize an interchange automorphism $\omega_{m+i,m+j}$ one uses an isotopy J_t on $\overline{V - (h_i \cup h_j)}$ such that J_1 fixes $\partial V - F$, interchanges the left attaching disk of h_i and the left attaching disk of h_j, interchanges the right attaching disk of h_i and the right attaching disk of h_j, interchanges α_i^0 and α_j^0, and interchanges α_i^1 and α_j^1 (we must assume here that the α_k^ℓ have been chosen "unknotted" and "unlinked" to ensure that isotopies interchanging them exist). One extends J_1 to a homeomorphism f of V which fixes $\partial V - F$ and realizes $\omega_{m+i,m+j}$ by extending J_1 so that it interchanges h_i and h_j.

To realize a flip automorphism σ_{m+j}, one uses an isotopy J_t on $\overline{V - h_j}$ such that J_1 fixes $\partial V - F$, interchanges the left attaching disk of h_j and the right attaching disk of h_j, and interchanges α_j^0 and α_j^1. One extends J_1 to a homeomorphism of V realizing σ_j by extending it to be a homeomorphism of h_j which interchanges its attaching disks.

We remark that these descriptions do not specify the slide homeomorphisms up to isotopy, since there are generally many choices of the isotopy J_t that achieve the specified conditions. The arcs around which the attaching disks slide in realizing the $\rho_{i,j}(x)$, $\lambda_{i,j}(x)$, and $\mu_{i,j}(x)$ need not be isotopic in ∂V, although they will be homotopic in V. Different choices will change the resulting homeomorphisms by Dehn twists about properly imbedded 2-disks. We refer the reader to section 2 of McCullough-Miller [**89**] for a more detailed discussion.

We are now ready to establish our main result for relative compression bodies. As mentioned in the introduction, Maskit [**77**] proved proposition 9.2.1 for relative compression bodies which are pared manifolds, and McCullough and Miller [**89**] proved it for compression bodies.

PROPOSITION 9.2.1. *Let $(V, \underline{\underline{v}})$ be an orientable relative compression body, such that the elements of $\underline{\underline{v}}$ are disjoint and incompressible. Let F be a compressible free side, such that each element of $\underline{\underline{v}}$ that meets F is an annulus. Then $\mathcal{R}(V, \underline{\underline{v}})$ has finite index in* $\mathrm{Out}(\pi_1(V), \pi_1(\underline{\underline{v}}))$.

We reiterate that example 9.0.2 shows that proposition 9.2.1 can fail without the assumption that every element of $\underline{\underline{v}}$ that meets F is an annulus.

PROOF OF PROPOSITION 9.2.1. Let $\{F_i\}$ be the constituents of V, if any. No F_i is a disk, since the elements of $\underline{\underline{v}}$ that meet F are incompressible annuli, and by the definition of relative compression body no F_i is a 2-sphere. Denote by A the union of the annuli in $|\underline{\underline{v}}|$ that meet F.

Select coordinates on V as a relative compression body so that each component $C \times \mathrm{I}$ of $\partial F_i \times \mathrm{I}$ is a collar neighborhood of a boundary component of an annulus A_C of A. If F_i is not an annulus, then $A_C \cap F_i$ is a collar neighborhood of C in F_i. If F_i is an annulus, then $A_C \cap F_i$ may be such a collar neighborhood for C, or it may happen that $\partial F_i \times \mathrm{I} \cup F_i$ is a single annulus component of $|\underline{\underline{v}}|$.

Let $\mathrm{Out}_0(\pi_1(V), \pi_1(\underline{\underline{v}}))$ be the subgroup of finite index in $\mathrm{Out}(\pi_1(V), \pi_1(\underline{\underline{v}}))$ consisting of the automorphisms which induce the trivial permutation on the (conjugacy classes of the) subgroups of $\pi_1(V)$ corresponding to the elements of $\underline{\underline{v}}$. Since V is aspherical, and the elements of $\underline{\underline{v}}$ are disjoint, any ϕ in $\mathrm{Out}_0(\pi_1(V), \pi_1(\underline{\underline{v}}))$ may be realized by a homotopy equivalence f that maps each element of $\underline{\underline{v}}$ to itself by a homotopy equivalence. By passing to a further finite index subgroup, still called $\mathrm{Out}_0(\pi_1(V), \pi_1(\underline{\underline{v}}))$, we may assume that f is homotopic to the identity map on each annular element of $\underline{\underline{v}}$. So, we may alter f so that it restricts to the identity map on each annulus in $|\underline{\underline{v}}|$. In particular, f preserves each component of ∂F_i.

We will first show that there is a subgroup $\mathrm{Out}_1(\pi_1(V), \pi_1(\underline{\underline{v}}))$ of finite index in $\mathrm{Out}_0(\pi_1(V), \pi_1(\underline{\underline{v}}))$ such that if ϕ lies in $\mathrm{Out}_1(\pi_1(V), \pi_1(\underline{\underline{v}}))$, then f is homotopic, relative to A, to an admissible homotopy equivalence that preserves each F_i, and restricts to a homeomorphism on F_i. Adjust f, relative to A, so that its restriction to F_i is transverse to the cocores of the 1-handles of V. The boundary (if any) of each F_i is preserved by f, so the preimage in F_i of the cocores consists of disjoint closed curves, but no arcs. Since F_i is incompressible, these closed curves are contractible in F_i. Using asphericity of V, f can be changed by homotopy relative to A so that $f(F_i)$ is disjoint from the cocores. Then, $f(F_i)$ lies in a product with one end equal to some F_j, so f may be changed further so that $f(F_i) \subseteq F_j$ for some j. If F_i has nonempty boundary, then $j = i$ since f preserves the boundary of F_i.

Let $f_i \colon F_i \to F_j$ be the restriction of f to F_i. Since $f_\#$ (and hence $(f_i)_\#$) is injective, the Baer-Nielsen Theorem 2.5.5 implies that f_i is properly homotopic to a finite covering map, so $f_\#(\pi_1(F_i))$ has finite index in $\pi_1(F_j)$. Since $f_\#(\pi_1(F_i))$ is a free factor of $\pi_1(V)$, lemma 9.1.1 implies that it is a free factor of $\pi_1(F_j)$. Since it also has finite index, $f_\#(\pi_1(F_i)) = \pi_1(F_j)$, so f_i is properly homotopic to homeomorphism. From now on, we assume that each f_i is a homeomorphism. Since the subgroups corresponding to different F_j are not conjugate in $\pi_1(V)$, the permutation of the F_i is uniquely determined by the homotopy class of f. So by passing to a subgroup $\mathrm{Out}_1(\pi_1(V), \pi_1(\underline{\underline{v}}))$ of finite index in $\mathrm{Out}_0(\pi_1(V), \pi_1(\underline{\underline{v}}))$, we may assume that f preserves each F_i.

Suppose B is an element of $\underline{\underline{v}}$ contained in the interior of some F_i (hence not one of the annuli of A). By asphericity and the definition of $\mathrm{Out}_0(\pi_1(V), \pi_1(\underline{\underline{v}}))$, $f_i(B)$ is homotopic in V into B. By theorem 3.1.2, the homotopy may be assumed to take place in $F_i \times \mathrm{I}$, and then by a further deformation to a homotopy into F_i, showing that $f_i(B)$ is homotopic in F_i into B. Using lemma 2.12.3, we may change f by homotopy to be admissible. Thus, if $\phi \in \mathrm{Out}_1(\pi_1(V), \pi_1(\underline{\underline{v}}))$, then f is homotopic, relative to A, to an admissible homotopy equivalence that preserves each F_i and restricts to a homeomorphism on each F_i.

Pass to a finite index subgroup $\mathrm{Out}_2(\pi_1(V), \pi_1(\underline{\underline{v}}))$ of $\mathrm{Out}_1(\pi_1(V), \pi_1(\underline{\underline{v}}))$ so that if $\phi \in \mathrm{Out}_2(\pi_1(V), \pi_1(\underline{\underline{v}}))$, then each f_i is orientation-preserving. We may also assume that $f(v) = v$. Since ϕ is induced by a homotopy equivalence f which preserves each F_i, ϕ takes G_i to a conjugate of itself for each $1 \leq i \leq m$. As explained at the start of this section, this implies that $f_\# \colon \pi_1(V, v) \to \pi_1(V, v)$ can be written as $\varphi_1\varphi_2$, where φ_1 is a product of generators of the form $\rho_{j,k}(x)$, $\lambda_{j,k}(x)$, $\mu_{j,k}(x)$, σ_j, and $\omega_{j,k}$, and $\varphi_2 = \prod_{i=1}^m \phi_i$ where each ϕ_i is a factor automorphism induced by an automorphism of G_i. In the case when $m = 0$, we take $\mathrm{Out}_2(\pi_1(V)) = \mathrm{Out}(\pi_1(V))$ and φ_2 to be the identity.

To finish the proof, we will construct an admissible homeomorphism gs that realizes ϕ. Recall that we showed that each slide automorphism, flip automorphism or interchange automorphism of the type used above can be realized by a homeomorphism which is the identity on $\partial V - F$. So there exists a homeomorphism g of V which is the identity on $\partial V - F$ and induces φ_1. Then $g^{-1}f$ induces φ_2. For each i, let ψ_i be the automorphism of G_i which induces ϕ_i, i.e. $\phi_i = (\psi_i)_i$. Since g is the identity on $\partial V - F$, each ψ_i must be induced by f_i, up to inner automorphism. Since f_i is orientation-preserving, it can be extended to a homeomorphism s_i of $\pi_1(V)$ which is the identity on $\partial V - (F \cup F_i)$ and induces ϕ_i on $\pi_1(V)$. (To construct s_i, extend f_i first to $F_i \times [0, 1/2]$ using an isotopy from f_i to a homeomorphism f_i' which fixes v_i and induces ψ_i on $\pi_1(F_i \times \{1/2\}, v_i)$, then extend to a homeomorphism on the rest of $F_i \times \mathrm{I}$ which fixes the attaching disks of the 1-handles of V, fixes β_i^1 if $i \geq 2$ and fixes each α_k^0, α_k^1 and β_j^0 if $i = 1$, then extend using the identity on the rest of V.) Let s be the composition $\prod_{i=1}^m s_i$; then

$s^{-1}g^{-1}f$ is the identity on $\partial V - F$ (so gs is admissible) and induces the identity automorphism on $\pi_1(V)$ (so gs realizes ϕ). □

The next proposition establishes our main result for small manifolds.

PROPOSITION 9.2.2. *Let* $(M, \underline{\underline{a}})$ *be a compact orientable irreducible 3-manifold with boundary pattern consisting of a (possibly empty) collection of disjoint incompressible submanifolds. If* $(M, \underline{\underline{a}})$ *is small, then* $\mathcal{R}(M, \underline{\underline{a}})$ *has finite index in* $\mathrm{Out}(\pi_1(M), \pi_1(\underline{\underline{a}}))$.

PROOF. When $(M, \underline{\underline{a}})$ is small, its fundamental group is a free product of the form $G_1 * G_2$, where corresponding to the types of small manifold defined in section 6.1 we have one of the following:

I. G_1 is either
   (a) the fundamental group of a nonorientable surface and contains $\pi_1(F_1)$ as a subgroup of index 2, or
   (b) infinite cyclic and contains $\pi_1(F_1)$ as a proper subgroup,

   and G_2 is infinite cyclic.

II. G_1 is either
   (a) the fundamental group of a nonorientable surface, and contains $\pi_1(F_1)$ as a subgroup of index at most 2, or
   (b) infinite cyclic and contains $\pi_1(F_1)$ as a proper subgroup, or
   (c) the fundamental group of the torus or Klein bottle, and $\pi_1(F_1)$ is infinite cyclic,

   and G_2 either equals $\pi_1(F_2)$, or satisfies (a), (b), or (c) with $\pi_1(F_2)$ in place of $\pi_1(F_1)$.

III. Either
   (a) $G_1 = \pi_1(F_1)$, and G_2 is infinite cyclic, or
   (b) G_1 and G_2 are both infinite cyclic and $\pi_1(F_1)$ and $\pi_1(F_2)$ are subgroups of (the same) finite index in G_1, or
   (c) G_1 is $\mathbb{Z} \times \mathbb{Z}$, G_2 is infinite cyclic, $\pi_1(F_1)$ is either G_1 or is a primitive infinite cyclic subgroup of G_1, and $\pi_1(F_2)$ is a primitive infinite cyclic subgroup of G_1.

As in the proof of proposition 9.2.1, $\mathrm{Out}(\pi_1(M), \pi_1(\underline{\underline{a}}))$ contains a subgroup $\mathrm{Out}_0(\pi_1(M), \pi_1(\underline{\underline{a}}))$ of finite index so that each ϕ in this subgroup is induced by a homotopy equivalence f which preserves each component of $|\underline{\underline{a}}|$ and restricts to the identity homeomorphism on the union A of the annuli in $|\underline{\underline{a}}|$ that meet F. From now on, we will consider only automorphisms that lie in this subgroup. Let V be a normally imbedded relative compression body neighborhood of F. We will examine each type of small manifold in turn.

Suppose M is of type I. Let W denote $\overline{M - V}$. Suppose first that M is of type Ia, so that W is a twisted I-bundle. Let F_0 be the zero section of W, and choose the basepoint of M to lie in F_0. Then $G_1 = \pi_1(F_0)$, and G_2 is infinite cyclic and is generated by an element ω represented by a loop which travels once over the 1-handle of V. Since the boundary (if any) of F_0 is homotopic into A, and f is the identity on A, we may assume that f fixes ∂F_0. In particular, the image under f of ∂F_0 is disjoint from the 1-handle of M. Using asphericity, f is homotopic, relative to A, so that $f(F_0) \subseteq W$ and hence so that $f(F_0) \subset F_0$ (since f is the identity on ∂F_0, this actually forces $f(F_0) = F_0$). By the Baer-Nielsen Theorem 2.5.5, $f|_{F_0} : F_0 \to F_0$ is homotopic to a covering map from F_0 to F_0. If $\partial F_0 \neq \emptyset$, then, since f fixes ∂F_0, this map is a homeomorphism. If $\partial F_0 = \emptyset$, then since $\pi_1(F_0)$ is indecomposable, lemma 9.1.1 shows that $f_{\#}(\pi_1(F_0)) = \pi_1(F_0)$, so again the covering map is a homeomorphism. Therefore $\phi(\pi_1(F_0))$ is conjugate to $\pi_1(F_0)$, and after conjugation ϕ preserves the index 2 subgroup $\pi_1(F_1) \subset \pi_1(F_0)$ which consists of all orientation-preserving loops.

Lemma 9.1.2 applies to show that there exist unique elements $\varphi_1 \in \mathrm{Aut}(G_1)$ and $\gamma \in \pi_1(F_0)$, so that, perhaps after composing by an inner automorphism, $\phi(g_1) = \varphi_1(g_1)$ for all $g_1 \in G_1 = \pi_1(F_0)$ and $\phi(\omega) = \gamma\omega^\epsilon$ where $\epsilon = \pm 1$. We define a homomorphism from $\mathrm{Out}_0(\pi_1(M), \pi_1(\underline{\underline{a}}))$ to $\mathbb{Z}/2$ by sending ϕ to ϵ and a second homomorphism from the kernel of the first to $\pi_1(F_0)/\pi_1(F_1) \cong \mathbb{Z}/2$ by sending ϕ to the coset of γ. The latter is a homomorphism since each ϕ preserves the subgroup $\pi_1(F_1)$ of $\pi_1(F_0)$. Let $\mathrm{Out}_1(\pi_1(M), \pi_1(\underline{\underline{a}}))$ be the kernel of this second homomorphism.

If $\phi \in \mathrm{Out}_1(\pi_1(M), \pi_1(\underline{\underline{a}}))$, then we may assume that there exists $\varphi_1 \in \mathrm{Aut}(G_1)$ and $\gamma \in \pi_1(F_1)$ such that $\phi(g_1) = \varphi_1(g_1)$ for all $g_1 \in \pi_1(F_0)$ and $\phi(\omega) = \gamma\omega$. Since φ_1 preserves the peripheral structure of F_0, it is realizable by a homeomorphism h_1 of F_0. Passing to another finite-index subgroup $\mathrm{Out}_2(\pi_1(M), \pi_1(\underline{\underline{a}}))$, we may assume that the basepoint-preserving lift of h_1 to F_1 is orientation-preserving. This ensures that there is an extension of h_1 to an orientation-preserving admissible homeomorphism h of M whose induced automorphism sends g_1 to $\varphi_1(g_1)$ and sends ω to ω. Let h' be an admissible homeomorphism which is the identity on W and slides one end of the handle around a loop in $F_1 \times \{1\}$ which represents γ. (The details in the construction of h' resemble those in the construction of the realization of a left slide automorphism.) Then $h'h$ is an admissible homeomorphism inducing ϕ. Therefore, $\mathrm{Out}_2(\pi_1(M), \pi_1(\underline{\underline{a}})) \subset \mathcal{R}(M, \underline{\underline{a}})$, so $\mathcal{R}(M, \underline{\underline{a}})$ has finite index in $\mathrm{Out}(\pi_1(M), \pi_1(\underline{\underline{a}}))$.

Now suppose that M is of type Ib, so that W is a solid torus. In this case $\pi_1(M)$ is free on two generators: ω_1, represented by a core curve of W, and ω_2, represented by a loop which travels once over the 1-handle attached to $F_1 \times \{1\}$, entering on an end we will call the left-hand attaching disk and leaving by the right-hand attaching disk. Now F is a torus with two holes whose boundary components are fixed by f (since they lie in A), hence are fixed up to conjugacy by ϕ. Each boundary component of F represents the same nonzero power ω_1^ℓ, so $\phi(\omega_1^\ell) = g\omega_1^\ell g^{-1}$ for some $g \in \pi_1(M)$. Since roots are unique in free groups, $\phi(\omega_1) = g\omega_1 g^{-1}$. By lemma 9.1.2 we may assume that ϕ fixes ω_1 and takes ω_2 to $\omega_1^n \omega_2^\epsilon$ where n is a well-defined integer and $\epsilon = \pm 1$. Define a homomorphism from $\mathrm{Out}_0(\pi_1(M), \pi_1(\underline{\underline{a}}))$ to $\mathbb{Z}/2$ by sending ϕ to ϵ. Define a second homomorphism from the kernel of the first to $\mathbb{Z}/\ell$ by sending ϕ to n. Suppose that ϕ is in the kernel $\mathrm{Out}_1(\pi_1(M), \pi_1(\underline{\underline{a}}))$ of the second homomorphism. We may assume that $\phi(\omega_1) = \omega_1$ and $\phi(\omega_2) = \omega_1^{r\ell}\omega_2$ for some r. In the portion of $F_1 \times \{1\}$ that lies in the boundary of W, we can choose a loop based at the left-hand attaching disk of the 1-handle that represents ℓ times the generator of $\pi_1(W)$. Sliding the left-hand attaching disk of the 1-handle r times around the reverse of this loop, we obtain a slide homeomorphism that induces ϕ. We may choose the slide homeomorphism to fix $\overline{\partial M - F}$. In particular, it will preserve $|\underline{\underline{a}}|$, so will be admissible. Therefore, $\mathrm{Out}_2(\pi_1(M), \pi_1(\underline{\underline{a}})) \subset \mathcal{R}(M, \underline{\underline{a}})$, so $\mathcal{R}(M, \underline{\underline{a}})$ has finite index in $\mathrm{Out}(\pi_1(M), \pi_1(\underline{\underline{a}}))$.

Assume that M is of type II, and write W_i for the component of $\overline{M-V}$ that contains F_i. When W_i is a solid torus or an I-bundle with one of its lids as a constituent, arguments similar to the type I case show that if $\phi \in \mathrm{Out}_0(\pi_1(M), \pi_1(\underline{\underline{a}}))$ then $\phi(G_i)$ is conjugate to either G_1 or G_2. When F_i is an annulus and W_i is an I-bundle over the torus or Klein bottle, G_i is a noncyclic indecomposable free factor of $G_1 * G_2$, so is conjugate to a unique indecomposable free factor of $G_1 * G_2$. In particular, if $\phi \in \mathrm{Out}_0(\pi_1(M), \pi_1(\underline{\underline{a}}))$, then $\phi(G_i)$ is again conjugate to either G_1 or G_2. Therefore, in all cases there exists a subgroup $\mathrm{Out}_1(\pi_1(M), \pi_1(\underline{\underline{a}}))$ of $\mathrm{Out}_0(\pi_1(M), \pi_1(\underline{\underline{a}}))$ of index at most 2 such that $\phi(G_i)$ is conjugate to G_i for each i.

We may apply lemma 9.1.2(i) to show each ϕ in $\mathrm{Out}_1(\pi_1(M), \pi_1(\underline{\underline{a}}))$ has a unique representative that preserves both G_i; denote its restriction to G_i by φ_i. If W_i is a solid torus or a twisted I-bundle with one of its lids as a constituent, one may use the methods of the type I case to show that there exists a finite index subgroup of $\mathrm{Out}_1(\pi_1(M), \pi_1(\underline{\underline{a}}))$, such that if ϕ lies in the subgroup, then there exists a homeomorphism h_i of W_i that preserves F_i, induces φ_i relative to a basepoint of W_i that lies in F_i, and extends to an admissible homeomorphism of $(M, \underline{\underline{a}})$ inducing the factor automorphism $(\varphi_i)_i$. If W_i is an I-bundle with F_i as lid and $\underline{\underline{w_i}}$ is its submanifold boundary pattern, then we may combine methods of type I case and the proof of proposition 9.2.1 to show that there exists a finite index subgroup of $\mathrm{Out}_1(\pi_1(M), \pi_1(\underline{\underline{a}}))$, such that if ϕ lies in the subgroup, then there exists an admissible homeomorphism h_i of $(W_i, \underline{\underline{w_i}})$ that preserves F_i, induces φ_i relative to a basepoint of W_i that lies in F_i, and extends to an admissible homeomorphism of $(M, \underline{\underline{a}})$ inducing the factor automorphism $(\varphi_i)_i$.

For the remaining possibilities, assume that F_i is an annulus and W_i is the I-bundle over the Klein bottle or torus. Suppose first that W_i is the I-bundle over the Klein bottle. Since $\mathrm{Out}(\pi_1(W_i))$ is finite, we may pass to another subgroup of finite index to assume that φ_i is inner. Since $\pi_1(\partial W_i)$ has index 2 in $\pi_1(W_i)$, we may further assume that φ_i is conjugation by an element in $\pi_1(\partial W_i)$. Then, there is a Dehn twist h_i about ∂W_i that induces φ_i, relative to a basepoint in F_i. Since h_i is the identity on ∂W_i, it extends to M using the identity on $M - W_1$. Suppose now that W_i is the I-bundle over the torus. If the component of ∂W_i that lies in ∂M does not contain any annuli of $|\underline{\underline{a}}|$ that are not homotopic to F_i, then by lemma 9.1.3(i) there is a finite index subgroup of $\mathrm{Aut}(\pi_1(W_i); \pi_1(F_i))$ generated by a Dehn twist in an annulus that meets ∂W_i in loops homotopic into F_i. So h_i may be chosen to preserve $|\underline{\underline{a}}|$ and to be the identity on F_i. Again, h_i extends using the identity to the rest of M. Finally, if ∂W_i does contain annuli of $\underline{\underline{a}}$ that are not homotopic to F_i, then lemma 9.1.3(ii) shows that we may pass to a finite-index subgroup to assume that φ_i is the identity automorphism. In this case, we just take h_i equal to the identity.

If ϕ lies in the intersection of all the relevant finite index subgroups, then $h_1 h_2$ is an admissible homeomorphism of M which induces ϕ. So we have shown that $\mathcal{R}(M, \underline{\underline{a}})$ has finite index in $\mathrm{Out}(\pi_1(M), \pi_1(\underline{\underline{a}}))$ if $(M, \underline{\underline{a}})$ is a small manifold of type II.

Finally, assume that M is of type III. Again, let W denote $\overline{M-V}$. We write $\pi_1(M) \cong G_1 * \mathbb{Z}$, where the second factor is generated by ω, which is represented by a loop that goes once over the 1-handle of V. As in the previous cases, we may assume, after passing to a finite index subgroup $\mathrm{Out}_1(\pi_1(M), \pi_1(\underline{\underline{a}}))$ of $\mathrm{Out}(\pi_1(M), \pi_1(\underline{\underline{a}}))$, that ϕ preserves G_1 up to conjugacy. Using lemma 9.1.2, we may assume that $\phi(g_1)=\varphi_1(g_1)$ and $\phi(\omega)=\gamma\omega^{\pm 1}$, where $\varphi_1 \in \mathrm{Aut}(G_1)$ and $\gamma \in G_1$.

Suppose M is of type IIIa. Since φ_1 preserves the peripheral structure of $\pi_1(F_1)$, there is a homeomorphism h_1 of F_1 that induces φ_1 on $\pi_1(F_1)$. There is a homeomorphism of M that induces φ_1 on $\pi_1(F_1)$ and sends ω to $\omega^{\pm 1}$; on $F_1 \times \mathrm{I}$ this is of the form $h_1 \times \mathrm{id}_I$ or $h_1 \times r$, where h_1 is a homeomorphism of F_1 that induces φ_1 and r is reflection in the I-fibers, depending on whether ω is sent to ω or ω^{-1}. Composing this with a homeomorphism which slides the left end of the 1-handle around a loop in $F_1 \times \{1\}$ representing γ if ω is sent to ω, or the right end of the 1-handle around a loop in $F_2 \times \{1\}$ representing γ if ω is sent to ω^{-1}, we obtain a homeomorphism inducing ϕ. It may be chosen to be admissible, since every element of $\underline{\underline{a}}$ is an annulus parallel to an annulus that meets F. In this case every element of $\mathrm{Out}_1(\pi_1(M), \pi_1(\underline{\underline{a}}))$ is realizable by an admissible homeomorphism.

If M is of type IIIb, then W is a solid torus and the argument is similar to the type Ib case. The surface F is a sphere with four holes, but otherwise there is little change.

If M is of type IIIc, then either part (i) or (ii) of lemma 9.1.3 shows that there is a subgroup of finite index in $\mathrm{Aut}(\pi_1(\mathrm{S}^1 \times \mathrm{S}^1 \times \mathrm{I}); \pi_1(F_1), \pi_1(F_2))$ realizable by homeomorphisms of $\mathrm{S}^1 \times \mathrm{S}^1 \times \mathrm{I}$ that preserve the elements of the boundary pattern and the attaching disks for the 1-handle. Passing to a subgroup of finite index in $\mathrm{Out}(\pi_1(M), \pi_1(\underline{\underline{a}}))$, we may assume that φ_1 lies in this subgroup. The composition of a homeomorphism realizing φ_1 and a Dehn twist about $\mathrm{S}^1 \times \mathrm{S}^1 \times \{1/2\}$ that sends ω to $\gamma\omega$ induces ϕ. $\square$

9.3. The infinite-index cases

In this section we complete the proof of Main Topological Theorem 1 by proving that $\mathcal{R}(M, \underline{\underline{a}})$ has infinite index in $\mathrm{Out}(\pi_1(M), \pi_1(\underline{\underline{a}}))$ whenever $(M, \underline{\underline{a}})$ is not a relative compression body and is not small. For this part of the proof we do not need the full strength of the hypotheses on $\underline{\underline{a}}$, we use only the assumption that no boundary circle of F is compressible. To indicate this weaker hypothesis, we will now denote the boundary pattern by $\underline{\underline{m}}$. At the end of this section, we discuss some further possible weakenings of this hypothesis.

PROPOSITION 9.3.1. *Let $(M, \underline{\underline{m}})$ be a compact orientable irreducible 3-manifold with boundary pattern consisting of a (possibly empty) collection of disjoint submanifolds. Suppose there exists a free side F which is compressible in M, but no boundary circle of F is contractible in M. If $(M, \underline{\underline{m}})$ is not a relative compression body and is not small, then $\mathcal{R}(M, \underline{\underline{m}})$ has infinite index in* $\mathrm{Out}(\pi_1(M), \pi_1(\underline{\underline{m}}))$.

Here is a sketch of the proof of proposition 9.3.1. Let $(V, \underline{\underline{v}})$ be a minimally imbedded relative compression body neighborhood of the compressible free side F of $(M, \underline{\underline{m}})$. For a loop α in M, we define a "wrapping" homotopy equivalence $h(\alpha)$ of $(M, \underline{\underline{m}})$ which takes a 1-handle of V and maps it to a loop that goes around α and then goes once over the handle. Provided that α itself does not go over the 1-handle, $h(\alpha)$ will be an admissible homotopy equivalence with admissible homotopy inverse $h(\alpha^{-1})$, where α^{-1} denotes the reverse path for α. To detect that the induced outer automorphism $h(\alpha)_\#$ is not realizable by an admissible homeomorphism, we select a certain loop β in F and observe that $h(\alpha)(\beta)$ is not freely homotopic to a loop disjoint from a properly imbedded constituent F_1 of V.

If there were an admissible homeomorphism inducing $h(\alpha)_\#$, then $h(\alpha)(\beta)$ would be freely homotopic into a free side of $(M, \underline{\underline{m}})$ and hence to be disjoint from F_1. To show that $\mathcal{R}(M, \underline{\underline{m}})$ has infinite index in $\mathrm{Out}(\pi_1(M), \pi_1(\underline{\underline{m}}))$, we find sequences $\{h(\alpha_r)\}$, so that for any $i \neq j$, $h(\alpha_i)_\# h(\alpha_j^{-1})_\#$ cannot be induced by an element of $\mathcal{H}(M, \underline{\underline{m}})$. This implies that the induced automorphisms $h(\alpha_r)_\#$ represent distinct cosets of $\mathcal{R}(M, \underline{\underline{m}})$ in $\mathrm{Out}(\pi_1(M), \pi_1(\underline{\underline{m}}))$. The proof breaks into various cases, according to the way in which the constituents of V separate M, and the nature of the complementary components.

PROOF. Let $(V, \underline{v})$ be a minimally imbedded relative compression body neighborhood of F, with constituents $F_1, \ldots, F_m$. Since F is compressible, $k \geq 1$, where k denotes the number of 1-handles of V. The assumption that the boundary circles of F are not contractible in M implies that no constituent F_i is simply connected. Since $(M, \underline{\underline{m}})$ is not itself a relative compression body, we may assume that either F_1 is nonseparating, or for the component M_1 of $\overline{M - V}$ that meets F_1, $\pi_1(F_1) \to \pi_1(M_1)$ is not surjective. We select the basepoint of M to lie in F_1. We will need rather precise descriptions of $\pi_1(M)$.

(i) When F_1 is separating, regard $\pi_1(M)$ as a free product with amalgamation $\pi_1(M_1) *_{\pi_1(F_1)} \pi_1(\overline{M - M_1})$. As explained on p. 187 of [**73**], every element of $\pi_1(M_1) *_{\pi_1(F_1)} \pi_1(\overline{M - M_1})$ can be expressed as a product $g_1 g_2 \cdots g_n$ where each g_i is in one of the factors $\pi_1(M_1)$ or $\pi_1(\overline{M - M_1})$, successive g_i's come from different factors, and if $n > 1$ then no g_i is in $\pi_1(F_1)$. Such a product is called *cyclically reduced* if $n = 1$ or if g_1 and g_n lie in different factors. Every element is conjugate to an element that can be written in cyclically reduced form. According to theorem IV.2.8 in [**73**], if $g_1 \cdots g_n$ is cyclically reduced, then any element in cyclically reduced form conjugate to $g_1 \cdots g_n$ can be obtained by cyclically permuting the g_i and then conjugating by an element of $\pi_1(F_1)$. In particular, an element written in cyclically reduced form which has length $n \geq 2$ cannot be conjugate into $\pi_1(M_1)$ or $\pi_1(\overline{M - M_1})$.

(ii) When F_1 is nonseparating, M is obtained from a manifold M_0 by identifying two copies of F_1 in ∂M_0. Under the quotient map from M_0 to M, one component V_0 of the preimage of V maps homeomorphically to V; the other is a copy F_0 of F_1. Let the copy of the basepoint of M that lies in V_0 be the basepoint of M_0, and choose a path τ in M_0 from that basepoint to the other copy of the basepoint of M, that lies in F_0. Now $\pi_1(F_1)$ sits naturally as a subgroup of $\pi_1(M_0)$, since there is a copy of F_1 in V_0 that contains the basepoint. Using the path τ, we regard $\pi_1(F_0)$ as a subgroup of $\pi_1(M_0)$. In M, τ becomes a closed loop, representing an element $t \in \pi_1(M)$. By Van Kampen's Theorem, $\pi_1(M)$ is an HNN extension $\pi_1(M_0) *_{\pi_1(F_1)}$, where $t \pi_1(F_1) t^{-1} = \pi_1(F_0)$. As explained on pp. 181-187 of [**73**], every element of $\pi_1(M_0) *_{\pi_1(F_1)}$ can be written in the form $g_0 t^{\epsilon_1} g_1 t^{\epsilon_2} g_2 \cdots g_{n-1} t^{\epsilon_n} g_n$ where each $g_i \in \pi_1(M_0)$ and each $\epsilon_i = \pm 1$. This element is called *reduced* if there is no consecutive subsequence $t^{-1} g_i t$ with $g_i \in \pi_1(F_0)$ and there is no consecutive subsequence $t g_i t^{-1}$ with $g_i \in \pi_1(F_1)$. It is called *cyclically reduced* if $g_n = 1$ and all its cyclic permutations ending in a power of t are reduced. According to theorem IV.2.5 of [**73**], if $g_0 t^{\epsilon_1} \cdots g_{n-1} t^{\epsilon_n}$ is cyclically reduced, then any conjugate element in cyclically reduced form also has exactly n appearances of $t^{\pm 1}$. In particular, a cyclically reduced element in which a power of t appears cannot be conjugate into $\pi_1(M_0)$

We illustrate the conjugacy condition in the HNN case with an example that will be useful in Case IIIb below. Note first that since F_1 is a constituent of V, we have $\pi_1(M_0)=\pi_1(F_1)*H$ where $\pi_1(F_0) \subseteq H$. In particular, $\pi_1(F_0)\cap\pi_1(F_1)=\{1\}$. Now, suppose that there exists a loop α in $\overline{M-V}$, based at the basepoint in F_1 and not (based) homotopic back into F_1. Writing $\alpha \simeq \tau^{-1}\tau\alpha\tau^{-1}\tau$ shows that α represents an element of the form $t^{-1}g_1t$, where $g_1 \notin \pi_1(F_0)$ since α is not homotopic into F_1. Now suppose that δ is a loop that represents an element $g_0 \in \pi_1(V) - \pi_1(F_1)$. Then the element represented by $\delta\alpha$ is $g_0t^{-1}g_1t$. Since $g_0 \notin \pi_1(F_1)$ and $g_1 \notin \pi_1(F_0)$, this element is cyclically reduced. Since it contains appearances of $t^{\pm 1}$, it is not conjugate into $\pi_1(M_0)$.

We will now detail the general construction of a wrapping homotopy equivalence. Typically, these will be used to wrap a 1-handle around some non-peripheral curve in M. Let D be a disk, let d_0 be the origin contained in D, and assume that $D \times \mathrm{I}$ is a 1-handle of V such that $\partial D \times \mathrm{I} \subset F$. Let α be a loop based at $\{d_0\} \times \{3/4\} \in D \times \{3/4\}$ such that α is disjoint from $D \times [0, 3/4)$. Construct a map $h(\alpha)\colon (M, \underline{\underline{m}}) \to (M, \underline{\underline{m}})$ which (1) is the identity outside $D \times \mathrm{I}$, (2) maps each $D \times \{t\}$ to itself for $t \in [0, 1/4] \cup [3/4, 1]$, and (3) collapses each $D \times \{t\}$ to $\{d_0\} \times \{t\}$ for $1/4 \le t \le 3/4$, and then maps the resulting interval $\{d_0\} \times [1/4, 3/4]$ around the path $(\{d_0\} \times [1/4, 3/4]) * \alpha$. Observe that if β is another loop based at $\{d_0\} \times \{3/4\} \in D \times \{3/4\}$ and disjoint from $D \times [0, 3/4)$, then $h(\alpha) \circ h(\beta)$ is admissibly homotopic to $h(\alpha\beta)$ relative to $\overline{M - D \times \mathrm{I}}$. In particular, $h(\alpha^{-1})$ is an admissible homotopy inverse to $h(\alpha)$.

In each case of the ensuing proof, we will construct a sequence $\{h(\alpha_r)\}$ whose induced automorphisms $h(\alpha_r)_\#$ will represent distinct cosets of $\mathcal{R}(M, \underline{\underline{m}})$ in $\mathrm{Out}(\pi_1(M), \pi_1(\underline{\underline{m}}))$. So we will need to show that if $i \neq j$, then $h(\alpha_i)_\# h(\alpha_j^{-1})_\#$ does not lie in $\mathcal{R}(M, \underline{\underline{m}})$. To prove that $h(\alpha_i)_\# h(\alpha_j^{-1})_\#$ does not lie in $\mathcal{R}(M, \underline{\underline{m}})$ we will find a loop ν in ∂M, whose image under $h(\alpha)$, where $\alpha = \alpha_i\alpha_j^{-1}$, is not peripheral, i. e. is not homotopic into ∂M. This loop ν will either be a loop β that intersects $\partial(D \times \{3/4\})$ in one point, or a loop of the form $\gamma\delta$ where γ and δ are arcs with endpoints in $\partial(D \times \{3/4\})$ and $\gamma\delta$ intersects $\partial(D \times \{3/4\})$ transversely in two points. Now $h(\alpha)(\ell)$ is homotopic to a loop obtained from ℓ by inserting α or α^{-1} at each (transverse) crossing of ℓ over $\partial(D \times \{3/4\})$. Thus $h(\alpha)(\ell)$ will be freely homotopic to $\alpha\beta$, in the first case, and to $\gamma\alpha\delta\alpha^{-1}$ in the second. In each case, the loops will be chosen so that $h(\alpha)(\ell)$ is not freely homotopic to a loop which is disjoint from F_1. If there were an admissible homeomorphism f realizing ϕ, then $f(\ell)$ would lie in a free side of $(M, \underline{\underline{m}})$ and hence would be disjoint from F_1.

The element of $\pi_1(M)$ represented by a loop whose name is a Greek letter will be denoted by the corresponding Roman letter, thus α represents the element a and so on. We must have $m \geq 1$, since otherwise M is a handlebody with $\underline{\underline{m}} = \emptyset$, and hence a compression body. In all the cases given below, the notation is as in (i) and (ii) above. In particular, M_1 denotes the component of $\overline{M - V}$ that contains F_1, and M_0 denotes the manifold obtained by splitting M along F_1, when F_1 is separating.

The argument is simplest when there are at least as many 1-handles as constituents, i. e. when $k \geq m$.

Case I: $k \geq m$.

In this case, the wrapping homotopy equivalences will affect a 1-handle $D \times \mathrm{I}$ in V for which $D \times \{0\}$ does not separate V. We may choose the compression body structure on V so that both ends of $D \times \mathrm{I}$ lie in $F_1 \times \{1\}$. Fix a loop β in F that intersects $\partial(D \times \{3/4\})$ transversely in a single point, oriented so that it passes over $D \times \mathrm{I}$ from $D \times \{0\}$ to $D \times \{1\}$. There are four subcases:

Case Ia: F_1 does not separate M.

Let τ be a loop in M based in $D \times \{3/4\}$ and disjoint from $D \times [0, 3/4)$, which represents the element t in the HNN decomposition $\pi_1(M) = \pi_1(M_0) *_{\pi_1(F_1)}$ described in (ii) above. Then $h(\tau^i\tau^{-j})(\beta)$ represents an element conjugate to bt^{i-j} with $b \in \pi_1(M_0)$. This is cyclically reduced, so when $i \neq j$ it is not conjugate into $\pi_1(M_0)$. Therefore $h(\tau^{i-j})(\beta)$ is not homotopic to a loop disjoint from F_1. (For this case, one can avoid the HNN theory by simply observing that under the homomorphism from $\pi_1(M_0)*_{\pi_1(F_1)}$ to $\mathbb{Z}$ that sends t to 1 and all elements of $\pi_1(M_0)$ to 0, bt^{i-j} maps to $i - j$.) So, $h(\tau^{i-j})$ does not lie in $\mathcal{R}(M, \underline{\underline{m}})$ and thus $\mathcal{R}(M, \underline{\underline{m}})$ has infinite index in $\mathrm{Out}(\pi_1(M), \pi_1(\underline{\underline{m}}))$.

Case Ib: V has only one constituent ($m = 1$), and $\pi_1(F_1)$ has infinite index in $\pi_1(M_1)$.

Since $m = 1$, F_1 is separating. Choose loops $\{\alpha_r \mid 1 \leq r\}$ that represent infinitely many distinct cosets of $\pi_1(F_1)$ in $\pi_1(M_1)$. Then $h(\alpha_i\alpha_j^{-1})(\beta)$ represents an element conjugate to $(a_ia_j^{-1})b$, with $a_ia_j^{-1}$ in $\pi_1(M_1) - \pi(F_1)$ and $b \in \pi_1(\overline{M - M_1}) - \pi_1(F_1)$. Provided that $i \neq j$, this element is cyclically reduced and has length 2 with respect to the amalgamated free product decomposition as described in (i) above, so is not conjugate to an element of $\pi_1(M_1)$ or $\pi_1(\overline{M - M_1})$. Therefore $h(\alpha_i\alpha_j^{-1})(\beta)$ is not freely homotopic to a loop disjoint from F_1, which as we have seen shows that the $h(\alpha_i)_\#$ represent distinct cosets of $\mathcal{R}(M, \underline{\underline{m}})$ in $\mathrm{Out}(\pi_1(M), \pi_1(\underline{\underline{m}}))$.

Case Ic: V has only one constituent ($m = 1$), and $\pi_1(F_1)$ has finite index in $\pi_1(M_1)$.

Since $\pi_1(F_1)$ has finite index in $\pi_1(M_1)$, the Finite Index Theorem 2.1.1 shows that either M_1 is an I-bundle with F_1 as a lid, or M_1 is a solid torus and F_1 is an annulus. Since $(M, \underline{\underline{m}})$ is not small, there must be a second 1-handle in V. Let α be the product of a loop α' such that $a' \in \pi_1(M_1) - \pi_1(F_1)$ and a loop α'' in V that goes once over the second 1-handle in V. Then $h(\alpha^r)(\beta)$ represents an element conjugate to $(a'a'')^r b$ with $a' \in \pi_1(M_1) - \pi_1(F_1)$, and a'' and $a''b$ in $\pi_1(V) - \pi_1(F_1)$. When $r > 0$, a cyclically reduced form for this element in the free product with amalgamation $\pi_1(M) = \pi_1(V) *_{\pi_1(F_1)} \pi_1(M_1)$ is $a' \cdot a'' \cdot a' \cdot \ldots \cdot a' \cdot a''b$, which has length at least 2. So when $r > 0$, $h(\alpha^r)(\beta)$ is not freely homotopic to a loop disjoint from F_1. Again the $h(\alpha^r)_\#$ represent distinct cosets of $\mathcal{R}(M, \underline{\underline{m}})$ in $\mathrm{Out}(\pi_1(M), \pi_1(\underline{\underline{m}}))$.

Case Id: V has at least two constituents ($m \geq 2$), and F_1 separates M.

Since F_1 separates, our choice of F_1 implies that $\pi_1(F_1) \to \pi_1(M_1)$ is not surjective. Let α be the product of a loop α' such that $a' \in \pi_1(M_1) - \pi_1(F_1)$ and a loop α'' such that a'' is a nontrivial element of $\pi_1(F_2)$. An argument similar to the one in Case Ic then applies to complete the proof that $\mathcal{R}(M, \underline{\underline{m}})$ has infinite index in $\mathrm{Out}(\pi_1(M), \pi_1(\underline{\underline{m}}))$.

For the remaining cases, we may assume that every imbedded 2-disk separates V, that is, that $k = m - 1$. We choose the compression body structure with exactly

$m-1$ 1-handles $D_i \times I$, $1 \le i \le m-1$, so that $D_i \times \{0\}$ is attached to $F_i \times \{1\}$ and $D_i \times \{1\}$ is attached to $F_{i+1} \times \{1\}$. Arcs in V from the basepoint in F_1 to these handles are used to regard loops based at points in the handles as elements of $\pi_1(M)$.

For the remaining cases, we may assume that every imbedded 2-disk separates V, that is, that $k = m-1$. We choose the compression body structure with exactly $m-1$ 1-handles $D_i \times I$, $1 \le i \le m-1$, so that $D_i \times \{0\}$ is attached to $F_i \times \{1\}$ and $D_i \times \{1\}$ is attached to $F_{i+1} \times \{1\}$, except that when $m \ge 2$, $D_2 \times \{0\}$ is attached to $F_3 \times \{1\}$ and $D_2 \times \{1\}$ is attached to $F_2 \times \{1\}$. Arcs in V from the basepoint in F_1 to these handles are used to regard loops based at points in the handles as elements of $\pi_1(M)$.

Case II: $k = m-1$ and $m \ge 3$.

Again there is a natural subdivision into subcases.

Case IIa: F_1 does not separate M.

We choose the compression body structure and notation so that F_1 and F_2 lie in the same component of $\overline{M - V}$. Choose a loop τ based at $D_2 \times \{3/4\}$ and disjoint from $D_2 \times [0, 3/4)$, which represents the HNN generator t of $\pi_1(M) = \pi_1(M_0)*_{\pi_1(F_1)}$. The homotopy equivalences $h(\tau^r)$ are defined using the 1-handle $D_2 \times I$. Choose an arc γ with endpoints in $D_2 \times \{3/4\}$ which is disjoint from $D_2 \times [0, 3/4)$ and which represents a nontrivial element $g \in \pi_1(F_1)$. Choose an arc δ with endpoints in $D_2 \times \{3/4\}$ which is disjoint from $D_2 \times (3/4, 1]$ and which represents a nontrivial element $d \in \pi_1(F_3)$, and so that $\gamma\delta$ is a loop. Now $h(\tau^r)(\gamma\delta)$ represents an element conjugate to $gt^{-r}dt^r$. Also, $g \notin \pi_1(F_0)$ (because $\pi_1(F_1) \cap \pi_1(F_0) = \{1\}$) and $d \notin \pi_1(F_1)$, so when $r < 0$ this element is cyclically reduced and hence is not conjugate into $\pi_1(M_0)$. So for $r < 0$, $h(\tau^r)_\#$ is not induced by any admissible homeomorphism, showing that they represent distinct cosets of $\mathcal{R}(M, \underline{\underline{m}})$ in $\mathrm{Out}(\pi_1(M), \pi_1(\underline{\underline{m}}))$.

Case IIb: F_1 separates M.

Since $\pi_1(F_1)$ must be a proper subgroup of $\pi_1(M_1)$, we can choose a loop α based at $D_2 \times \{3/4\}$ representing an element of the form $a_1 a_2$ with $a_1 \in \pi_1(M_1) - \pi_1(F_1)$ and $a_2 \in \pi_1(F_2) - \{1\}$. Let $\gamma\delta$ be as in Case IIa. Then $h(\alpha^r)(\gamma\delta)$ represents an element conjugate to $g(a_1a_2)^{-r}d(a_1a_2)^r$. Written in normal form, this is

$$ga_2^{-1} \cdot a_1^{-1} \cdot a_2^{-1} \cdot \ldots \cdot a_1^{-1} \cdot d \cdot a_1 \cdot a_2 \cdot \ldots \cdot a_1$$

with $a_1^{\pm 1}$ in $\pi_1(M_1) - \pi_1(F_1)$, and ga_2^{-1}, d, and $a_2^{\pm 1}$ in $\pi_1(\overline{M - M_1}) - \pi_1(F_1)$, when $r > 0$. As before, this shows that if $r > 0$, then $h(\alpha^r)_\#$ is not induced by an admissible homeomorphism.

We have now treated all cases for which $k \ge m$ or $m > 2$, so from now on we may assume that $k = 1$ and $m = 2$. We next handle the cases in which the normalizer of $\pi_1(F_1)$ has infinite index in $\pi_1(M_1)$.

Case III: $k = 1$, $m = 2$ and the normalizer of $\pi_1(F_1)$ has infinite index in $\pi_1(M_1)$.

Choose elements $a_r \in \pi_1(M_1)$ representing distinct cosets of the normalizer of $\pi_1(F_1)$ in $\pi_1(M_1)$. For each $a_i a_j^{-1}$ with $i \ne j$, choose $g_{i,j}$ in $\pi_1(F_1)$ so that $(a_i a_j^{-1}) g_{i,j} (a_i a_j^{-1})^{-1} \in \pi_1(M_1) - \pi_1(F_1)$. Let $\gamma_{i,j}$ be an arc in F with endpoints in $\partial(D_1 \times \{3/4\})$ and disjoint from $D_1 \times (3/4, 1]$, which represents $g_{i,j}$, and let δ be an arc in F with endpoints in $\partial(D_1 \times \{3/4\})$ and disjoint from $D_1 \times [0, 3/4)$, representing a nontrivial element of $\pi_1(F_2)$, such that $\gamma_{i,j}\delta$ is a loop intersecting $\partial(D_1 \times \{1/4\})$ in two points. Then $h(\alpha_i\alpha_j^{-1})(\gamma_{i,j}\delta)$ represents an element conjugate to $(a_i a_j^{-1})g_{i,j}(a_i a_j^{-1})^{-1}d$.

Case IIIa: F_1 separates M.

Since $(a_i a_j^{-1})g_{i,j}(a_i a_j^{-1})^{-1} \in \pi_1(M_1) - \pi_1(F_1)$ and $d \in \pi_1(\overline{M - M_1}) - \pi_1(F_1)$, $h(\alpha_i\alpha_j^{-1})(\gamma_{i,j}\delta)$ is not homotopic to a loop disjoint from F_1. As usual, this shows that the $h(\alpha_i)_\#$ represent distinct cosets of $\mathcal{R}(M, \underline{\underline{m}})$ in $\mathrm{Out}(\pi_1(M), \pi_1(\underline{\underline{m}}))$.

Case IIIb: F_1 does not separate M.

Notice that $(a_i a_j^{-1})g_{i,j}(a_i a_j^{-1})^{-1}$ can be represented by a loop in M based in F_1 and contained in $\pi_1(\overline{M - V})$, and not based homotopic into F_1. As explained in the example after (ii) above, this shows that $\delta(a_i a_j^{-1})g_{i,j}(a_i a_j^{-1})^{-1}$ represents an element of the form $g_0 t^{-1} g_1 t$ where $g_0 \notin \pi_1(F_1)$ and $g_1 \notin \pi_1(F_0)$, and this element cannot be conjugate into $\pi_1(M_0)$. So again $\mathcal{R}(M, \underline{\underline{m}})$ has infinite index in $\mathrm{Out}(\pi_1(M), \pi_1(\underline{\underline{m}}))$.

The remaining case requires more elaborate preparation.

Case IV: $k = 1$, $m = 2$ and for each component G of the frontier of V in M which is contained in a component W of $\overline{M - V}$, the normalizer of $\pi_1(G)$ in $\pi_1(W)$ has finite index.

Since M is not small, the Finite Index Theorem 2.1.1 ensures that F_1 may be chosen so that $\pi_1(F_1)$ has infinite index in $\pi_1(M_1)$. We will use the following result, which is lemma 1.6 of [**117**].

LEMMA 9.3.2. *Let M be an orientable irreducible 3-manifold and S an incompressible component of ∂M. Suppose there is a subgroup H of finite index in $\pi_1(S)$ which is normal in a subgroup H' of $\pi_1(M)$. Then either*

(i) *$H' \subset \pi_1(S)$, or*

(ii) *$\pi_1(S)$ is of index 2 in $\pi_1(M)$, $S = \partial M$, and M is compact and homeomorphic to the I-bundle over a nonorientable surface.*

We first claim that F_1 is an annulus contained in a torus boundary component of M_1. Let N be the covering of M_1 corresponding to the normalizer of $\pi_1(F_1)$. Then N is compact and there is an exact sequence

$$1 \to \pi_1(F_1) \to \pi_1(N) \to Q \to 1$$

with Q infinite. According to theorem 11.1 of [**51**], one of the following must hold:

(1) $\pi_1(F_1) \cong \mathbb{Z}$,

(2) $\pi_1(F_1) \ncong \mathbb{Z}$ and N is a fiber bundle over S^1 with fiber a compact manifold,

(3) $\pi_1(F_1) \ncong \mathbb{Z}$ and N is the union of two twisted I-bundles along their lids, or

(4) N contains a two-sided projective plane.

Suppose for contradiction that case (1) does not hold. Case (4) cannot hold since N is orientable. In cases (2) and (3), each component of ∂N and hence of ∂M_1 is an incompressible torus. Since $\pi_1(F_1) \neq \mathbb{Z}$ and F_1 is incompressible, F_1 must be a torus boundary component of M_1. By lemma 9.3.2, $\pi_1(F_1)$ has index 2 in $\pi_1(M_1)$, which would be a contradiction. We conclude that case (1) holds, so F_1 is an annulus. By corollary 12.8 of [**51**], N is Seifert fibered so ∂N and hence ∂M_1 consist of tori. This proves the claim.

Let T be the boundary torus of M_1 which contains F_1. Since $\pi_1(F_1)$ has infinite index in $\pi_1(M_1)$, M_1 is not a solid torus and therefore T is incompressible. If possible, choose F_1 so that the normalizer of $\pi_1(T)$ in $\pi_1(M_1)$ has infinite index. By a trick which involves altering the boundary pattern, the remaining possibilities will be reduced to cases already considered.

Case IVa: The normalizer of $\pi_1(T)$ in $\pi_1(M_1)$ has infinite index in $\pi_1(M_1)$.

Let $\underline{\underline{m}}'$ be the boundary pattern on M obtained by removing from $\underline{\underline{m}}$ all elements of $\underline{\underline{m}}$ that meet T. If F_2 does not meet T, then $(M, \underline{\underline{m}}')$ has $F \cup (\partial F_1 \times \mathrm{I}) \cup (T \cap \partial M)$ as a free side. If F_2 does not lie in T, then $(M, \underline{\underline{m}}')$ has $F \cup (\partial F_1 \times \mathrm{I}) \cup (\partial F_2 \times I) \cup (T \cap \partial M)$ as a free side. In either case, $V \cup (T \times \mathrm{I})$ is a minimally imbedded relative compression body neighborhood of this free side where $T \times \mathrm{I}$ is a collar neighborhood of T in M_1. Either case Ib (if $F_2 \subset T$) or case IIIa or IIIb (if $F_2 \not\subset T$) applies to show that there are infinitely many admissible homotopy equivalences $h(\alpha_i)$ of $(M, \underline{\underline{m}}')$, such that $h(\alpha_i \alpha_j^{-1})_\#$ is not induced by an admissible homeomorphism of $(M, \underline{\underline{m}}')$, and hence not by an admissible homeomorphism of $(M, \underline{\underline{m}})$. Moreover, we may assume that the 1-handle of V' is the same as the 1-handle of V, so that the $h(\alpha_i)$ are the identity on $\overline{\partial M - F}$. This shows that each $h(\alpha_i)_\#$ lies in $\mathrm{Out}(\pi_1(M), \pi_1(\underline{\underline{m}}))$. Therefore the $h(\alpha_i)_\#$ represent distinct cosets of $\mathcal{R}(M, \underline{\underline{m}})$ in $\mathrm{Out}(\pi_1(M), \pi_1(\underline{\underline{m}}))$.

Case IVb: The normalizer of $\pi_1(T)$ in $\pi_1(M_1)$ has finite index in $\pi_1(M_1)$.

By lemma 9.3.2, M_1 is either $T \times \mathrm{I}$ or the I-bundle over the Klein bottle.

Suppose for contradiction that F_1 separates M, and let M_2 the the component of $\overline{M - V}$ that contains F_2. If the index of $\pi_1(F_2)$ in $\pi_1(M_2)$ is finite, then either M_2 is an I-bundle with F_2 as its lid, or F_2 is an annulus and M_2 is a solid torus. If the index is infinite, then since we are not in case IVa, we have as for F_1 that F_2 is an annulus and M_2 is either $T \times \mathrm{I}$ or the I-bundle over the Klein bottle. In any of these cases, M would be small. So F_1 and F_2 must be nonseparating annuli which meet M_1. They must lie in the same boundary torus T of M_1, otherwise M_1 must be $T \times \mathrm{I}$ and M is small of type IIIc.

Let the components of $\overline{T - (F_1 \cup F_2)}$ be A and B, and form a new boundary pattern $\underline{\underline{m}}'$ on M by removing from $\underline{\underline{m}}$ all elements that meet A. There are boundary circles C_i of F_i so that $C_i \times \mathrm{I}$ meets A, so $F \cup (C_1 \times \mathrm{I}) \cup (C_2 \times \mathrm{I}) \cup A$ is a free side of $(M, \underline{\underline{m}}')$. A minimally imbedded compression body neighborhood of this free side has exactly one 1-handle, and one constituent which is an annulus (the frontier of a regular neighborhood of $F_1 \cup A \cup F_2$ in M_1). Case Ib applies to $(M, \underline{\underline{m}}')$, and exactly as in case IVa this shows that $\mathcal{R}(M, \underline{\underline{m}})$ has infinite index in $\mathrm{Out}(\pi_1(M), \pi_1(\underline{\underline{m}}))$. □

Finally, we note that the case where components of ∂F are allowed to be compressible can undoubtedly be handled using methods similar to those in the proof of proposition 9.3.1. Since this would require the definition of additional small manifolds (for example, a twisted I-bundle with boundary pattern consisting of its sides together with a collection of disjoint disks in its lid), we have restricted to the case when ∂F is incompressible.

CHAPTER 10

The Case When the Boundary Pattern Is Useful

The main goal of this chapter is to prove Main Topological Theorem 2, stated in chapter 8. Roughly speaking, it says that for a 3-manifold with a boundary pattern whose completion is useful, the subgroup of automorphisms of the fundamental group that can be realized by homeomorphisms has finite index in the group of all automorphisms if and only if the components of the characteristic submanifold that meet the boundary of the 3-manifold are not too complicated. The characteristic submanifold is taken with respect to the completed boundary pattern.

For the index of the realizable subgroup to be finite, all the Seifert-fibered components of the characteristic submanifold that meet the boundary must be from a certain list that fiber over surfaces of very small genus, and have few exceptional fibers. However, the way in which these components meet the boundary can also affect the index. This is illustrated in examples 1.4.3 and 1.4.4 from chapter 1. These examples have homeomorphic characteristic submanifolds (the product of circle and a disk with two holes), but in example 1.4.3 the characteristic submanifold meets the boundary in two tori, and the index is infinite, while in example 1.4.4 it meets the boundary in a single torus, and the index is finite. Thus both the topological types of these components and their intersection with the boundary must be taken into account. The I-bundle components of the characteristic submanifold are similarly constrained.

The first stage of the proof of Main Topological Theorem 2 is theorem 10.1.1. It states that the realizable automorphisms have finite index if and only if for each component of the characteristic submanifold, the subgroup of a certain group of admissible self-homotopy-equivalences realizable by homeomorphisms has finite index. Its proof is complicated, and will be sketched in detail in section 10.1 after the necessary notation has been set up. To complete the proof of Main Topological Theorem 2, we must determine the fibered manifolds for which the realizable subgroup in theorem 10.1.1 has finite index. The case of I-fibered manifolds is handled first, in section 10.2, then the Seifert-fibered case, in section 10.3. In section 10.4, these elements are assembled into the formal proof of Main Topological Theorem 2.

We now set up some basic notation, and use it to give a more precise rendering of the previous outline. Let M be a compact orientable irreducible 3-manifold with a (possibly empty) boundary pattern $\underline{\underline{m}}$ whose completion $\overline{\underline{\underline{m}}}$ is useful and nonempty. This excludes the possibility that $(M, \underline{\underline{m}})$ is one of the exceptional fibered manifolds (EF1)-(EF5) or (ESF), defined in section 2.6. Throughout this section, we work under the strong assumption that the elements of $\underline{\underline{m}}$ are *disjoint;* consequently, unless $(M, \underline{\underline{m}}) = (D^3, \underline{\underline{d}})$ where $\underline{\underline{d}}$ consists of a single 2-disk, no element of $\overline{\underline{\underline{m}}}$ is simply-connected, since an element of $\overline{\underline{\underline{m}}}$ that met a disk of $\overline{\underline{\underline{m}}}$ would be compressible and $\overline{\underline{\underline{m}}}$ could not be useful.

In section 10.1, we will define a subgroup $\operatorname{Out}_2(\pi_1(M), \pi_1(\underline{\underline{m}}))$ which has finite index in $\operatorname{Out}(\pi_1(M), \pi_1(\underline{\underline{m}}))$. An admissible homotopy equivalence that induces an automorphism in $\operatorname{Out}_2(\pi_1(M), \pi_1(\underline{\underline{m}}))$ and preserves the characteristic submanifold of $(M, \overline{\underline{\underline{m}}})$ must preserve each component V_i of this characteristic submanifold, in fact it must preserve each component of the frontier of each V_i. Restricting such a homotopy equivalence to the V_i produces a homotopy equivalence on each V_i, which lies in a certain group $\mathcal{E}(V_i, \underline{\underline{v_i}}', \underline{\underline{v_i}}'')$ of homotopy equivalences. This restriction is well-defined on elements of $\operatorname{Out}_2(\pi_1(M), \pi_1(\underline{\underline{m}}))$, so this process defines a homomorphism Ψ from $\operatorname{Out}_2(\pi_1(M), \pi_1(\underline{\underline{m}}))$ to a product $\prod_{i=1}^n \mathcal{E}(V_i, \underline{\underline{v_i}}', \underline{\underline{v_i}}'')$.

Each group $\mathcal{E}(V_i, \underline{\underline{v_i}}', \underline{\underline{v_i}}'')$ contains a subgroup $\mathcal{H}(V_i, \underline{\underline{v_i}}', \underline{\underline{v_i}}'')$ consisting of the classes which are realizable by orientation-preserving homeomorphisms. The crucial property of Ψ is that (apart from one exceptional type of manifold) up to finite index the subgroup $\mathcal{R}(M, \underline{\underline{m}})$ is the preimage $\Psi^{-1}(\prod_{i=1}^n \mathcal{H}(V_i, \underline{\underline{v_i}}', \underline{\underline{v_i}}''))$. This leads to the main result of section 10.1, theorem 10.1.1, which states that $\mathcal{R}(M, \underline{\underline{m}})$ has finite index in $\operatorname{Out}(\pi_1(M), \pi_1(\underline{\underline{m}}))$ precisely when each $\mathcal{H}(V_i, \underline{\underline{v_i}}', \underline{\underline{v_i}}'')$ has finite index in $\mathcal{E}(V_i, \underline{\underline{v_i}}', \underline{\underline{v_i}}'')$.

In section 10.2, the I-fibered manifolds for which $\mathcal{H}(V_i, \underline{\underline{v_i}}', \underline{\underline{v_i}}'')$ has finite index are determined. Actually, this section treats all fibered manifolds which are bundles, so applies to the Seifert-fibered manifolds with no exceptional fibers (apart from the exceptional case of $\mathrm{S}^1 \times \mathrm{S}^1 \times \mathrm{I}$ with boundary pattern $\overline{\underline{\underline{\emptyset}}}$), as well as all I-bundles. For bundles, proposition 10.2.2 shows that the realization question can be translated directly into a realization question for the base surface $(B, \underline{\underline{b}})$. One simply needs to tell when the realizable subgroup $\mathcal{H}(B, \underline{\underline{b}})$ has finite index in the group of admissible homotopy equivalences $\mathcal{E}(B, \underline{\underline{b}})$. For this two-dimensional problem, lemmas 10.2.3 and 10.2.4 treat the finite- and infinite-index cases respectively.

The remaining Seifert-fibered manifolds are treated in section 10.3. Specialized arguments are used for the cases when V is a solid torus, or fibers over the disk with two exceptional orbits of type $(2, 1)$ (the latter is a Seifert fibering on the I-bundle over the Klein bottle). Lemma 10.3.6 addresses the remaining cases for which the sum of the number of exceptional fibers and the rank of $H_1(B)$ is 2; provided that the boundary patterns are suitably restricted, these comprise most of the cases when the index is finite. Lemma 10.3.7 treats the cases when the sum is at least 3. There is exactly one such manifold where the index is finite, the manifold that fibers over the 2-disk with three exceptional fibers of type $(2, 1)$, and has empty boundary pattern. For the other cases, as well as the infinite-index cases in lemma 10.3.6 that are not covered by proposition 10.2.2, homotopy equivalences that we call "sweeps" are used to produce representatives for infinitely many cosets of the subgroup of realizable automorphisms.

In the remainder of this introductory section, we develop some notation that will be used throughout this chapter. Let Σ be the characteristic submanifold of $(M, \overline{\underline{\underline{m}}})$, and let $\widehat{\underline{\underline{\sigma}}}$ be the *proper* boundary pattern on Σ as a submanifold of $(M, \overline{\underline{\underline{m}}})$. Thus $\widehat{\underline{\underline{\sigma}}}$ consists of the components of the intersections of Σ with each element of $\overline{\underline{\underline{m}}}$, together with the set of components of the frontier of Σ, which we denote by $\underline{\underline{\sigma}}''$. Let $(V_1, \widehat{\underline{\underline{v_1}}}), \ldots, (V_n, \widehat{\underline{\underline{v_n}}})$ be the components of $(\Sigma, \widehat{\underline{\underline{\sigma}}})$, and denote the set of components of the frontier of V_i by $\underline{\underline{v_i}}''$.

Let $\underline{\underline{\sigma}}$ and $\underline{\underline{v_i}}$ be the elements of $\widehat{\underline{\underline{\sigma}}}$ and $\widehat{\underline{\underline{v_i}}}$ that lie in $|\underline{\underline{m}}|$; that is, these are the submanifold boundary patterns on Σ and V_i as submanifolds of $(M, \underline{\underline{m}})$. By lemma 2.10.8, each annulus and torus in $\underline{\underline{m}}$ must be an element of $\underline{\underline{\sigma}}$.

Let $\widetilde{\underline{\underline{m}}} \cup \underline{\underline{\sigma}}''$ be the proper boundary pattern on $\overline{M - \Sigma}$ as a submanifold of $(M, \underline{\underline{m}})$, that is, the set $\widetilde{\underline{\underline{m}}} = \{\text{components of } F \cap \overline{M - \Sigma} \mid F \in \underline{\underline{m}}\}$, together with the set $\underline{\underline{\sigma}}''$ of components of the frontier of $\overline{M - \Sigma}$. For each component S_j of $\overline{M - \Sigma}$, let $\underline{\underline{s_j}}$ denote the set of elements of $\widetilde{\underline{\underline{m}}} \cup \underline{\underline{\sigma}}''$ that lie in S_j; this is the proper boundary pattern on S_j.

For each i, if $(V_i, \widehat{\underline{\underline{v_i}}})$ is Seifert-fibered let $\underline{\underline{v_i}}'$ denote the elements of $\underline{\underline{v_i}}$ that are not annuli belonging to $\underline{\underline{m}}$, while if $(V_i, \widehat{\underline{\underline{v_i}}})$ is an I-bundle let $\underline{\underline{v_i}}' = \underline{\underline{v_i}}$. Let $\underline{\underline{\sigma}}'$ denote $\bigcup_{i=1}^{n} \underline{\underline{v_i}}'$.

Let $(V, \underline{\underline{v}}' \cup \underline{\underline{v}}'')$ be one of the $(V_i, \underline{\underline{v_i}}' \cup \underline{\underline{v_i}}'')$. Define $\mathcal{E}(V, \underline{\underline{v}}', \underline{\underline{v}}'')$ to be the group of path components of the space of admissible homotopy equivalences from $(V, \underline{\underline{v}}' \cup \underline{\underline{v}}'')$ to $(V, \underline{\underline{v}}' \cup \underline{\underline{v}}'')$ which preserve each element of $\underline{\underline{v}}' \cup \underline{\underline{v}}''$ and whose restriction to each element of $\underline{\underline{v}}''$ is a homeomorphism which is isotopic to the identity. Define $\mathcal{H}(V, \underline{\underline{v}}', \underline{\underline{v}}'')$ to be the subgroup of $\mathcal{E}(V, \underline{\underline{v}}', \underline{\underline{v}}'')$ consisting of all classes which contain orientation-preserving homeomorphisms.

10.1. The homomorphism Ψ

The goal of this section is to prove the following theorem, which reduces the finite index realization problem in the useful case to the analysis of homotopy equivalences of I-bundles and Seifert-fibered manifolds.

THEOREM 10.1.1. *Let M be a compact orientable irreducible 3-manifold, with a boundary pattern $\underline{\underline{m}}$ whose completion is useful and nonempty. Assume that the elements of $\underline{\underline{m}}$ are disjoint. Then $\mathcal{R}(M, \underline{\underline{m}})$ has finite index in* $\mathrm{Out}(\pi_1(M), \pi_1(\underline{\underline{m}}))$ *if and only if for each component V_i of the characteristic submanifold of $(M, \overline{\underline{\underline{m}}})$, the subgroup $\mathcal{H}(V_i, \underline{\underline{v_i}}', \underline{\underline{v_i}}'')$ of elements of $\mathcal{E}(V_i, \underline{\underline{v_i}}', \underline{\underline{v_i}}'')$ realizable by orientation-preserving homeomorphisms has finite index in $\mathcal{E}(V_i, \underline{\underline{v_i}}', \underline{\underline{v_i}}'')$.*

To prove the theorem, we will define a subgroup $\mathrm{Out}_2(\pi_1(M), \pi_1(\underline{\underline{m}}))$ of finite index in $\mathrm{Out}(\pi_1(M), \pi_1(\underline{\underline{m}}))$, and construct a homomorphism Ψ from $\mathrm{Out}_2(\pi_1(M), \pi_1(\underline{\underline{m}}))$ to $\prod_{i=1}^{n} \mathcal{E}(V_i, \underline{\underline{v_i}}', \underline{\underline{v_i}}'')$. Defining Ψ on a subgroup of finite index allows numerous simplifications, for example it allows one to work using homotopy equivalences that preserve each component of Σ. The basic idea in defining Ψ is to take a homotopy equivalence that induces the automorphism, deform it so that it preserves Σ, and then take its restrictions to the components of Σ as the coordinates in $\prod_{i=1}^{n} \mathcal{E}(V_i, \underline{\underline{v_i}}', \underline{\underline{v_i}}'')$. Roughly speaking, an automorphism can be realized by an orientation-preserving admissible homeomorphism of $(M, \underline{\underline{m}})$ if and only if each of these restrictions is admissibly homotopic to a homeomorphism, so one can think of Ψ as isolating the parts of the automorphism that might fail to be realizable by homeomorphisms and allowing us to test them individually.

The difficulty in defining Ψ is that an automorphism determines a homotopy equivalence only up to homotopy, not up to admissible homotopy. Indeed, there are many admissible homotopy equivalences of the $(V_i, \underline{\underline{v_i}})$ that are homotopic but not admissibly homotopic to the identity. For example, we will see in the proof of proposition 10.2.1 that the "sweep" construction mentioned above produces numerous such examples. Consequently, a homomorphism from a finite-index subgroup of

$\mathrm{Out}(\pi_1(M), \pi_1(\underline{\underline{m}}))$ to $\prod_{i=1}^{n} \mathcal{E}(V_i, \underline{\underline{v_i}} \cup \underline{\underline{v_i}}'')$ would not be well-defined. Fortunately, by using $\underline{\underline{v_i}}' \cup \underline{\underline{v_i}}''$ rather than $\underline{\underline{v_i}} \cup \underline{\underline{v_i}}''$, we can define Ψ.

Here is a step-by-step summary of how we proceed.

1. In lemma 10.1.2, we find a subgroup $\mathrm{Out}_1(\pi_1(M), \pi_1(\underline{\underline{m}}))$ of finite index in $\mathrm{Out}(\pi_1(M), \pi_1(\underline{\underline{m}}))$, whose elements are realizable by admissible homotopy equivalences satisfying four properties called (C1)-(C4). In particular they preserve each component of Σ and $\overline{M - \Sigma}$, and restrict to an orientation-preserving homeomorphism on each component of the latter.

2. To examine the difference between homotopy and admissible homotopy, we introduce the trace. The trace is defined for an element F of $\underline{\underline{m}}$ with respect to a homotopy between two admissible homotopy equivalences, each of which preserves F. As shown in lemma 10.1.5, the trace for F measures the inability to deform the homotopy to one that preserves F. Lemma 10.1.6 tells when nontrivial traces can arise. Its proof shows that apart from one exceptional case, they can arise only when F is an annulus in a Seifert-fibered component V of Σ, and the two homotopy equivalences differ by a "sweep" that moves F around a loop in V. Sweeps are certain homotopy equivalences that will be defined precisely in section 10.2.

3. We prove proposition 10.1.4, which says that if two homotopy equivalences satisfying the conditions of lemma 10.1.2 induce the same outer automorphism on $\pi_1(M)$, then their restrictions to $\overline{M - \Sigma}$ and Σ are admissibly homotopic, provided that we use the boundary pattern $\underline{\underline{\sigma}}' \cup \underline{\underline{\sigma}}''$ on Σ.

4. By virtue of proposition 10.1.4, the restriction homomorphism that takes an element of $\mathrm{Out}_1(\pi_1(M), \pi_1(\underline{\underline{m}}))$, represents it by an admissible homotopy equivalence satisfying the conditions of lemma 10.1.2, and restricts it to an element of the group of homotopy equivalences on each component of the characteristic submanifold and each component of its complement is well-defined. We show in lemma 10.1.7 that the restrictions to components of $\overline{M - \Sigma}$ form a finite group. Passing to the kernel of these restrictions produces a finite-index subgroup $\mathrm{Out}_2(\pi_1(M), \pi_1(\underline{\underline{m}}))$ of $\mathrm{Out}_1(\pi_1(M), \pi_1(\underline{\underline{m}}))$. Roughly speaking, these are the automorphisms that are trivial on $\overline{M - V}$. The homomorphism Ψ is just the restriction homomorphism, restricted to this subgroup of automorphisms.

5. After seeing in lemma 10.1.8 that (apart from an exceptional case) Ψ is surjective, we show that $\Psi^{-1}(\prod_{i=1}^{n} \mathcal{H}(V_i, \underline{\underline{v_i}}', \underline{\underline{v_i}}''))$ is closely related to the realizable subgroup $\mathcal{R}(M, \underline{\underline{m}})$ of $\mathrm{Out}(\pi_1(M), \pi_1(\underline{\underline{m}}))$. The precise statement is theorem 10.1.9.

6. Theorem 10.1.1 is deduced from theorem 10.1.9.

From now until the deduction of theorem 10.1.1 from theorem 10.1.9, we will assume that M is not the 3-ball. Also, as we have indicated, there are some *exceptional cases,* for which certain of the results in this section do not hold. At various times, we will exclude the following possibilities for $(M, \underline{\underline{m}})$:

(E1) $M = S^1 \times S^1 \times I$, $\underline{\underline{m}}$ contains an annulus, and $(M, \underline{\underline{m}})$ can be admissibly Seifert-fibered (i. e. all annuli in $\underline{\underline{m}}$ are homotopic in M).

(E2) $(M, \underline{\underline{m}})$ is an admissibly fibered twisted I-bundle.

Note that for both of these exceptional cases, the characteristic submanifold of $(M, \overline{\underline{\underline{m}}})$ is all of M.

Our first lemma uses the Classification Theorem 2.11.1 to realize automorphisms in $\mathrm{Out}(\pi_1(M), \pi_1(\underline{\underline{m}}))$ by admissible homotopy equivalences that preserve Σ and $\overline{M - \Sigma}$.

LEMMA 10.1.2. $\mathrm{Out}(\pi_1(M), \pi_1(\underline{\underline{m}}))$ *has a subgroup* $\mathrm{Out}_1(\pi_1(M), \pi_1(\underline{\underline{m}}))$ *of finite index in which each element can be induced by an admissible homotopy equivalence* $f\colon (M, \underline{\underline{m}}) \to (M, \underline{\underline{m}})$ *satisfying the following conditions:*

- (C1) *f preserves Σ and $\overline{M - \Sigma}$, its restriction to $(\overline{M - \Sigma}, \underline{\underline{\widetilde{m}}} \cup \underline{\underline{\sigma}}'')$ is an admissible homeomorphism, and its restriction to $(\Sigma, \underline{\underline{\sigma}} \cup \underline{\underline{\sigma}}'')$ is an admissible homotopy equivalence.*
- (C2) *f preserves each component of Σ and each component of $\overline{M - \Sigma}$.*
- (C3) *f preserves each element of $\underline{\underline{m}}$ and each element of $\underline{\underline{\sigma}}''$.*
- (C4) *The restrictions of f to the components of $\overline{M - \Sigma}$ are orientation-preserving homeomorphisms, and the restriction of f to each element of $\underline{\underline{\sigma}}''$ that is a square or an annulus is admissibly isotopic to the identity.*

There are many places in this chapter where we use the assumption that the elements of $\underline{\underline{m}}$ are disjoint, but lemma 10.1.2 appears to be one of the most crucial, since as the next example shows, it is false without this assumption.

EXAMPLE 10.1.3. *A 3-manifold $(M, \underline{\underline{m}})$ such that no subgroup of finite index in* $\mathrm{Out}(\pi_1(M), \pi_1(\underline{\underline{m}}))$ *can be realized by admissible homotopy equivalences.*

Let $M = F \times S^1$ where F is a closed surface of genus 2 with one boundary component. Let $\underline{\underline{f}}$ be a complete boundary pattern on F consisting of arcs, and let $\underline{\underline{m}} = \{k \times S^1 \mid k \in \underline{\underline{f}}\}$. So $\underline{\underline{m}}$ consists of annuli, each of whose fundamental groups is generated by a loop that generates the center of $\pi_1(M) = \pi_1(F) \times \mathbb{Z}$. Since the center is characteristic, $\mathrm{Out}(\pi_1(M), \pi_1(\underline{\underline{m}})) = \mathrm{Out}(\pi_1(M))$. However, the subgroup of $\mathrm{Out}(\pi_1(M))$ realizable by homotopy equivalences that preserve ∂M has infinite index (example 9.0.2 displays an infinite set of distinct coset representatives for the realizable subgroup.)

PROOF OF LEMMA 10.1.2. Any element of $\mathrm{Out}(\pi_1(M), \pi_1(\underline{\underline{m}}))$ can be induced by a homotopy equivalence f of M, since M is aspherical, and since the elements of $\underline{\underline{m}}$ are disjoint, f may be chosen to be admissible for $(M, \underline{\underline{m}})$. By definition, each element of $\mathrm{Out}(\pi_1(M), \pi_1(\underline{\underline{m}}))$ permutes the conjugacy classes of the subgroups of $\pi_1(M)$ corresponding to the elements of $\underline{\underline{m}}$, so there is a subgroup $\mathrm{Out}_0(\pi_1(M), \pi_1(\underline{\underline{m}}))$ of finite index in $\mathrm{Out}(\pi_1(M), \pi_1(\underline{\underline{m}}))$ consisting of automorphisms that fix the conjugacy classes of these subgroups. Suppose f induces an automorphism in $\mathrm{Out}_0(\pi_1(M), \pi_1(\underline{\underline{m}}))$. Since M is aspherical, and the elements of $\underline{\underline{m}}$ are incompressible and aspherical, any two elements of $\underline{\underline{m}}$ which determine the same conjugacy class in $\pi_1(M)$ are homotopic in M. Since the elements are also disjoint, f may be changed by homotopy to take each element F of $\underline{\underline{m}}$ to itself, and each $(f|_F)_\#\colon \pi_1(F) \to \pi_1(F)$ will be an isomorphism. Since the elements of $\underline{\underline{m}}$ are aspherical and disjoint, $f\colon (M, |\underline{\underline{m}}|) \to (M, |\underline{\underline{m}}|)$ is a homotopy equivalence of pairs, and hence is an admissible homotopy equivalence. By the Classification Theorem 2.11.1, f can be changed by admissible homotopy so that it satisfies (C1).

Let $\mathcal{E}_0(M,\underline{\underline{m}})$ denote the group of admissible homotopy classes of admissible homotopy equivalences of $(M,\underline{\underline{m}})$ that send each element of $\underline{\underline{m}}$ to itself. Sending an admissible homotopy class to its induced automorphism defines a homomorphism

$$\mathcal{E}_0(M,\underline{\underline{m}}) \longrightarrow \mathrm{Out}_0(\pi_1(M),\pi_1(\underline{\underline{m}}))\ .$$

So far, we have shown that this homomorphism is surjective, and that any element of $\mathcal{E}_0(M,\underline{\underline{m}})$ contains a homotopy equivalence which satisfies (C1).

Suppose two admissible homotopy equivalences satisfying (C1) and preserving the elements of $\underline{\underline{m}}$ are admissibly homotopic. By the Homotopy Splitting Theorem 2.11.4, we may assume that for all $t \in I$ the homotopy satisfies $H_t^{-1}(\Sigma)=\Sigma$. It follows that $H_t^{-1}(\overline{M-\Sigma})=\overline{M-\Sigma}$. Thus, taking an element of $\mathcal{E}_0(M,\underline{\underline{m}})$, selecting a representative that satisfies (C1), and restricting it to components of Σ and $\overline{M-\Sigma}$ gives restrictions that are well-defined up to admissible homotopy on each component of $(\Sigma,\underline{\underline{\sigma}}\cup\underline{\underline{\sigma}}'')$ and $(\overline{M-\Sigma},\underline{\underline{\widetilde{m}}}\cup\underline{\underline{\sigma}}'')$. This shows that the permutations induced on components of Σ, components of $\overline{M-\Sigma}$, components of $\Sigma\cap\overline{M-\Sigma}$, and elements of $\underline{\underline{m}}$ by elements of $\mathcal{E}_0(M,\underline{\underline{m}})$ are well-defined. Therefore, there is a subgroup $\mathcal{E}_1(M,\underline{\underline{m}})$ of finite index in $\mathcal{E}_0(M,\underline{\underline{m}})$ realizable by admissible homotopy equivalences which satisfy (C1)-(C3).

Define a homomorphism from $\mathcal{E}_1(M,\underline{\underline{m}})$ to a product of m copies of $\mathbb{Z}/2$, where m is the number of components of $\overline{M-\Sigma}$, by taking -1 in each coordinate corresponding to a component on which the restriction of f is not admissibly homotopic to an orientation-preserving homeomorphism. Each element in the kernel is representable by an f satisfying (C1)-(C3) whose restriction to each component of $\overline{M-\Sigma}$ is orientation-preserving. Replace $\mathcal{E}_1(M,\underline{\underline{m}})$ by this kernel. The mapping class groups of the square and the annulus are finite, so after replacing $\mathcal{E}_1(M,\underline{\underline{m}})$ by a subgroup of finite index, we may assume that the restriction to each element of $\underline{\underline{\sigma}}''$ that is a square or an annulus is admissibly isotopic to the identity. Letting $\mathrm{Out}_1(\pi_1(M),\pi_1(\underline{\underline{m}}))$ be the image of $\mathcal{E}_1(M,\underline{\underline{m}})$ in $\mathrm{Out}_0(\pi_1(M),\pi_1(\underline{\underline{m}}))$ gives the desired subgroup of finite index in $\mathrm{Out}(\pi_1(M),\pi_1(\underline{\underline{m}}))$. □

The next proposition is the step that necessitates the definition of $\underline{\underline{\sigma}}'$ and its ensuing complications. The "sweep" homotopy equivalences that we will use in sections 10.2 and 10.3 show that the proposition would fail drastically if one were to use $\underline{\underline{\sigma}}$ instead of $\underline{\underline{\sigma}}'$ in its statement.

PROPOSITION 10.1.4. *Suppose that two admissible homotopy equivalences f and f' of $(M,\underline{\underline{m}})$ satisfy conditions (C1)-(C4), and induce the same automorphism in $\mathrm{Out}_1(\pi_1(M),\pi_1(\underline{\underline{m}}))$. If $(M,\underline{\underline{m}})$ is not an exceptional case (E2), then there exists a homotopy from f to f' which preserves Σ and $\overline{M-\Sigma}$ and restricts to admissible homotopies on $(\overline{M-\Sigma},\underline{\underline{\widetilde{m}}}\cup\underline{\underline{\sigma}}'')$ and on $(\Sigma,\underline{\underline{\sigma}}'\cup\underline{\underline{\sigma}}'')$.*

We will first motivate the proof of proposition 10.1.4, and present some of the ideas used in it. The condition that f and f' induce the same outer automorphism tells us that there is a homotopy from f to f', but the homotopy need not be admissible. To keep track of where an element of $\underline{\underline{m}}$ moves during such a homotopy, we use the *trace*. The trace is defined for an element F of $\underline{\underline{m}}$ with respect to a homotopy between two admissible homotopy equivalences, each of which preserves F. It measures the inability to deform the homotopy to one that preserves F. If the homotopy is between two admissible homotopy equivalences that satisfy the conditions (C1)-(C4), and it does not admit a deformation to a homotopy that preserves F, then lemma 10.1.6 below shows that either $(M, \underline{\underline{m}})$ is an exceptional manifold of type (E2), or F is an annulus of $\underline{\underline{m}}$ that lies in a Seifert-fibered component of Σ. To exclude such annuli, we have to work with $\underline{\underline{\sigma}}'$ rather than $\underline{\underline{\sigma}}$ in proposition 10.1.4. In fact, homotopies with nontrivial trace occur in the proof of proposition 10.1.4. These homotopies move annuli of $\underline{\underline{m}}$ around nontrivial paths in Seifert-fibered components of Σ, and the homotopy equivalence which is the end result of such a homotopy is called a "sweep". In many cases, sweeps are not admissibly homotopic to homeomorphisms, and they will be used in the cases when the realizable automorphisms have infinite index in a way somewhat analogous to the way that wrapping homotopy equivalences were used in chapter 9. Although sweeps appear in the proof of proposition 10.1.4, we delay their general definition until section 10.2 (just before lemma 10.2.4) so as not to run too far afield of the development of Ψ, which is the central theme of the current section.

To define the trace, consider an element $F \in \underline{\underline{m}}$ and a homotopy between two admissible homotopy equivalences f and f' that preserve F. Fix a basepoint p in the interior of F. Change f and f' by admissible homotopy, so that they both preserve p. During the homotopy from f to f', p then moves around a loop representing an element α of $\pi_1(M, p)$ (i. e. the restriction of the homotopy to $\{p\} \times \mathrm{I}$ is a loop based at p which represents the element α).

The element α lies in the normalizer of $\pi_1(F, p)$ in $\pi_1(M, p)$. For suppose that β is a loop at p representing an element of $\pi_1(F, p)$. The restrictions of the homotopy to the loops $\{p\} \times [0, t] * (\beta \times \{t\}) * \overline{\{p\} \times [0, t]}$ give a path homotopy from $f \circ \beta$ to $\alpha * (f' \circ \beta) * \alpha^{-1}$. Since $f_\#$ and $f'_\#$ preserve the subgroup $\pi_1(F, p)$ of $\pi_1(M, p)$, it follows that α is in the normalizer.

A different choice of admissible homotopies making f and f' preserve p will change α by (pre- and post-) multiplication by elements of $\pi_1(F, p)$. Strictly speaking, then, the *trace* of F during the homotopy is defined to be the double coset $\pi_1(F, p)\, \alpha\, \pi_1(F, p)$ in the normalizer of $\pi_1(F, p)$ in $\pi_1(M, p)$, although in practice we will just refer to α as the trace. In particular, we say that the trace *lies in* $\pi_1(F)$ to mean that this double coset equals $\pi_1(F, p)$.

A key property of the trace is that if one changes the homotopy by a deformation which restricts to an admissible homotopy on $M \times \partial \mathrm{I}$, then the trace will be unchanged. For the endpoints of $\{p\} \times \mathrm{I}$ move only within F, so α is changed only by pre- and post-multiplication by elements of $\pi_1(F, p)$.

Here is the main property of the trace that we will use.

LEMMA 10.1.5. *Let $F \in \underline{\underline{m}}$, and let f and f' be admissible homotopy equivalences of $(M, \underline{\underline{m}})$ that preserve F. Suppose that $H\colon M \times I \to M$ is a homotopy from f to f', not necessarily admissible. Let N be an open neighborhood of F. If the trace of F during H lies in $\pi_1(F)$, then there is a deformation of H relative to $(\overline{M - N} \times I) \cup (M \times \partial I)$ so that $H(F \times I) \subset F$.*

PROOF. Suppose that the trace α of the homotopy H lies in $\pi_1(F, p)$. The first step will be to observe that there is a deformation of H (all deformations of homotopies will be relative to $(\overline{M - N} \times \mathrm{I}) \cup (M \times \partial \mathrm{I})$) to a product of homotopies $K * L * K'$, where K and K' are admissible homotopies fixed on $\overline{M - N} \times \mathrm{I}$, and each L_t fixes p.

Note first that any given homotopy K starting at f can become the first part of H, and analogously, any homotopy K' ending at f' can become its final part. For one can use first a deformation of H to $C * H$, where C is the constant homotopy that equals f at every step, then a deformation of C to $K * \overline{K}$ to change $C * H$ to $K * (\overline{K} * L)$, where as usual $\overline{K}$ denotes the reverse homotopy defined by $\overline{K}(x,t) = K(x, 1-t)$. Using this principle, we may change H to $K*(\overline{K}*H*\overline{K'})*K'$, where K is an admissible homotopy fixed on $\overline{M - N}$ that first moves $f(p)$ to p, then moves p around a loop in F that represents α, and where K' moves p to $f'(p)$. Then, the trace of $\overline{K} * H * \overline{K'}$ will be trivial, as an element of $\pi_1(F)$. We will now use the homotopy extension property to deform $\overline{K} * H * \overline{K'}$ to a homotopy L that fixes p at each level. Starting with $\overline{K} * H * \overline{K'}$ on $(M \times \mathrm{I}) \times \{0\}$, construct a partial homotopy on the subspace $(M \times \mathrm{I} \times \{0\}) \cup (M \times \partial\mathrm{I} \times \mathrm{I}) \cup (\overline{M - N} \times \mathrm{I} \times \mathrm{I}) \cup (\{p\} \times \mathrm{I} \times \mathrm{I})$ of $(M \times \mathrm{I}) \times \mathrm{I}$ as follows. On all of these except $\{p\} \times \mathrm{I} \times \mathrm{I}$ it will agree with $\overline{K} * H * \overline{K'}$ at each time, while on $\{p\} \times \mathrm{I} \times \mathrm{I}$ a contraction of the trace of $\overline{K} * H * \overline{K'}$ to the constant loop at p is used. Applying the Homotopy Extension Property, this partial homotopy extends to a homotopy $(M \times \mathrm{I}) \times \mathrm{I} \to M$, whose restriction to $(M \times \mathrm{I}) \times \{1\}$ is L.

We have a deformation of the original H to $K * L * K'$. Since K and K' are admissible, and hence preserve F, it is now sufficient to find a deformation of L to a homotopy that preserves F. Let g and g' be the starting and ending maps of L. Since each L_t fixes p, $g_\#$ and $g'_\#$ are the same automorphism on $\pi_1(M,p)$, and hence on the subgroup $\pi_1(F,p)$. Since F is aspherical, this implies that g and g' are homotopic on F. Now select a homotopy h' between the restrictions of g to F and g' to F. Using an extension L' of h' to an admissible homotopy of g, relative to p and fixed on $\overline{M - N}$, deform L to $L' * (\overline{L'} * L)$. Since L' preserves F, we need only show that there is a deformation of $\overline{L'} * L$ to a homotopy that preserves F. That is, we may assume that for our original L from g to g', each L_t fixes p, and g and g' agree on F.

Since M is Haken, it is aspherical, and we can now use a standard technique to make the homotopy agree with g on F at each level. It is convenient to regard L as a map from $M \times \mathrm{I}$ to $M \times \mathrm{I}$, by using L_t to send $M \times \{t\}$ to $M \times \{t\}$. We will change L only on $\overline{M - N} \times \mathrm{I}$, using the homotopy extension property to extend homotopies defined on subsets of $F \times \mathrm{I}$. Regard F as triangulated with p as a vertex, and let T be a maximal tree in the 1-skeleton $F^{(1)}$. Since p is a deformation retract of T, the restriction of L to $T \times \mathrm{I}$ is homotopic to the identity, so we may assume that each L_t is the identity on T. If σ is a 1-simplex of F not in T, then the restriction of L to $\sigma \times \mathrm{I}$, together with the map defined on $\sigma \times \mathrm{I}$ using $g|_\sigma$ at every level, define a map from the 2-sphere into $M \times \mathrm{I}$. Since $\pi_2(M \times \mathrm{I}) = 0$, these two maps are homotopic relative to $\sigma \times \partial\mathrm{I} \cup \partial\sigma \times \mathrm{I}$, so we may assume that each L_t agrees with g on σ. Repeating for all 1-simplices of F not in T, we may assume that each L_t agrees with g on $F^{(1)}$. Since $\pi_3(M \times \mathrm{I}) = 0$, a similar argument using the 2-simplices of F allows us to make each L_t agree with g on all of F. □

Lemma 10.1.5 shows that the traces of the homotopy at the elements of $\underline{\underline{m}}$ represent the only obstruction to making the homotopy admissible. When the trace does not lie in $\pi_1(F,p)$, there might still exist an admissible homotopy from f to f', but there is no deformation of the given homotopy relative to $M \times \partial\mathrm{I}$ to a map taking p into F, so certainly no deformation that makes it preserve F.

In our applications of lemma 10.1.5, we will be working with boundary patterns whose elements are disjoint. Then, if all the relevant traces are trivial, lemma 10.1.5 can be applied repeatedly to produce an admissible homotopy from f to f'.

Next, we will analyze more carefully what happens when the trace of a homotopy between two of the homotopy equivalences produced by lemma 10.1.2 does not lie in $\pi_1(F)$.

LEMMA 10.1.6. *Let F be an element of $\underline{\underline{m}}$ and let α be the trace of F during a homotopy between two admissible homotopy equivalences f and f' which satisfy conditions (C1)-(C4). Suppose α does not lie in $\pi_1(F)$. Then F is contained in a component $(V, \widehat{\underline{\underline{v}}})$ of the characteristic submanifold of $(M, \overline{\underline{\underline{m}}})$, and $\alpha \in \pi_1(V, p)$. Moreover, either*

(i) *F is an annulus and $(V, \widehat{\underline{\underline{v}}})$ is Seifert-fibered, or*
(ii) *$(M, \underline{\underline{m}})$ is an exceptional case (E2).*

PROOF. Consider any essential closed curve in F passing through p. The images of this curve during the homotopy from f to f' form an admissible singular annulus in $(M, \overline{\underline{\underline{m}}})$, with both ends in F. Since α does not lie in $\pi_1(F, p)$, this annulus is essential. By the enclosing property of the characteristic submanifold, these annuli are admissibly homotopic into Σ. Therefore every essential closed curve in F is homotopic in F into $\Sigma \cap F$. Since F is not simply-connected, there is at least one essential closed curve in F, so $\Sigma \cap F$ is nonempty. Also, since the frontier of Σ is essential, the boundary circles of $\Sigma \cap F$ are essential in F. Since F is a 2-manifold and is not simply-connected, it follows that some component F_1 of $\Sigma \cap F$ is a deformation retract of F. Since Σ is admissibly fibered, and F_1 is an element of the submanifold boundary pattern $\underline{\underline{\sigma}}$ and is not simply-connected, F_1 is either a torus or annulus, or is a lid of an I-bundle component of $(\Sigma, \widehat{\underline{\underline{\sigma}}})$.

Let V be the component of Σ which contains F_1. We will show that $f(F_1) \subseteq F_1$ and $f'(F_1) \subseteq F_1$. Since f and f' satisfy (C1)-(C4), each carries V into V and F into F, so also carries $V \cap F$ into $V \cap F$. Now if F is an annulus, then $F \subset V$ by lemma 2.10.8. In this case, $F_1 = F$ so f and f' preserve it. If F_1 is not an annulus, then only one component of $V \cap F$ can be a deformation retract of F (all other components would be boundary-parallel annuli), so again f and f' preserve F_1.

To complete the proof, it suffices to show that if $(V, \widehat{\underline{\underline{v}}})$ is an I-bundle, or if F_1 is a torus and $(V, \widehat{\underline{\underline{v}}})$ is Seifert-fibered, then $(M, \underline{\underline{m}})$ is an exceptional case (E2).

We have seen that there are essential annuli containing α which are homotopic into V. Therefore $\alpha \in \pi_1(V, p)$, so α is an element of $\pi_1(V) - \pi_1(F_1)$ which normalizes $\pi_1(F_1)$.

Assume for now that $(V, \widehat{\underline{\underline{v}}})$ is an I-bundle and that F_1 is a side of $(V, \widehat{\underline{\underline{v}}})$. Then $F = F_1$, since F_1 is an annulus, and $(V, \widehat{\underline{\underline{v}}})$ must be the I-bundle over the Möbius band, since otherwise $\pi_1(F_1, p)$ equals its own normalizer in $\pi_1(V, p)$. By lemma 2.10.8, Σ and hence V contain a regular neighborhood of F. Since $(V, \widehat{\underline{\underline{v}}})$ is admissibly imbedded in $(M, \overline{\underline{\underline{m}}})$, its lid must also be an element of $\overline{\underline{\underline{m}}}$, and hence $(V, \widehat{\underline{\underline{v}}}) = (M, \overline{\underline{\underline{m}}})$, so $(M, \underline{\underline{m}})$ is an exceptional case (E2).

Assume now that $(V, \widehat{\underline{\underline{v}}})$ is an I-bundle and that F is a lid of $(V, \widehat{\underline{\underline{v}}})$. Since α is not in $\pi_1(F, p)$, $(V, \widehat{\underline{\underline{v}}})$ must be twisted. To show that $(M, \underline{\underline{m}})$ is an exceptional case (E2), it remains to argue that $M = V$.

Regard $(V, \widehat{\underline{\underline{v}}})$ as the quotient of $F \times \mathrm{I}$ by an involution of the form $\tau \times \rho$, where τ is a free orientation-reversing involution of F and $\rho(t) = 1 - t$. We will show that f and $\tau f'$ are homotopic maps from F to F. Let $q\colon F \times \mathrm{I} \to V$ be the quotient map. We choose coordinates on $F \subset V$ so that $q(x, 0) = x$ and $q(x, 1) = \tau(x)$. Now $f\colon F \to F$ admits two lifts to the covering $q\colon F \times \mathrm{I} \to V$. One, say $\widetilde{f}$, carries F

to $F \times \{0\}$, and has the form $\widetilde{f}(x) = (f(x), 0)$. The other carries F to $F \times \{1\}$, sending x to $(\tau f(x), 1)$. The two lifts of f' have the same form. The restriction to F of the homotopy from f to f' is an essential map $(F \times \mathrm{I}, F \times \partial\mathrm{I}) \to (M, \underline{\underline{m}})$ so by the Extended Enclosing Theorem 2.9.2 it is admissibly homotopic into Σ and hence into V. Therefore there is a map $h\colon F \times \mathrm{I} \to V$ which restricts to $f|_F$ and $f'|_F$ on $F \times \{0\}$ and $F \times \{1\}$ respectively, and whose trace at F does not lie in $\pi_1(F)$. Let $\widetilde{h}\colon F \times \mathrm{I} \to F \times \mathrm{I}$ be the lift of h for which $\widetilde{h}(F \times \{0\}) = F \times \{0\}$. Since the trace of the homotopy at F does not lie in $\pi_1(F)$, $\widetilde{h}(F \times \{1\}) = F \times \{1\}$. Thus $\widetilde{h}(x, 0) = (f(x), 0)$ and $\widetilde{h}(x, 1) = (\tau f'(x), 1)$. If $p\colon F \times \mathrm{I} \to F \times \{0\}$ denotes the projection, then $qp\widetilde{h}$ is a homotopy from f to $\tau f'$ as maps from F to F.

Suppose for contradiction that $M \neq V$, and let A be a component of the frontier of V, so A is a side of $(V, \widehat{\underline{\underline{v}}})$. It meets two boundary circles C_1 and C_2 of F, and $\tau(C_1) = C_2$ since they cover the same boundary circle of F/τ. By condition (C4), f and f' restrict to the identity on A, so each preserves C_1 and preserves C_2. Since f and $\tau f'$ are homotopic as maps into F, C_1 is homotopic to $\tau f'(C_1) = C_2$ in F. Therefore F must be an annulus, so $(V, \widetilde{\underline{\underline{v}}})$ is the I-bundle over the Möbius band. By lemma 2.10.8, Σ and hence V contain a regular neighborhood of F. So $\partial V \subseteq \partial M$, which implies that $M = V$ and M is an exceptional case of type (E2).

Now assume that $(V, \underline{\underline{v}})$ is Seifert-fibered and that F_1 is a torus. Since $\pi_1(F)$ does not equal its normalizer in $\pi_1(V)$, lemma 9.3.2 shows that V must be homeomorphic to the I-bundle over the Klein bottle. So $F = \partial V$. Consequently, the frontier of V is empty, and $M = V$. Moreover, $\underline{\underline{v}} = \overline{\underline{\underline{\emptyset}}}$, so $(M, \underline{\underline{m}})$ is an exceptional case (E2). □

With the properties of the trace developed, we can now proceed with proposition 10.1.4.

PROOF OF PROPOSITION 10.1.4. By hypothesis, f and f' induce the same outer automorphism so there is a homotopy from f to f'. We will improve it to one which preserves Σ and $\overline{M - \Sigma}$ and restricts to admissible homotopies on $(\overline{M - \Sigma}, \widetilde{\underline{\underline{m}}} \cup \underline{\underline{\sigma}}'')$ and on $(\Sigma, \underline{\underline{\sigma}}' \cup \underline{\underline{\sigma}}'')$.

Suppose F is an element of $\underline{\underline{m}}$ such that the trace of the homotopy at F is nontrivial. Since $(M, \underline{\underline{m}})$ is not an exceptional case (E2), lemma 10.1.6 shows that F is an annulus that lies in a Seifert-fibered component V of Σ. By lemma 2.10.8, V contains a regular neighborhood of F in ∂M, so F is disjoint from the elements of $\underline{\underline{v}}' \cup \underline{\underline{v}}''$. By lemma 10.1.6, the trace at F lies in $\pi_1(V)$. It can be represented by a loop in V based at p that misses the exceptional fibers. Define a partial homotopy on F starting at f; on F it first shrinks $f(F)$ down to the fiber that contains p, then moves this fiber through a path of fibers in such a way that p moves around the trace, returning to the original fiber. Also, outside a regular neighborhood of F, define the partial homotopy to equal f at every level. By the homotopy extension property, this partial homotopy extends to a homotopy K_F of M. Assuming that the regular neighborhood of F is sufficiently small, each level of K_F preserves Σ and agrees with f on $|\underline{\underline{m}} - \{F\}| \cup \overline{M - \Sigma}$.

Let K be the product of the homotopies K_F for all F for which the trace of the original homotopy was nontrivial. Deform the original homotopy (relative to $M \times \partial\mathrm{I}$) to a product $K * K'$, where K' starts by doing the reverse of K, then does the original homotopy. For every element of $\underline{\underline{m}}$, the trace during K' is trivial, so by lemma 10.1.5 there is a deformation of K' to a homotopy which is admissible for

$(M, \underline{\underline{m}})$. By the Homotopy Splitting Theorem 2.11.4, we may then deform K' so that each level, K' preserves Σ and $\overline{M - \Sigma}$. Then, $K * K'$ satisfies the conclusion of proposition 10.1.4. □

Recall that $S_1, \ldots, S_m$ are the components of $\overline{M - \Sigma}$, with their proper boundary patterns s_j as submanifolds of $(M, \underline{\underline{m}})$. We will now define a homomorphism

$$\Phi\colon \operatorname{Out}_1(\pi_1(M), \pi_1(\underline{\underline{m}})) \to \prod_{j=1}^{m} \mathcal{E}(S_j, \underline{\underline{s_j}})$$

where $\mathcal{E}(S_j, \underline{\underline{s_j}})$ denotes the group of path components of admissible homotopy equivalences of $(S_j, \underline{\underline{s_j}})$. If $M = \Sigma$, then $\prod_{j=1}^{m} \mathcal{E}(S_j, \underline{\underline{s_j}})$ is the trivial group and Φ is the zero homomorphism. Otherwise, given an element ϕ of $\operatorname{Out}_1(\pi_1(M), \pi_1(\underline{\underline{m}}))$, use lemma 10.1.2 to choose an admissible homotopy equivalence inducing ϕ and satisfying (C1)-(C4). By proposition 10.1.4, the restriction of this homotopy equivalence to each $(S_j, \underline{\underline{s_j}})$ is a well-defined element of $\mathcal{E}(S_j, \underline{\underline{s_j}})$. The element with these coordinates is defined to be $\Phi(\phi)$.

LEMMA 10.1.7. *The image of Φ is finite.*

PROOF. Since the homotopy equivalences used to define Φ satisfy condition (C4), the coordinate of $\Phi(\phi)$ in each $\mathcal{E}(S_j, \underline{\underline{s_j}})$ lies in the subgroup $\mathcal{H}(S_j, \underline{\underline{s_j}})$ of elements that are realizable by orientation-preserving admissible homeomorphisms.

As explained in section 2.9, each $(S_j, \underline{\underline{s_j}})$ is either simple or is an I-bundle over a square, annulus, or torus. If $(S_j, \underline{\underline{s_j}})$ is simple, then by the Finite Mapping Class Group Theorem 2.11.2, $\mathcal{H}(S_j, \underline{\underline{s_j}})$ is finite. The same is true if $(S_j, \underline{\underline{s_j}})$ is an I-bundle over a square or annulus, so it remains only to consider the case when $(S_j, \underline{\underline{s_j}})$ is an I-bundle over a torus, say $(S_j, \underline{\underline{s_j}}) = (T \times \mathrm{I}, \underline{\underline{\emptyset}})$. Such a component cannot meet ∂M, since if so it would have to contain a torus component of ∂M (since it has the boundary pattern $\underline{\underline{\emptyset}}$, but by lemma 2.10.8, every torus boundary component must lie in Σ.

Let V_1 and V_2 be the components of Σ that meet S_j (possibly $V_1 = V_2$); they are Seifert-fibered since they meet $\overline{M - \Sigma}$ in tori, whereas the frontier of any I-bundle component of Σ must consist of annuli and squares. Suppose f is an admissible homotopy equivalence which satisfies conditions (C1)-(C4). Then f preserves V_1 and V_2 and restricts to a homeomorphism on each component of their frontiers. Neither $(V_i, \underline{\underline{\widehat{v_i}}})$ is $(\mathrm{S}^1 \times \mathrm{S}^1 \times \mathrm{I}, \underline{\underline{\emptyset}})$, for if so, then either $V_1 = V_2$, which would imply that M is closed (it would be a torus bundle over the circle), or $V_1 \neq V_2$, in which case one of them could be refibered so that the fibering on $V_1 \cup V_2$ extended over S_j violating the maximality of Σ. So by the Fiber-preserving Self-map Theorem 2.8.6, f must preserve the fiber of each $(V_i, \underline{\underline{\widehat{v_i}}})$ up to homotopy. The fibers of V_1 and V_2 must represent linearly independent elements of $\pi_1(T)$, since otherwise their Seifert fiberings would extend over S_j. The only elements of $\operatorname{Out}(\pi_1(T)) \cong \mathrm{GL}(2, \mathbb{Z})$ which have determinant 1 and preserve two linearly independent elements are the identity I_2 and $-I_2$. Therefore (since f is orientation-preserving on $(S_j, \underline{\underline{s_j}})$) there are at most two possibilities for the restriction of f to S_j; it can be admissibly isotopic to the identity or to a homeomorphism preserving the levels $T \times \{t\}$ and sending each element of $\pi_1(S_j)$ to its negative. □

Define $\mathrm{Out}_2(\pi_1(M), \pi_1(\underline{\underline{m}}))$ to be the kernel of Φ. Lemma 10.1.7 shows that $\mathrm{Out}_2(\pi_1(M), \pi_1(\underline{\underline{m}}))$ has finite index in $\mathrm{Out}(\pi_1(M), \pi_1(\underline{\underline{m}}))$. It consists of the elements of $\mathrm{Out}(\pi_1(M), \pi_1(\underline{\underline{m}}))$ induced by admissible homotopy equivalences which satisfy (C1)-(C4) and whose restrictions to $\overline{M - \Sigma}$ are admissibly homotopic to the identity. Provided that $(M, \underline{\underline{m}})$ is not an exceptional case (E2), proposition 10.1.4 shows that restricting such a homotopy equivalence to Σ determines a well-defined homomorphism

$$\Psi\colon \mathrm{Out}_2(\pi_1(M), \pi_1(\underline{\underline{m}})) \to \prod_{i=1}^{n} \mathcal{E}(V_i, \underline{\underline{v_i}}', \underline{\underline{v_i}}'') \,.$$

LEMMA 10.1.8. *Assume that $(M, \underline{\underline{m}})$ is not an exceptional case (E1) or (E2). Then Ψ is surjective.*

PROOF. An element of $\prod_{i=1}^{n} \mathcal{E}(V_i, \underline{\underline{v_i}}', \underline{\underline{v_i}}'')$ can be represented by an admissible homotopy equivalence of $(\Sigma, \underline{\underline{\sigma}}' \cup \underline{\underline{\sigma}}'')$ which is the identity on $|\underline{\underline{\sigma}}''|$. This extends using the identity map on $\overline{M - \Sigma}$ to a homotopy equivalence f of M which satisfies (C1)-(C4), except that it may fail to be admissible, by failing to preserve annuli of $\underline{\underline{m}}$ that lie in Seifert-fibered components of Σ. We will show that it can be selected to preserve these annuli as well. Then, its induced automorphism is an element of $\mathrm{Out}_2(\pi_1(M), \pi_1(\underline{\underline{m}}))$ which Ψ maps to the given element of $\prod_{i=1}^{n} \mathcal{E}(V_i, \underline{\underline{v_i}}', \underline{\underline{v_i}}'')$.

Each element of $\underline{\underline{m}}$ that is not an annulus in a Seifert-fibered component of Σ is actually mapped to itself by f, since it is contained in $|\underline{\underline{\sigma}}'| \cup |\underline{\underline{\widetilde{m}}}|$. Now consider an annulus F of $\underline{\underline{m}}$ contained in a Seifert-fibered component $(V, \underline{\underline{\widehat{v}}})$. By lemma 2.10.8, V contains a regular neighborhood of F.

Suppose that $M \neq V$, so the frontier of V is nonempty. Since f is isotopic to the identity on each component of the frontier of Σ, the fiber of V is preserved up to homotopy. Since the core circle of the annulus is a fiber, f is homotopic, without changing it on $\overline{M - \Sigma}$ or the rest of $|\underline{\underline{m}}|$, to a map preserving F.

From now on, assume that $M = V$, so $\underline{\underline{v}}'' = \underline{\underline{\emptyset}}$. If the Fiber-preserving Self-map Theorem 2.8.6 applies to $(V, \underline{\underline{v}}')$, then the unoriented fiber of V is preserved up to homotopy by f, so f may be changed by homotopy so that it preserves all annuli of $\underline{\underline{m}}$. If the Fiber-preserving Self-map Theorem 2.8.6 does not apply to $(V, \underline{\underline{v}}')$, then since $(V, \underline{\underline{v}}')$ is Seifert-fibered, either $(V, \overline{\underline{\underline{v}}'})$ does not have useful boundary pattern, or $(V, \overline{\underline{\underline{v}}'}) = (\mathrm{S}^1 \times \mathrm{S}^1 \times \mathrm{I}, \underline{\underline{\emptyset}})$.

Suppose that $(V, \overline{\underline{\underline{v}}'})$ does not have useful boundary pattern. Since $(V, \underline{\underline{\widehat{v}}})$ is Seifert-fibered, and $\underline{\underline{v}}''$ is empty, no element of $\underline{\underline{v}}'$ can be an annulus (for since $M = V$, it would be an annulus of $\underline{\underline{m}}$, so by definition would not be in $\underline{\underline{v}}'$). So $\overline{\underline{\underline{v}}'}$ can only fail to be useful when V has a compressible boundary torus. Since V is irreducible, this implies that V is a solid torus, so $(M, \underline{\underline{m}})$ is a fibered solid torus and $\underline{\underline{v}}'$ is empty. Now $\mathcal{E}(V, \underline{\underline{\emptyset}}, \underline{\underline{\emptyset}})$ has two elements, classified by their effect on $\pi_1(V)$, and each element is realizable by an admissible homotopy equivalence which preserves each element of $\underline{\underline{m}}$. (The nontrivial element is induced by the homeomorphism from $\mathrm{S}^1 \times D^2$ to $\mathrm{S}^1 \times D^2$ which sends (θ, z) to $(\overline{\theta}, \overline{z})$, where the bars denote complex conjugation. Since this preserves the unoriented fiber up to homotopy, it is homotopic to a homotopy equivalence that preserves each annulus of $\underline{\underline{m}}$.) Therefore $\Psi\colon \mathrm{Out}_2(\pi_1(M), \pi_1(\underline{\underline{m}})) \to \mathcal{E}(V, \underline{\underline{\emptyset}}, \underline{\underline{\emptyset}})$ is an isomorphism of groups of order two.

Finally, suppose that $(V,\overline{\underline{\underline{v}}}')=(\mathrm{S}^1\times\mathrm{S}^1\times\mathrm{I},\overline{\underline{\underline{\emptyset}}})$. Since $(M,\underline{\underline{m}})$ is not an exceptional case of type (E1), $\underline{\underline{m}}$ cannot contain annuli. Therefore $\underline{\underline{m}}=\underline{\underline{v}}'$, which is empty or consists of one or both boundary components of V. So we have $\mathrm{Out}_2(\pi_1(M),\pi_1(\underline{\underline{m}}))\cong \mathrm{GL}(2,\mathbb{Z})$, $\mathcal{E}(V,\underline{\underline{v}}',\underline{\underline{\emptyset}})\cong \mathrm{GL}(2,\mathbb{Z})$, and $\Psi\colon \mathrm{GL}(2,\mathbb{Z})\to\mathcal{E}(V,\underline{\underline{v}}',\underline{\underline{\emptyset}})$ is an isomorphism. □

The following theorem details the properties of Ψ. From it, we will deduce theorem 10.1.1.

THEOREM 10.1.9. *Let M be a compact, orientable, irreducible 3-manifold, with a boundary pattern $\underline{\underline{m}}$ whose completion is useful and nonempty. Assume that the elements of $\underline{\underline{m}}$ are disjoint and that $(M,\underline{\underline{m}})$ is not an exceptional case (E1) or (E2), and let ϕ be an element of $\mathrm{Out}_2(\pi_1(M),\pi_1(\underline{\underline{m}}))$. Let $\mathcal{H}(V_i,\underline{\underline{v_i}}',\underline{\underline{v_i}}'')$ be the subgroup of elements of $\mathcal{E}(V_i,\underline{\underline{v_i}}',\underline{\underline{v_i}}'')$ realizable by orientation-preserving homeomorphisms.*

(i) *If ϕ can be induced by an orientation-preserving homeomorphism that preserves each element of $\underline{\underline{m}}$, then $\Psi(\phi)$ lies in $\prod_{i=1}^n\mathcal{H}(V_i,\underline{\underline{v_i}}',\underline{\underline{v_i}}'')$.*

(ii) *If $\Psi(\phi)$ lies in $\prod_{i=1}^n\mathcal{H}(V_i,\underline{\underline{v_i}}',\underline{\underline{v_i}}'')$, then ϕ can be induced by an orientation-preserving homeomorphism which preserves Σ and $\overline{M-\Sigma}$, restricts to the identity map on $\overline{M-\Sigma}$, and preserves each element of $\underline{\underline{m}}$ that is not an annulus in a Seifert-fibered component of Σ.*

PROOF. To prove part (i), suppose $\phi\in\mathrm{Out}_2(\pi_1(M),\pi_1(\underline{\underline{m}}))$. By definition, $\Psi(\phi)$ is obtained as the restriction of a homotopy equivalence f inducing ϕ, which satisfies (C1)-(C4). Suppose ϕ can be induced by an orientation-preserving homeomorphism h of M that preserves each element of $\underline{\underline{m}}$. Since h is a homeomorphism, it is admissible for $(M,\overline{\underline{\underline{m}}})$. Because the characteristic submanifold of $(M,\overline{\underline{\underline{m}}})$ is unique up to admissible isotopy, we may deform h by admissible isotopy to assume that it preserves Σ. Since $\phi\in\mathrm{Out}_2(\pi_1(M),\pi_1(\underline{\underline{m}}))$, the kernel of Φ, the restriction of h to $\overline{M-\Sigma}$ is admissibly homotopic to the identity. Using lemma 2.12.1, we may assume that the admissible homotopy is an isotopy on the frontier of Σ. Therefore it extends to an admissible homotopy of M which is an isotopy on Σ. The homeomorphism resulting from this homotopy restricts to show that $\Psi(\phi)$ lies in $\prod_{i=1}^n\mathcal{H}(V_i,\underline{\underline{v_i}}',\underline{\underline{v_i}}'')$.

For (ii), suppose that $\Psi(\phi)$ lies in $\prod_{i=1}^n\mathcal{H}(V_i,\underline{\underline{v_i}}',\underline{\underline{v_i}}'')$. By definition of Ψ, there exists an admissible homotopy equivalence f of $(M,\underline{\underline{m}})$ which satisfies (C1)-(C4) and restricts to representatives of the coordinates of $\Psi(\phi)$ on the V_i. Since ϕ lies in $\mathrm{Out}_2(\pi_1(M),\pi_1(\underline{\underline{m}}))$, the restriction of f to $\overline{M-\Sigma}$ is admissibly homotopic to the identity. Since $\Psi(\phi)$ lies in $\prod_{i=1}^n\mathcal{H}(V_i,\underline{\underline{v_i}}',\underline{\underline{v_i}}'')$, the restriction of f to each $(V_i,\underline{\underline{v_i}}'\cup\underline{\underline{v_i}}'')$ is admissibly homotopic to an orientation-preserving homeomorphism. Extend such a homotopy to $\overline{M-\Sigma}$ to get a homotopy from f to a map g, which preserves each element of $\underline{\underline{m}}$ that is not an annulus in a Seifert-fibered component of Σ. The restriction of g to $\overline{M-\Sigma}$ is admissibly homotopic to the identity. Again using lemma 2.12.1, we may assume that the homotopy is an isotopy on $|\underline{\underline{\sigma}}''|$. Such a homotopy extends to an isotopy on $\overline{M-\Sigma}$, fixed outside a neighborhood of the frontier of Σ, which results in a homeomorphism h which is the identity on $\overline{M-\Sigma}$ and preserves each element of $\underline{\underline{m}}$ which is not an annulus in a Seifert-fibered component of Σ. □

Using theorem 10.1.9, we can now prove theorem 10.1.1.

PROOF OF THEOREM 10.1.1. Assume for now that M is not a 3-ball or $\mathrm{S}^1 \times \mathrm{S}^1 \times \mathrm{I}$ (in particular, $(M, \underline{\underline{m}})$ is not an exceptional case (E1)), and is not an exceptional case (E2).

If $\mathcal{R}(M, \underline{\underline{m}})$ has finite index in $\mathrm{Out}(\pi_1(M), \pi_1(\underline{\underline{m}}))$, then a finite-index subgroup of $\mathrm{Out}_2(\pi_1(M), \pi_1(\underline{\underline{m}}))$ can be realized by orientation-preserving homeomorphisms that preserve each element of $\underline{\underline{m}}$. By lemma 10.1.8, Ψ is surjective, so theorem 10.1.9(i) shows that $\prod_{i=1}^{n} \mathcal{H}(V_i, \underline{\underline{v_i}}', \underline{\underline{v_i}}'')$ has finite index in $\prod_{i=1}^{n} \mathcal{E}(V_i, \underline{\underline{v_i}}', \underline{\underline{v_i}}'')$.

For the converse, we assume that $\prod_{i=1}^{n} \mathcal{H}(V_i, \underline{\underline{v_i}}', \underline{\underline{v_i}}'')$ has finite index in $\prod_{i=1}^{n} \mathcal{E}(V_i, \underline{\underline{v_i}}', \underline{\underline{v_i}}'')$. We will show that the automorphisms in a subgroup of finite index in $\mathrm{Out}_2(\pi_1(M), \pi_1(\underline{\underline{m}}))$ (and hence of finite index in $\mathrm{Out}(\pi_1(M), \pi_1(\underline{\underline{m}}))$) can be realized by admissible orientation-preserving homeomorphisms of $(M, \underline{\underline{m}})$.

By theorem 10.1.9(ii), each element in a finite-index subgroup of $\mathrm{Out}_2(\pi_1(M), \pi_1(\underline{\underline{m}}))$ can be realized by an orientation-preserving homeomorphism h that preserves Σ and restricts to the identity map on $\overline{M - \Sigma}$, and preserves each element of $\underline{\underline{m}}$ that is not an annulus in a Seifert-fibered component of Σ.

Since M is not $\mathrm{S}^1 \times \mathrm{S}^1 \times \mathrm{I}$, no torus of ∂M is homotopic into another torus boundary component. This implies that the permutation that h induces on torus boundary components of V depends only on its induced outer automorphism. So by passing to a smaller finite-index subgroup of $\mathrm{Out}_2(\pi_1(M), \pi_1(\underline{\underline{m}}))$, we may assume that h preserves each torus boundary component of M.

Suppose that $M \neq \Sigma$. Then h is the identity on the frontier of Σ. On any component of $\Sigma \cap \partial M$ that is an annulus A, h fixes the boundary circles of A, so h can be changed by isotopy in a neighborhood of A so that it preserves each fiber in A, and in particular it preserves any annulus of $\underline{\underline{m}}$ that lies in A. On any torus component of $\Sigma \cap \partial M$, h preserves the oriented fiber up to homotopy, and is orientation-preserving. So h can be changed by isotopy in a neighborhood of the torus boundary components to take each fiber in $\Sigma \cap \partial M$ to itself, and hence preserve each element of $\underline{\underline{m}}$ in these torus boundary components.

Suppose that $M = \Sigma$. By the Fiber-preserving Self-map Theorem 2.8.6, h preserves the unoriented fiber of M up to homotopy. By passing to a smaller subgroup in $\mathrm{Out}_2(\pi_1(M), \pi_1(\underline{\underline{m}}))$ we may assume that h preserves the oriented fiber up to homotopy. Therefore it is isotopic to a homeomorphism that preserves each fiber in each boundary component, hence each element of $\underline{\underline{m}}$.

It remains to examine the cases when M is a 3-ball or $\mathrm{S}^1 \times \mathrm{S}^1 \times \mathrm{I}$. If M is the 3-ball, then $\mathrm{Out}(\pi_1(M), \pi_1(\underline{\underline{m}}))$ and each $\mathcal{E}(V_i, \underline{\underline{v_i}}', \underline{\underline{v_i}}'')$ is finite, so both indices in the theorem are finite.

Assume now that M is homeomorphic to $\mathrm{S}^1 \times \mathrm{S}^1 \times \mathrm{I}$.

Suppose first that $\underline{\underline{m}}$ contains no annuli, that is, $\overline{\underline{\underline{m}}} = \underline{\underline{\emptyset}}$. Then $V = M$, $\underline{\underline{v}}' = \underline{\underline{m}}$, and $\underline{\underline{v}}'' = \underline{\underline{\emptyset}}$. In this case, $\mathcal{E}(V, \underline{\underline{v}}', \underline{\underline{\emptyset}}) \cong \mathrm{Out}(\pi_1(M), \pi_1(\underline{\underline{m}}))$ and $\mathcal{H}(V, \underline{\underline{v}}', \underline{\underline{\emptyset}}) \cong \mathcal{R}(M, \underline{\underline{m}})$, so the theorem is a tautology. In fact, $\mathcal{E}(V, \underline{\underline{v}}', \underline{\underline{\emptyset}}) \cong \mathrm{GL}(2, \mathbb{Z})$, while $\mathcal{H}(V, \underline{\underline{v}}', \underline{\underline{\emptyset}})$ corresponds to $\mathrm{GL}(2, \mathbb{Z})$, if $\underline{\underline{v}}'$ is empty, and to $\mathrm{SL}(2, \mathbb{Z})$ if not (for if $\underline{\underline{v}}'$ is nonempty, the homeomorphism cannot interchange the boundary components, and being orientation-preserving it must induce an element of $\mathrm{SL}(2, \mathbb{Z})$). So the indices are finite.

Suppose $\underline{\underline{m}}$ contains annuli, not all of which are pairwise homotopic. Then Σ has two components, isotopic to $\mathrm{S}^1 \times \mathrm{S}^1 \times [0, 1/3]$ and to $\mathrm{S}^1 \times \mathrm{S}^1 \times [2/3, 1]$. For each of these components, $\underline{\underline{v}}' = \underline{\underline{\emptyset}}$ and $\underline{\underline{v}}''$ is a single boundary component, so each

$\mathcal{E}(V, \underline{\underline{v}}', \underline{\underline{v}}'')$ is the trivial group. On the other hand, lemma 9.1.3(ii) shows that $\mathrm{Out}(\pi_1(M), \pi_1(\underline{\underline{m}}))$ is finite. So the theorem holds in this case.

Finally, suppose that $(M, \underline{\underline{m}})$ contains annuli, all of which are pairwise homotopic (that is, $(M, \underline{\underline{m}})$ is an exceptional case (E1)). Then $M = V$, $\underline{\underline{v}}'' = \underline{\underline{\emptyset}}$ and $\overline{\underline{\underline{v}}'} = \overline{\underline{\underline{\emptyset}}}$. Since $\underline{\underline{m}}$ contains at least one annulus, lemma 9.1.3(i) shows that $\mathrm{Out}(\pi_1(M), \pi_1(\underline{\underline{m}}))$ has an infinite cyclic subgroup of finite index, generated by a Dehn twist homeomorphism which preserves each element of $\underline{\underline{m}}$. So $\mathcal{R}(M, \underline{\underline{m}})$ has finite index in $\mathrm{Out}(\pi_1(M), \pi_1(\underline{\underline{m}}))$. Now, $\underline{\underline{v}}'$ is either empty or is one of the boundary components of M, so as in the cases when $\overline{\underline{\underline{m}}} = \overline{\underline{\underline{\emptyset}}}$, $\mathcal{H}(V, \underline{\underline{v}}', \underline{\underline{\emptyset}})$ has index at most 2 in $\mathcal{E}(V, \underline{\underline{v}}', \underline{\underline{\emptyset}})$. So the theorem holds (somewhat by accident) in this case as well.

The remaining case is when $(M, \underline{\underline{m}})$ is an exceptional case (E2), that is, when $(M, \overline{\underline{\underline{m}}})$ is an admissibly fibered twisted I-bundle over a nonorientable surface $(N, \underline{\underline{n}})$. We have $V = M$, $\underline{\underline{v}}' = \underline{\underline{m}}$, and $\underline{\underline{v}}'' = \underline{\underline{\emptyset}}$. By lemma 10.1.2, the elements in a finite-index subgroup of $\mathrm{Out}(\pi_1(M), \pi_1(\underline{\underline{m}}))$ can be induced by admissible homotopy equivalences of $(M, \underline{\underline{m}})$.

Suppose first that the lid of $(M, \overline{\underline{\underline{m}}})$ is not contained in $\underline{\underline{m}}$. Since the elements of $\underline{\underline{m}}$ are disjoint, $\underline{\underline{m}}$ consists exactly of the sides of $(M, \overline{\underline{\underline{m}}})$, and $\underline{\underline{n}}$ consists of the boundary components of N. Lemma 2.11.3 shows that any admissible homotopy equivalence of $(M, \underline{\underline{m}})$ is admissibly homotopic to a homeomorphism, so $\mathcal{R}(M, \underline{\underline{m}})$ has finite index in $\mathrm{Out}(\pi_1(M), \pi_1(\underline{\underline{m}}))$. Lemma 2.11.3 also shows that $\mathcal{H}(V, \underline{\underline{v}}', \underline{\underline{\emptyset}})$ equals $\mathcal{E}(V, \underline{\underline{v}}', \underline{\underline{\emptyset}})$, so the theorem holds in this case.

Assume now that the lid of $(M, \overline{\underline{\underline{m}}})$ is contained in $\underline{\underline{m}}$. Since the elements of $\underline{\underline{m}}$ are disjoint, the lid is the only element of $\underline{\underline{m}}$, and $\underline{\underline{n}} = \underline{\underline{\emptyset}}$. We will use the following commutative diagram, in which the vertical maps are inclusions, and the second horizontal map of each row takes homotopy classes to their induced outer automorphisms.

$$\begin{array}{ccccc}
\mathcal{H}(V, \underline{\underline{v}}', \underline{\underline{\emptyset}}) & \xrightarrow{=} & \mathcal{H}(M, \underline{\underline{m}}) & \longrightarrow & \mathcal{R}(M, \underline{\underline{m}}) \\
\downarrow & & \downarrow & & \downarrow \\
\mathcal{E}(V, \underline{\underline{v}}', \underline{\underline{\emptyset}}) & \xrightarrow{=} & \mathcal{E}(M, \underline{\underline{m}}) & \longrightarrow & \mathrm{Out}(\pi_1(M), \pi_1(\underline{\underline{m}}))
\end{array}$$

Suppose first that $\mathcal{H}(V, \underline{\underline{v}}', \underline{\underline{\emptyset}})$ has finite index in $\mathcal{E}(V, \underline{\underline{v}}', \underline{\underline{\emptyset}})$. Since the homomorphism in the bottom row has image of finite index, the diagram shows that $R(M, \underline{\underline{m}})$ has finite index in $\mathrm{Out}(\pi_1(M), \pi_1(\underline{\underline{m}}))$.

Conversely, assume $\mathcal{R}(M, \underline{\underline{m}})$ has finite index in $\mathrm{Out}(\pi_1(M), \pi_1(\underline{\underline{m}}))$. We claim that $\mathcal{H}(M, \underline{\underline{m}})$ is the full preimage of $\mathcal{R}(M, \underline{\underline{m}})$. Let f be an admissible homotopy equivalence of $(M, \underline{\underline{m}})$ for which $f_\#$ lies in $\mathcal{R}(M, \underline{\underline{m}})$. Then there is an admissible homeomorphism h of $(M, \underline{\underline{m}})$ with $h_\# = f_\#$, so $(fh^{-1})_\#$ is the identity automorphism of $\pi_1(M)$. Let $p\colon M \to N$ be the projection, and fix an admissible section $s\colon (N, \underline{\underline{\emptyset}}) \to (M, \underline{\underline{m}})$. Then $p(fh^{-1})s$ induces the identity outer automorphism on $\pi_1(N)$, and since N is aspherical, $p(fh^{-1})s$ is homotopic to the identity on N. By part (iv) of proposition 10.2.2 below, it follows that fh^{-1} is admissibly homotopic to a homeomorphism of M, and hence that f is. This proves the claim. Now, the diagram implies that $\mathcal{H}(V, \underline{\underline{v}}', \underline{\underline{\emptyset}})$ has finite index in $\mathcal{E}(V, \underline{\underline{v}}', \underline{\underline{\emptyset}})$. □

10.2. Realizing homotopy equivalences of I-bundles

In this section and the next, we examine the groups $\mathcal{E}(V, \underline{\underline{v}}', \underline{\underline{v}}'')$ that appear in theorem 10.1.1. The main results tell exactly when $\mathcal{H}(V, \underline{\underline{v}}', \underline{\underline{v}}'')$ has finite index in $\mathcal{E}(V, \underline{\underline{v}}', \underline{\underline{v}}'')$. Throughout this section we assume that $(V, \overline{\underline{\underline{v}}})$ is an I-bundle or S^1-bundle. The remaining Seifert-fibered cases are examined in section 10.3.

PROPOSITION 10.2.1. *Let $(V, \underline{\underline{v}})$ be an irreducible 3-manifold with boundary pattern whose completion is useful and nonempty, and let $(B, \underline{\underline{b}})$ be a connected 2-manifold with boundary pattern. Suppose that either*

(a) *there is an admissible I-fibering of $(V, \overline{\underline{\underline{v}}})$ over $(B, \overline{\underline{\underline{b}}})$, with projection $p \colon V \to B$ for which $\{p^{-1}(k) \mid k \in \underline{\underline{b}}\} = \{G \in \underline{\underline{v}} \mid G$ is not a lid of $(V, \overline{\underline{\underline{v}}})\}$, or*

(b) *$\partial V \neq \emptyset$, $(V, \overline{\underline{\underline{v}}}) \neq (S^1 \times S^1 \times I, \overline{\underline{\underline{\emptyset}}})$, and there is an admissible Seifert fibering $p \colon (V, \underline{\underline{v}}) \to (B, \underline{\underline{b}})$ with no exceptional fibers.*

Let $\underline{\underline{v}}'$ and $\underline{\underline{v}}''$ be two boundary patterns on V with $\underline{\underline{v}}' \cup \underline{\underline{v}}'' = \underline{\underline{v}}$ and which do not share any common elements. In the I-bundle case, assume that no element of $\underline{\underline{v}}''$ is a lid. Then:

(1) *If $|\underline{\underline{b}}| = \partial B$, then the index of $\mathcal{H}(V, \underline{\underline{v}}', \underline{\underline{v}}'')$ in $\mathcal{E}(V, \underline{\underline{v}}', \underline{\underline{v}}'')$ is finite.*

(2) *If $|\underline{\underline{b}}| \neq \partial B$, then the index is finite if and only if one of the following holds:*
   - (i) *B is a disk.*
   - (ii) *B is an annulus or Möbius band, and each boundary circle of B contains at most one component of $|\underline{\underline{b}}|$ that is an arc.*
   - (iii) *B is a disk with two holes, and $|\underline{\underline{b}}|$ consists of two boundary circles.*
   - (iv) *B is a torus with one hole and $|\underline{\underline{b}}|$ is empty.*

To prove proposition 10.2.1, we first show in proposition 10.2.2 that an admissible homotopy equivalence of $(V, \underline{\underline{v}})$ is admissibly homotopic to a homeomorphism if and only if a corresponding admissible homotopy equivalence of the base surface (obtained by using a section of the bundle, applying the homotopy equivalence of V, and then projecting) is admissibly homotopic to a homeomorphism. With some additional argument, which constitutes the actual proof of proposition 10.2.1, proposition 10.2.2 implies that $\mathcal{H}(V, \underline{\underline{v}}', \underline{\underline{v}}'')$ has finite index in $\mathcal{E}(V, \underline{\underline{v}}', \underline{\underline{v}}'')$ if and only if $\mathcal{H}(B, \underline{\underline{b}})$ has finite index in $\mathcal{E}(B, \underline{\underline{b}})$, where $\mathcal{H}(B, \underline{\underline{b}})$ is, as usual, the subgroup realizable by orientation-preserving homeomorphisms. So we are reduced to determining the $(B, \underline{\underline{b}})$ for which the index is finite (in particular, the Baer-Nielsen theorem says the index is finite whenever $|\underline{\underline{b}}| = \partial B$). This is accomplished in two lemmas: the finite index cases are given in lemma 10.2.3 and the infinite index cases in lemma 10.2.4. These lemmas will also be used in section 10.3. In order to state and prove lemma 10.2.4, we formally define sweeps of 2-manifolds, the homotopy equivalences which are the analogue of the wrapping homotopy equivalences of 3-manifolds $h(\alpha)$ used in chapter 9. To deduce proposition 10.2.1, we must also define sweeps of fibered 3-manifolds, which are fiber-preserving homotopy equivalences which project to sweeps of their base surfaces. They too play a major role in section 10.3. The proof of proposition 10.2.1 closes the section.

PROPOSITION 10.2.2. *Let $(V, \underline{\underline{v}})$, $(B, \underline{\underline{b}})$, and $p \colon V \to B$ be as in the statement of proposition 10.2.1, and fix an admissible section $s \colon (B, \underline{\underline{b}}) \to (V, \underline{\underline{v}})$ for p. Suppose that $f \colon (V, \underline{\underline{v}}) \to (V, \underline{\underline{v}})$ is an admissible homotopy equivalence. If $(V, \overline{\underline{\underline{v}}})$ is an I-bundle, then assume that f takes each element of $\underline{\underline{v}}$ that is a lid to a lid. Then*

(i) *f is admissibly homotopic to a fiber-preserving map,*

(ii) *pfs is an admissible homotopy equivalence of $(B, \underline{\underline{b}})$,*

(iii) *sending* f *to* pfs *defines a homomorphism* $P\colon \mathcal{E}(V,\underline{\underline{v}}) \to \mathcal{E}(B,\underline{\underline{b}})$, *and*
(iv) f *is admissibly homotopic to a homeomorphism if and only if* pfs *is admissibly homotopic to a homeomorphism.*
(v) *Suppose that* $(V,\overline{\underline{\underline{v}}})$ *is an* I*-bundle. If some lid of* $(V,\overline{\underline{\underline{v}}})$ *is not an element of* $\underline{\underline{v}}$, *then* P *is injective. If every lid of* $(V,\overline{\underline{\underline{v}}})$ *is an element of* $\underline{\underline{v}}$, *then* P *has kernel of order* 2 *generated by a homeomorphism* r *which is reflection in the* I*-fibers, except in the case when* $(V,\underline{\underline{v}})$ *is the I-bundle over the Möbius band and* $\underline{\underline{v}}$ *consists of the lid alone, in which case* P *is injective.*

Note that in the Seifert-fibered case, the hypothesis of no exceptional fibers guarantees that the section s exists. Part (v) is not needed in our present work, but it has been included as potentially useful information about P. In a remark at the end of this section, we explain that P need not be surjective, but it always has image of finite index.

PROOF. Suppose that $(V,\overline{\underline{\underline{v}}})$ is Seifert-fibered or is an I-bundle having every lid in $\underline{\underline{v}}$. By the Fiber-preserving Self-map Theorem 2.8.6, we may assume that f is fiber-preserving. If $(V,\overline{\underline{\underline{v}}})$ is an I-bundle but not every lid is in $\underline{\underline{v}}$, then $\underline{\underline{b}}$ must be a complete boundary pattern on B (otherwise some element of $\overline{\underline{\underline{v}}}$ would contain a lid as a proper subset). So, in this case, lemma 2.11.3 implies that f is admissibly homotopic to a fiber-preserving homeomorphism. This completes the proof of statement (i).

Let f be an admissible homotopy equivalence of $(V,\underline{\underline{v}})$. By statement (i), we may assume that f is fiber-preserving and has a fiber-preserving admissible homotopy inverse g. Since f and g are fiber-preserving, $pfspgs{=}pfgs$. Since fg is admissibly homotopic to the identity map, so is $pfgs$. Thus, pgs is an admissible homotopy inverse for pfs. Statement (ii) follows immediately.

The function P which takes the admissible homotopy class of f in $\mathcal{E}(V,\underline{\underline{v}})$ to the admissible homotopy class of pfs in $\mathcal{E}(B,\underline{\underline{b}})$ is well-defined. Given any two elements of $\mathcal{E}(V,\underline{\underline{v}})$, statement (i) implies that we may choose fiber-preserving representatives f and g. In this case, P is a homomorphism, since $pfspgs{=}pfgs$, and this verifies statement (iii).

Assume that f is admissibly homotopic to a homeomorphism g. Since g is a homeomorphism, it must also be admissible for $(V,\overline{\underline{\underline{v}}})$. Suppose $(V,\overline{\underline{\underline{v}}})$ is an I-bundle. We claim that g must carry lids of $(V,\overline{\underline{\underline{v}}})$ to lids. By hypothesis, g takes each element of $\underline{\underline{v}}$ that is a lid to a lid. If some lid is not an element of $\underline{\underline{v}}$, then as noted above, $\underline{\underline{b}}$ must be complete, so all sides of $(V,\overline{\underline{\underline{v}}})$ are elements of $\underline{\underline{v}}$. Since g must take elements of $\overline{\underline{\underline{v}}}-\underline{\underline{v}}$ to elements of $\overline{\underline{\underline{v}}}-\underline{\underline{v}}$, it must take lids to lids.

In either the I-fibered or Seifert-fibered case, we may now apply the Unique Fibering Theorem 2.8.1 to assume that g is fiber-preserving. If F is the admissible homotopy from f to g, then pFs is an admissible homotopy on $(B,\underline{\underline{b}})$ from pfs to the homeomorphism pgs.

For the converse direction of (iv), assume that pfs is admissibly homotopic to a homeomorphism. We have seen that f is admissibly homotopic to a fiber-preserving map, so assume that f preserves fibers.

Consider first the case when $(V,\underline{\underline{v}})$ is an I-bundle. We may assume that the image of s does not meet the lids, so for each $b\in B$, the fiber $p^{-1}(b)$ is the union of two arcs that meet at the point $s(b)$. Since f preserves fibers, we may change it by admissible fiber-preserving homotopy so that it preserves the image of s and is a homeomorphism on each fiber, linear on each of the two arcs in the fiber (in the case that the I-bundle is twisted, the fibers may be assumed to carry a linear structure, since the covering transformation for the 2-fold covering by a product I-bundle may be chosen to be reflection in the I-fibers). Then, an admissible homotopy from pfs to a homeomorphism of $(B, \underline{\underline{b}})$ defines a homotopy on $s(B)$ from the restriction of f to $s(B)$ to a homeomorphism of $s(B)$, and this can be extended linearly on the two arcs in each fiber to produce a fiber-preserving homotopy from f to a homeomorphism of $(V, \underline{\underline{v}})$. This completes the proof of (iv) in the I-bundle case.

We will now prove (v), before returning to complete the proof of (iv). In case pfs is admissibly homotopic to 1_B, the previous construction produces an admissible homotopy from f to a homeomorphism which is the identity on $s(B)$ and is a linear homeomorphism on the complementary intervals, that is, either to the identity or to a reflection r.

Suppose first that the I-bundle is a product. If both lids are in $\underline{\underline{v}}$, then they are interchanged by r, and r cannot be admissibly homotopic to the identity. If exactly one lid is in $\underline{\underline{v}}$, then r is not admissible, so the previous isotopy of f produced the identity homeomorphism. If neither lid is in $\underline{\underline{v}}$, then r is admissibly homotopic to the identity. These cases are as given in (v).

Suppose now that the I-bundle is twisted. If the lid is not in $\underline{\underline{v}}$, then r is admissibly homotopic to the identity. Suppose that the lid is in $\underline{\underline{v}}$. Now $(V, \underline{\underline{v}})$ is doubly covered by a product I-bundle $(\widetilde{B} \times \mathrm{I}, \underline{\underline{w}})$, and r lifts to a reflection $\widetilde{r}$ on $\widetilde{B} \times \mathrm{I}$, which interchanges its lids. Suppose that r is admissibly homotopic to the identity. The homotopy lifts to an admissible homotopy from $\widetilde{r}$ to a covering transformation, necessarily nontrivial since the lids are interchanged. It follows that the covering transformation induces the identity automorphism on $\pi_1(\widetilde{B} \times \mathrm{I})$. The Lefschetz fixed-point formula implies that $H_1(\widetilde{B} \times \mathrm{I})$ has rank 1, so $\widetilde{B}$ is an annulus and B is a Möbius band. In this case, if the side of $(V, \underline{\underline{v}})$ is also in $\underline{\underline{v}}$, then r is orientation-reversing on the torus ∂V but would be homotopic preserving ∂V to the identity, which is orientation-preserving on ∂V. Therefore we must have $\underline{\underline{v}}$ consisting of the lid alone. In this case, r is indeed admissibly homotopic to the identity. To see this, regard the Möbius band $s(B)$ as an I-bundle over its center circle C. There is an admissible homotopy from r to the identity which preserves $s(B)$ and at all times carries each fiber of V homeomorphically to some fiber. The first stage of the homotopy is an isotopy preserving $s(B)$ and moving a point on C once around C. At the end of this isotopy, each I-fiber of $s(B)$ over C is reflected across C, but on the I-fibers of V that meet C, the map is now the identity. The second stage is a homotopy which preserves $s(B)$ and each I-fiber of $s(B)$, but changes the reflection on each fiber of $s(B)$ over C to the identity homeomorphism. This completes the proof of (v).

It remains to prove (iv) when $(V, \underline{\underline{v}})$ is Seifert-fibered. Since it has no exceptional fibers, $p \colon V \to B$ is an S^1-bundle. Since f is a homotopy equivalence, its restriction to each fiber must have degree ± 1 as a map from the circle to the circle.

Assume first that B is orientable, so that V is a product S^1-bundle. Using the section s, we may change f by a fiber-preserving admissible homotopy to a map g whose restriction to each fiber is a homeomorphism. (Explicitly, the bundle is a product bundle so its universal covering can be given as $\alpha \times \beta \colon \widetilde{B} \times \mathbb{R} \to B \times \mathrm{S}^1$, where $\beta(r) = e^{2\pi i r}$ and a lift $\widetilde{s} \colon \widetilde{B} \to \widetilde{B} \times \mathbb{R}$ of s has the form $\widetilde{s}(x) = (x, 0)$. Let $\widetilde{f} \colon \widetilde{B} \times \mathbb{R} \to \widetilde{B} \times \mathbb{R}$ be a lift of f. For each $(x, n) \in \widetilde{B} \times \mathbb{Z}$, we have $\widetilde{f}(x, n) = (y, t)$

and $\widetilde{f}(x, n+1) = (y, t+1)$ for some $(y,t) \in \widetilde{B} \times \mathbb{R}$. Therefore we can change $\widetilde{f}$ on $\{x\} \times [n, n+1]$, relative to $\{x\} \times \{n, n+1\}$, using a straight-line homotopy in the $\mathbb{R}$ factor of $\widetilde{B} \times \mathbb{R}$. The result is a map $\widetilde{g}$ which is a linear homeomorphism from $\{x\} \times [n, n+1]$ to $\{y\} \times [t, t+1]$. These homotopies fit together on all of $\widetilde{B} \times \mathbb{R}$ to give an equivariant homotopy from $\widetilde{f}$ to $\widetilde{g}$ which induces a homotopy from f to a map g on V, such that $pf = pg$ and the restriction of g to any fiber is a homeomorphism to another fiber.) The admissible homotopy from pfs to a homeomorphism can now be used to change the B-coordinate of g, while keeping the S^1-coordinate unchanged. (Explicitly, if we write $g(x,t) = (g_B(x,t), g_S(x,t))$, then since g is fiber-preserving we have $g_B(x,t) = pgs(x)$. If h_t is an admissible homotopy from $pgs = pfs$ to a homeomorphism, the homotopy g_t is defined by $g_t(x,t) = (h_t(x), g_S(x,t))$.) The result is an admissible fiber-preserving homotopy from g to a homeomorphism of $(V, \underline{\underline{v}})$.

If B is nonorientable, the orientable double covering map $\widehat{B} \to B$ lifts to a double covering $\widehat{B} \times \mathrm{S}^1 \to V$ and the section $s\colon B \to V$ lifts to an equivariant section from $\widehat{B}$ to $\widehat{B} \times \mathrm{S}^1$. Since pfs is homotopic to a homeomorphism, f lifts to $\widehat{f}\colon \widehat{B} \times \mathrm{S}^1 \to \widehat{B} \times \mathrm{S}^1$. The homotopy of $\widehat{f}$ constructed above is equivariant, so it induces the desired homotopy of f on V. This completes the proof of (iv) in the case that $(V, \underline{\underline{v}})$ is a S^1-bundle. □

As we have noted, proposition 10.2.2 will reduce proposition 10.2.1 to a matter of determining when the index of $\mathcal{H}(B, \underline{\underline{b}})$ in $\mathcal{E}(B, \underline{\underline{b}})$ is finite. In the next lemma, we treat the finite-index cases. Note that these include all cases when B is a closed surface, since then $|\underline{\underline{b}}| = \partial B$.

LEMMA 10.2.3. *Let $(B, \underline{\underline{b}})$ be a connected 2-manifold with boundary pattern. If $|\underline{\underline{b}}| = \partial B$, then the index of $\mathcal{H}(B, \underline{\underline{b}})$ in $\mathcal{E}(B, \underline{\underline{b}})$ is finite. Also, the index is finite if any of the following holds.*

(i) *B is a disk.*
(ii) *B is an annulus or Möbius band, and each boundary circle of B contains at most one component of $|\underline{\underline{b}}|$ that is an arc.*
(iii) *B is a disk with two holes, and $|\underline{\underline{b}}|$ consists of two boundary circles.*
(iv) *B is a torus with one hole and $|\underline{\underline{b}}|$ is empty.*

PROOF. If $|\underline{\underline{b}}| = \partial B$, then by the Baer-Nielsen Theorem 2.5.5, $\mathcal{H}(B, \underline{\underline{b}}) = \mathcal{E}(B, \underline{\underline{b}})$. From now on, assume that $|\underline{\underline{b}}| \neq \partial B$. Since $\underline{\underline{b}}$ consists of arcs and circles, we may pass to a subgroup of finite index in $\mathcal{E}(B, \underline{\underline{b}})$ for which each homotopy class can be represented by a homotopy equivalence f which is the identity on $|\underline{\underline{b}}|$. In the cases when $\pi_1(B)$ is infinite cyclic, we may also assume that f induces the identity automorphism on $\pi_1(B)$. It suffices to show that for the manifolds listed in lemma 10.2.3, any such f is admissibly homotopic to a homeomorphism.

When B is a disk, f may be changed by admissible homotopy to be the identity on all of ∂B. Then, the Alexander construction produces an admissible homotopy to the identity.

Suppose that B is a Möbius band and at most one component of $|\underline{\underline{b}}|$ is an arc. Since f is assumed to induce the identity automorphism on $\pi_1(B)$, f is admissibly homotopic to a map which is the identity on ∂B. Let α be an essential arc in B. Examining the universal cover of B shows that f is homotopic relative to ∂B to be the identity on α. Then, all components of the (transverse) preimage of α other than α itself are inessential circles, so f is further homotopic so that $f^{-1}(\alpha)=\alpha$. The Alexander trick applied to the disk that results from cutting B along α now gives an admissible homotopy from f to the identity.

When B is an annulus having at most one arc of $|\underline{\underline{b}}|$ in each boundary circle, we may again assume that f is the identity on ∂B. Again select an essential arc α, and let h be a Dehn twist about the center circle of B. For some value of n, $h^{-n}f(\alpha)$ is homotopic relative to its endpoints to α. Similarly to the Möbius band case, this implies that $h^{-n}f$ is admissibly homotopic to the identity, so f is admissibly homotopic to h^n.

Suppose that B is a disk with two holes and $|\underline{\underline{b}}|$ consists of two boundary circles. Let α and β be loops based at a basepoint v_0 in the interior of V, such that α and β are freely homotopic to the circles of $|\underline{\underline{b}}|$, and $\alpha\beta$ is freely homotopic to the other boundary circle C. We may assume that f fixes the basepoint v_0 in the interior of B. Since f is assumed to be the identity on $|\underline{\underline{b}}|$, its induced automorphism must send α to some conjugate $\gamma\alpha\gamma^{-1}$ and β to some $\delta\beta\delta^{-1}$. Changing f by admissible homotopy fixing $|\underline{\underline{b}}|$ but moving the basepoint, we may assume that $\gamma=1$. Since the images must generate the free group $\pi_1(V,v_0)$, $\delta\beta\delta^{-1}$ must be of the form $\alpha^n\beta\alpha^{-n}$ so the induced automorphism carries $\alpha\beta$ to $\alpha^n\alpha\beta\alpha^{-n}$. This implies that the image of C under f is freely homotopic to C, so f is admissibly homotopic to a map that preserves all of ∂B. By the Baer-Nielsen Theorem 2.5.5, this map is admissibly homotopic to a homeomorphism.

Finally, suppose that B is a torus with one hole and $|\underline{\underline{b}}|$ is empty. Every automorphism of $\pi_1(B)$ takes the element represented by the boundary circle C to itself or its inverse up to conjugacy, since each Nielsen generator [**99**] of the automorphism group of a free group on two generators takes the commutator of the generators to a conjugate of itself or its inverse. Again, this implies that f is homotopic to a map that takes C to C, and the Baer-Nielsen Theorem 2.5.5 shows that f is admissibly homotopic to a homeomorphism. □

As explained above, to handle the infinite-index cases, we will define a general kind of admissible homotopy equivalence of a surface, which is the 2-dimensional analogue of the wrapping homotopy equivalences defined in chapter 9 using 1-handles of M. Start with a 2-manifold $(B,\underline{\underline{b}})$ and suppose that γ is a properly imbedded arc in B whose endpoints lie in $\partial B - |\underline{\underline{b}}|$. Let $\gamma\times[-1,1]$ be a product neighborhood of $\gamma=\gamma\times\{0\}$ with $\gamma\times[-1,1]\cap\partial B=\partial\gamma\times I\subseteq\partial B-|\underline{\underline{b}}|$ (i. e. a 2-dimensional 1-handle with cocore γ). Suppose further that there is a loop α in the interior of B, based at a point $b_0\times\{1/2\}\in\gamma\times\{1/2\}$, and disjoint from $\gamma\times(-1/2,1/2)$. Such a loop is called *compatible* with γ (this depends on the choice of product neighborhood $\gamma\times[-1,1]$, a choice that will be understood as having been made in our arguments). Define an admissible homotopy equivalence $h(\gamma,\alpha)$ of $(B,\underline{\underline{b}})$ as follows. It will fix all points outside $\gamma\times(-1,1)$ (in particular, it is the identity on $|\underline{\underline{b}}|$). Map each $\gamma\times\{t\}$ to $\gamma\times\{t\}$ in such a way that for $-1/2\le t\le 1/2$, $\gamma\times\{t\}$ is collapsed to $b_0\times\{t\}$, then map the arc $b_0\times[-1/2,1/2]$ around the path product of $b_0\times[-1/2,1/2]$ and α. The admissible homotopy inverse of $h(\gamma,\alpha)$ is $h(\gamma,\alpha^{-1})$. We call $h(\gamma,\alpha)$ a homotopy equivalence obtained by *sweeping* γ *around* α, or just a *sweep*. When γ is the frontier of a regular neighborhood of a component G of $|\underline{\underline{b}}|$, we may also say $h(\gamma,\alpha)$ is obtained by *sweeping* G *around* α. In chapter 11, we will show explicitly how a sweep can be written as a product of the homotopy equivalences called Dehn flips by Johannson [**58**].

LEMMA 10.2.4. *Let* $(B, \underline{\underline{b}})$ *be a connected 2-manifold with boundary pattern. If* $|\underline{\underline{b}}| \neq \partial B$ *and* $(B, \underline{\underline{b}})$ *is not one of the cases listed in lemma 10.2.3, then there exist a properly embedded arc* $\gamma \in B$ *(whose endpoints lie in* $\partial B - |b|$*) and an orientation-preserving loop* α *in the interior of* B *such that no non-zero power of the sweep* $h(\gamma, \alpha)$ *of* $(B, \underline{\underline{b}})$ *is admissibly homotopic to a homeomorphism. Consequently,* $\mathcal{H}(B, \underline{\underline{b}})$ *has infinite index in* $\mathcal{E}(B, \underline{\underline{b}})$.

PROOF. Suppose first that B is an annulus or Möbius band. By hypothesis there must be a component C of ∂B which contains at least two arc components β_1 and β_2 of $|\underline{\underline{b}}|$. Let β' and β'' denote the components of $\overline{C - (\beta_1 \cup \beta_2)}$, so that $C = \beta_1 \cup \beta' \cup \beta_2 \cup \beta''$. Let γ be the frontier of a small regular neighborhood of β_1 in B. Taking α to be parallel to the center circle, if B is an annulus, or the square of the center circle, if B is a Möbius band, we observe that no nonzero power of $h(\gamma, \alpha)$ carries both β' and β'' to arcs which are admissibly homotopic into ∂B, so no nonzero power of $h(\gamma, \alpha)$ can be admissibly homotopic to a homeomorphism of $(B, \underline{\underline{b}})$.

Suppose that B is not a disk, annulus, or Möbius band, and some component β of $|\underline{\underline{b}}|$ is an arc. Let γ be the frontier of a regular neighborhood of β, and let C be the boundary circle of B that contains β. Since B is not a disk, annulus, or Möbius band, there exists an orientation-preserving loop α_0 based in β, no power of which is freely homotopic into C. For some loop α freely homotopic to α_0, there is a homotopy equivalence $h(\gamma, \alpha)$ obtained by sweeping γ around α. Then, $h(\gamma, \alpha)^n$ carries C to the loop $\alpha_0^n\, C\, \alpha_0^{-n}$, and for $n \neq 0$ this loop is not homotopic keeping its basepoint in β to a loop in C. Consequently, for $n \neq 0$, $h(\gamma, \alpha)$ is not admissibly homotopic to a homeomorphism.

From now on we will assume that all components of $|\underline{\underline{b}}|$ are boundary circles.

Suppose B is a disk with two holes. As in the proof of lemma 10.2.3, choose generators α and β of $\pi_1(B)$, each homotopic to a boundary circle, so that $\alpha\beta$ is freely homotopic to the third boundary circle. Since $(B, \underline{\underline{b}})$ is not one of the cases in lemma 10.2.3, we may assume that the boundary circles corresponding to β and $\alpha\beta$ are not components of $|\underline{\underline{b}}|$. Let γ be an arc connecting these two circles. For $n \neq 0$, $h(\gamma, \alpha)^n$ carries β to $\beta\alpha^n$. This is homotopic to a boundary circle only when $n = 1$, but in that case $h(\gamma, \alpha)$ carries $\alpha\beta$ to $\alpha\beta\alpha$, which is not freely homotopic to a boundary circle. Thus no nonzero power of $h(\gamma, \alpha)$ is homotopic to a map which preserves the boundary of B.

The remaining case is that B is not closed and not a disk, Möbius band, annulus, torus with one hole, or disk with two holes, and $|\underline{\underline{b}}|$ consists of boundary circles, and there is at least one boundary circle C not contained in $|\underline{\underline{b}}|$. Suppose first that B is nonorientable. Choose free generators $v_1, \ldots, v_g, c_1, \ldots, c_r$ for $\pi_1(B)$, where each v_i runs through a crosscap and each c_j encircles a boundary component, so that C is represented by $c = \prod_{i=1}^g v_i^2 \prod_{j=1}^r c_j$. Since B is nonorientable, $g \geq 1$, and if $r = 0$ (i. e. if B has only one boundary component) then $g \geq 2$ since B is not a Möbius band. We may choose loops representing the generators so that there is a proper arc γ, with $\gamma \cap \partial B \subset C$, which crosses a loop representing v_1 in one point and does not meet the other generating loops. The element $\prod_{i=2}^g v_i^2 \prod_{j=1}^r c_j$ is represented by an orientation-preserving loop α compatible with γ. Then, $h(\gamma, \alpha)^n$ carries v_1 to $v_1 \left(\prod_{i=2}^g v_i^2 \prod_{j=1}^r c_j\right)^n$ and carries c to

$$\left(v_1 (\prod_{i=2}^g v_i^2 \prod_{j=1}^r c_j)^n\right)^2 \prod_{i=2}^g v_i^2 \prod_{j=1}^r c_j,$$

so the image of C is not homotopic into ∂B.

Suppose that B is orientable and not planar. We can choose free generators $a_1, b_1, a_2, \ldots, b_g, c_1, \ldots, c_r$ for $\pi_1(B)$ so that C is homotopic to $\prod_{i=1}^{g}[a_i, b_i] \prod_{j=1}^{r} c_j$, with $g \geq 1$, and if $r=0$ then $g \geq 2$. This time use an arc γ dual to a_1 and a loop α representing $\prod_{i=2}^{g}[a_i, b_i] \prod_{j=1}^{r} c_j$.

Finally, if B is planar then it has at least four boundary components, so $\pi_1(B)$ is generated by $c_1, \ldots, c_r$ where C is homotopic to $\prod_{j=1}^{r} c_j$ and $r \geq 3$. Choose an arc γ that separates B into an annulus B_1, with $\pi_1(B_1)$ generated by c_1, and a disk-with-holes B_2 with $\pi_1(B_2)$ generated by $c_2, \ldots, c_r$. If α represents c_2, then $h(\gamma, \alpha)^n$ carries C to $c_2^n c_1 c_2^{1-n} \prod_{j=3}^{r} c_j$, which is homotopic into ∂B only when $n=0$. $\square$

To deduce proposition 10.2.1, we introduce 3-dimensional sweeps, which are fiber-preserving admissible homotopy equivalences $H(\gamma, \alpha)$ of $(V, \underline{\underline{v}})$ which project to sweeps of $(B, \underline{\underline{b}})$. Actually, proposition 10.2.1 can be proved without using 3-dimensional sweeps, by developing more properties of the homomorphism defined in proposition 10.2.2. But 3-dimensional sweeps will be used extensively in the next section, so we introduce them here. In the definition, we allow $(V, \underline{\underline{v}})$ to be Seifert-fibered, since this generality will be needed later.

To fix notation, let $p\colon V \to B$ be an admissible fibering of $(V, \underline{\underline{v}})$ as an I-bundle or Seifert-fibered space over $(B, \underline{\underline{b}})$. Suppose that α and γ are an arc and a loop in B as in the definition of 2-dimensional sweep, and assume further that

(i) α is orientation-preserving, and
(ii) if $(V, \underline{\underline{v}})$ is Seifert-fibered, then α and $\gamma \times [-1, 1]$ are disjoint from the image of the exceptional fibers.

With these assumptions, there is a neighborhood N of $\gamma \times [-1, 1] \cup \alpha$ such that

(1) $\overline{N}$ is disjoint from $|\underline{\underline{b}}|$,
(2) $h(\gamma, \alpha)^{-1}(\{x\}) = \{x\}$ for $x \in V - N$, and
(3) $p^{-1}(N)$ is a product $N \times \mathrm{I}$ or $N \times \mathrm{S}^1$ whose fibering agrees with the fibering of V.

Using such a neighborhood, we can define a fiber-preserving lift of $h(\gamma, \alpha)$ by $H(\gamma, \alpha)(x, t) = (h(\gamma, \alpha)(x), t)$ on $p^{-1}(N)$, while $H(\gamma, \alpha)$ will be the identity map outside of $p^{-1}(N)$. Observe that $H(\gamma, \alpha)$ is fiber-preserving, and is an admissible homotopy equivalence (an admissible homotopy from $h(\gamma, \alpha)h(\gamma, \overline{\alpha})$ to the identity of B, supported on N, lifts to an admissible homotopy from $H(\gamma, \alpha)H(\gamma, \overline{\alpha})$ to the identity of V). We caution that the admissible homotopy class of $H(\gamma, \alpha)$ is not necessarily uniquely defined, since we have made a choice of product structure for $p^{-1}(N)$. In practice, we simply make a choice and have $H(\gamma, \alpha)$ available for use; no uniqueness is ever needed.

If $(V, \underline{\underline{v}})$ is an I-bundle, then $H(\gamma, \alpha)$ preserves each lid and is the identity on the sides of $(V, \underline{\underline{v}})$. If $(V, \underline{\underline{v}})$ is Seifert-fibered, then it is the identity on all of $|\underline{\underline{v}}|$. We call $H(\gamma, \alpha)$ a homotopy equivalence obtained by *sweeping* $p^{-1}(\gamma)$ *around* α, or just a *sweep,* and when $p^{-1}(\gamma)$ is the frontier of a regular neighborhood of an annulus component G of $|\underline{\underline{v}}|$, we may say it is obtained by *sweeping* G *around* α. We also may speak of a sweep of $p^{-1}(\gamma)$ around a loop α in V; this means a sweep around the loop in B that is the image of α (providing, of course, that this projected loop meets the requirements for sweeping to be defined).

We can now deduce proposition 10.2.1.

PROOF OF PROPOSITION 10.2.1. Let $s\colon B \to V$ be a section for p, whose image is disjoint from the lids of $(V, \underline{\underline{v}})$ if $(V, \underline{\underline{v}})$ is an I-bundle. By proposition 10.2.2, we can define a homomorphism $P\colon \mathcal{E}(V, \underline{\underline{v}}', \underline{\underline{v}}'') \to \mathcal{E}(B, \underline{\underline{b}})$ by sending $\langle f \rangle$ to $\langle pfs \rangle$.

In the cases listed in lemma 10.2.3, $\mathcal{H}(B, \underline{\underline{b}})$ has finite index in $\mathcal{E}(B, \underline{\underline{b}})$ so proposition 10.2.2 shows that $\mathcal{H}(V, \underline{\underline{v}}', \underline{\underline{v}}'')$ has finite index in $\mathcal{E}(V, \underline{\underline{v}}', \underline{\underline{v}}'')$.

In the cases of lemma 10.2.4, let $h(\gamma, \alpha)$ be a sweep as given in that lemma, and let $H(\gamma, \alpha)$ be a lift of $h(\gamma, \alpha)$ to a sweep of $(V, \overline{\underline{\underline{v}}})$, so that $pH(\gamma, \alpha)s = h(\gamma, \alpha)$. For $n \neq 0$, $h(\gamma, \alpha)^n$ is not admissibly homotopic to a homeomorphism of $(B, \underline{\underline{b}})$, so proposition 10.2.2 implies that $H(\gamma, \alpha)^n$ is also not admissibly homotopic to a homeomorphism. Therefore the latter represent distinct cosets of $\mathcal{H}(V, \underline{\underline{v}}', \underline{\underline{v}}'')$ in $\mathcal{E}(V, \underline{\underline{v}}', \underline{\underline{v}}'')$, and the index is infinite. □

Remark: The homomorphism P in proposition 10.2.2 need not be surjective. For example, suppose that B is obtained by removing an open disk from a Möbius band. Its fundamental group is free on generators v_1 and c_1, where v_1 is represented by the center circle of the Möbius band and c_1 by an orientation-preserving loop that encircles the boundary of the removed disk. There is a homotopy equivalence g from B to B whose induced automorphism interchanges v_1 and c_1. If V is the S^1-bundle over B, then g is not the image of a homotopy equivalence of V under $P\colon \mathcal{E}(V, \underline{\underline{\emptyset}}, \underline{\underline{\emptyset}}) \to \mathcal{E}(B, \underline{\underline{\emptyset}})$. For according to section 2.1, $\pi_1(V)$ has a presentation

$$\langle t, v_1, c_1 \mid v_1 t v_1^{-1} = t^{-1}, [c_1, t] = 1 \rangle \ .$$

One can check that t generates the unique maximal infinite cyclic normal subgroup of $\pi_1(V)$. Consequently any automorphism α of $\pi_1(E)$ must preserve this subgroup, so must take t to either t or t^{-1}. But if α interchanges v_1 and c_1, it cannot also preserve the relations in the group. So g is not in the image of P. In all cases, the image of P has finite index, since any homotopy equivalence of $(B, \underline{\underline{b}})$ that preserves the orientation homomorphism $\pi_1(B) \to \mathbb{Z}/2$ does lift to a homotopy equivalence of $(V, \underline{\underline{v}})$.

10.3. Realizing homotopy equivalences of Seifert-fibered manifolds

In this section, we complete the determination of when the groups $\mathcal{H}(V, \underline{\underline{v}}', \underline{\underline{v}}'')$ have finite index in $\mathcal{E}(V, \underline{\underline{v}}', \underline{\underline{v}}'')$.

PROPOSITION 10.3.1. *Suppose that $(V, \underline{\underline{v}}' \cup \underline{\underline{v}}'')$ is admissibly Seifert-fibered, and the completion of $\underline{\underline{v}}' \cup \underline{\underline{v}}''$ is useful. If $|\underline{\underline{v}}' \cup \underline{\underline{v}}''| = \partial V$, then the index of $\mathcal{H}(V, \underline{\underline{v}}', \underline{\underline{v}}'')$ in $\mathcal{E}(V, \underline{\underline{v}}', \underline{\underline{v}}'')$ is finite. If $|\underline{\underline{v}}' \cup \underline{\underline{v}}''| \neq \partial V$, then the index is finite if and only if one of the following holds.*

- (i) *V is a solid torus.*
- (ii) *V is either $S^1 \times S^1 \times I$ or the I-bundle over the Klein bottle, and each boundary component of V contains at most one component of $|\underline{\underline{v}}' \cup \underline{\underline{v}}''|$ that is an annulus.*
- (iii) *V is fibered over the annulus with one exceptional fiber, and no component of $|\underline{\underline{v}}' \cup \underline{\underline{v}}''|$ is an annulus.*
- (iv) *V is fibered over the disk with two holes with no exceptional fiber, and $|\underline{\underline{v}}' \cup \underline{\underline{v}}''|$ consists of two boundary tori of V.*
- (v) *V is fibered over the disk with two exceptional fibers, over the Möbius band with one exceptional fiber, or over the torus with one hole with no exceptional fiber, and $|\underline{\underline{v}}' \cup \underline{\underline{v}}''|$ is empty.*
- (vi) *V is fibered over the disk with three exceptional fibers of type $(2,1)$, and $|\underline{\underline{v}}' \cup \underline{\underline{v}}''|$ is empty.*

The proof of proposition 10.3.1 constitutes the remainder of this section. As in the proof of proposition 10.2.1, the finite-index cases are examined on a case-by-case basis, and the infinite-index cases are detected using the sweeps $H(\gamma,\alpha)$ defined in section 10.2. To demarcate the major case divisions, we organize the proof as a series of lemmas.

PROOF OF PROPOSITION 10.3.1. When $|\underline{\underline{v}}'\cup\underline{\underline{v}}''|=\partial V$, Waldhausen's Theorem 2.5.6 shows immediately that $\mathcal{H}(V,\underline{\underline{v}}',\underline{\underline{v}}'') = \mathcal{E}(V,\underline{\underline{v}}',\underline{\underline{v}}'')$. From now on, we will assume that $|\underline{\underline{v}}'\cup\underline{\underline{v}}''| \neq \partial V$.

We first treat the cases (i)–(vi) in which $|\underline{\underline{v}}'\cup\underline{\underline{v}}''| \neq \partial V$ and yet $\mathcal{H}(V,\underline{\underline{v}}',\underline{\underline{v}}'')$ has finite index in $\mathcal{E}(V,\underline{\underline{v}}',\underline{\underline{v}}'')$. The following lemma handles case (i).

LEMMA 10.3.2. *Suppose that V is a solid torus and $\underline{\underline{v}}'\cup\underline{\underline{v}}''$ is any boundary pattern that consists of incompressible annuli. Then $\mathcal{E}(V,\underline{\underline{v}}',\underline{\underline{v}}'')$ is finite.*

PROOF. Let $\mathcal{E}_0$ be the subgroup of index at most 2 in $\mathcal{E}(V,\underline{\underline{v}}',\underline{\underline{v}}'')$ consisting of the elements that induce the identity automorphism on $\pi_1(V)\cong\mathbb{Z}$. Since V is aspherical, these are homotopic to the identity map, and are admissibly homotopic to maps which are the identity on $|\underline{\underline{v}}'\cup\underline{\underline{v}}''|$ (since by definition each homotopy equivalence in $\mathcal{E}(V,\underline{\underline{v}}',\underline{\underline{v}}'')$ preserves each element of $\underline{\underline{v}}'\cup\underline{\underline{v}}''$).

If $\underline{\underline{v}}'\cup\underline{\underline{v}}''$ is empty, then $\mathcal{E}_0$ is trivial so $\mathcal{E}(V,\underline{\underline{v}}',\underline{\underline{v}}'')$ is finite. So we will assume that $|\underline{\underline{v}}'\cup\underline{\underline{v}}''|$ is nonempty and not equal to ∂V. Let A_0, $A_1,\ldots$, A_n $(n\geq 0)$ be the components of $|\underline{\underline{v}}'\cup\underline{\underline{v}}''|$. There exists $N>0$, such that each $\pi_1(A_i)$ has index N in $\pi_1(V)$. To prove that $\mathcal{E}_0$ is finite, we will use the traces of a homotopy from f to the the identity map to define an element $T(f)\in(\mathbb{Z}/N)^n$, so that $T\colon\mathcal{E}_0\to(\mathbb{Z}/N)^n$ is an injective homomorphism.

Choose basepoints $a_i\in A_i$. Let f represent an element of $\mathcal{E}_0$, where f is the identity on each A_i. For any element of $\pi_1(V)$, there is a (nonadmissible) homotopy from the identity map to the identity map whose trace at a_0 is the given element (see section 10.1 for a discussion of the trace; in the present situation the trace is a well-defined element of $\pi_1(V,a_0)$, since the homotopy is between maps that fix a_0). This homotopy just rotates in the S^1-factor through the number of turns needed to achieve the given element of $\pi_1(V)$ as trace. Given a homotopy from f to the identity having trace m, we can follow it with a homotopy from the identity to the identity with trace $-m$ to obtain a homotopy H from f to the identity whose trace at a_0 is zero. By lemma 10.1.5, such a homotopy admits a deformation to a homotopy which is the identity on A_0 at each stage. For each $1\leq i\leq n$, define $t_i\in\pi_1(V)$ to be the trace of H at a_i. These t_i depend only on f and not on the choice of H. For given two such choices H and H', one can form a homotopy from the identity map to the identity map by taking the reverse of H followed by H'. Since this homotopy has trivial trace at a_0, it has trivial trace at each a_i (for any homotopy from the identity to the identity, the traces at different basepoints are freely homotopic, since if γ is a path between the basepoints then the restrictions of the homotopy to $\gamma(t)\times I$ determines a free homotopy).

Suppose f and f' are admissibly homotopic, and each restricts to the identity on $|\underline{\underline{v}}'\cup\underline{\underline{v}}''|$. During any admissible homotopy from f to f', each a_i must travel some number of times around A_i, so the values of t_i determined by f and f' differ by a multiple of N. That is, the residue class of t_i in $\mathbb{Z}/N$ is well-defined on elements of $\mathcal{E}_0$. Define a homomorphism $T\colon\mathcal{E}_0\to(\mathbb{Z}/N)^n$ by sending the admissible homotopy class of f to $(t_1,\ldots,t_n)$.

Consider an element in the kernel of T, and select a representative which is the identity on $|\underline{\underline{v}}' \cup \underline{\underline{v}}''|$. Then for this f each t_i is a multiple of N, say $t_i = k_i N$. There is an admissible isotopy $\{i_t\}$ of V which starts at the identity, is constant outside a regular neighborhood of A_i, and moves a_i once around A_i (hence N times around V). Let h_i be the ending homeomorphism of this isotopy. There is a homotopy from f to $h_1^{-k_1} h_2^{-k_2} \cdots h_n^{-k_n}$ which has trivial trace at each a_i, obtained by following the homotopy from f to the identity map with the isotopy $\{i_t^{-1}\}$ from the identity map to $h_1^{-k_1} h_2^{-k_2} \cdots h_n^{-k_n}$. Lemma 10.1.5 implies that this homotopy is is deformable to an admissible homotopy. But $h_1^{-k_1} h_2^{-k_2} \cdots h_n^{-k_n}$ is admissibly isotopic to the identity map. Therefore T is injective, so $\mathcal{E}_0$ is finite. □

We remark that the homomorphism T in the proof of lemma 10.3.2 is also surjective. For example, a homotopy equivalence f with $T(f) = (1, 0, \ldots, 0)$ is the end result of a homotopy starting at the identity that fixes all A_i except A_1, rotates the S^1-factor of A_1 so that a_1 travels once around the S^1-factor of $V = D^2 \times S^1$ (but only part of the way around the S^1-factor of A_1, since A_1 makes N turns around the S^1-factor of V) and then moves A_1 through V, keeping points within the D^2 fibers, until the identity map is achieved on A_1.

The manifolds in case (ii), and in (iii) when V is the I-bundle over the Klein bottle, are covered by the next two lemmas.

LEMMA 10.3.3. *Suppose that* $V = S^1 \times S^1 \times I$ *and* $|\underline{\underline{v}}' \cup \underline{\underline{v}}''| \neq \partial V$. *Then* $\mathcal{H}(V, \underline{\underline{v}}', \underline{\underline{v}}'')$ *has finite index in* $\mathcal{E}(V, \underline{\underline{v}}', \underline{\underline{v}}'')$ *if and only if each boundary component of* V *contains at most one component of* $|\underline{\underline{v}}' \cup \underline{\underline{v}}''|$ *which is an annulus.*

PROOF. Suppose first that $\underline{\underline{v}}' \cup \underline{\underline{v}}''$ contains an annulus. Then proposition 10.2.1 applies, and completes the proof. So assume that $\underline{\underline{v}}' \cup \underline{\underline{v}}''$ consists of tori. If $\underline{\underline{v}}''$ contains a boundary component of V then every element of $\mathcal{E}(V, \underline{\underline{v}}', \underline{\underline{v}}'')$ induces the identity automorphism on $\pi_1(V)$, so $\mathcal{E}(V, \underline{\underline{v}}', \underline{\underline{v}}'')$ is trivial. If $\underline{\underline{v}}''$ is empty, then $\mathcal{E}(V, \underline{\underline{v}}', \underline{\underline{\emptyset}}) \cong \mathrm{Out}(\pi_1(V), \pi_1(\underline{\underline{v}}')) \cong \mathrm{GL}(2, \mathbb{Z})$ and since every element of $\mathrm{SL}(2, \mathbb{Z})$ can be induced by an orientation-preserving homeomorphism, $\mathcal{H}(V, \underline{\underline{v}}', \underline{\underline{\emptyset}})$ has index at most 2 in $\mathcal{E}(V, \underline{\underline{v}}', \underline{\underline{\emptyset}})$ (it has index 1 if $\underline{\underline{v}}'$ is empty, and index 2 otherwise). □

LEMMA 10.3.4. *Suppose that* V *is the I-bundle over the Klein bottle, and* $|\underline{\underline{v}}' \cup \underline{\underline{v}}''| \neq \partial V$. *Then* $\mathcal{H}(V, \underline{\underline{v}}', \underline{\underline{v}}'')$ *has finite index in* $\mathcal{E}(V, \underline{\underline{v}}', \underline{\underline{v}}'')$ *if and only if at most one component of* $|\underline{\underline{v}}' \cup \underline{\underline{v}}''|$ *is an annulus.*

PROOF. Recall from lemma 2.8.5 that there are two Seifert fiberings of the I-bundle over the Klein bottle. Proposition 10.2.1 could be applied to the nonsingular fibering, as in lemma 10.3.3. But this is unnecessary, since the following argument needed for the singular fibering applies to the nonsingular fibering as well.

Since $\pi_1(V)$ is the fundamental group of the Klein bottle, which has only four outer automorphisms, we may pass to a subgroup $\mathcal{E}_0(V, \underline{\underline{v}}', \underline{\underline{v}}'')$ of finite index in $\mathcal{E}(V, \underline{\underline{v}}', \underline{\underline{v}}'')$ to assume that f is homotopic to the identity map. If $|\underline{\underline{v}}' \cup \underline{\underline{v}}''|$ is empty, then such a homotopy is automatically admissible, so $\mathcal{E}(V, \underline{\underline{v}}', \underline{\underline{v}}'')$ is finite and the lemma holds.

Suppose $|\underline{\underline{v}}' \cup \underline{\underline{v}}''|$ consists of a single annulus A. Using the trace, we will construct a certain subgroup $\mathcal{E}_1(V, \underline{\underline{v}}', \underline{\underline{v}}'')$ of index at most 2 in $\mathcal{E}_0(V, \underline{\underline{v}}', \underline{\underline{v}}'')$, then show that every element of $\mathcal{E}_1(V, \underline{\underline{v}}', \underline{\underline{v}}'')$ is admissibly homotopic to a homeomorphism.

We will define a homomorphism $T\colon \mathcal{E}_0(V,\underline{\underline{v}}',\underline{\underline{v}}'') \to \pi_1(V,v_0)/\pi_1(\partial V,v_0)$, where this quotient is a cyclic group of order 2. Fix a basepoint v_0 in the interior of an element of $\underline{\underline{v}}' \cup \underline{\underline{v}}''$. Let $f \in \mathcal{E}_0(V,\underline{\underline{v}}',\underline{\underline{v}}'')$. We may change f by an admissible isotopy so that $f(v_0) = v_0$. Choose a homotopy f_t from the identity map id_V to f, and let $\mathrm{tr}(f_t)$ denote the trace of f_t at v_0. Define $T(f)$ to be the element of $\pi_1(V,v_0)/\pi_1(\partial V,v_0)$ represented by $\mathrm{tr}(f_t)$.

If g_t is another homotopy from id_V to f, then g_t followed by the reverse of f_t is a homotopy from id_V to id_V, so its trace lies in the center of $\pi_1(V,v_0)$, which is contained in $\pi_1(\partial V,v_0)$. Since the trace of this homotopy is $\mathrm{tr}(g_t)\,\mathrm{tr}(f_t)^{-1}$, $T(f)$ is independent of the choice of homotopy f_t. It is also independent of the isotopy used to make $f(v_0) = v_0$, since the isotopy changing one choice to another has trace in $\pi_1(\partial V,v_0)$.

Finally, to see that T is a homomorphism, let $f_t\colon \mathrm{id}_V \simeq f$ and $g_t\colon \mathrm{id}_V \simeq g$ be homotopies. The path product $g_t \cdot (gf_t)$ is a homotopy from id_V to gf, whose trace at v_0 is $\mathrm{tr}(g_t)\,g(\mathrm{tr}(f_t))$. So $T(gf) = T(g)\,g_\#(T(f))$. Since $\pi_1(\partial V,v_0)$ is the unique rank 2 maximal abelian subgroup of $\pi_1(V,v_0)$, $g_\#(T(f))$ lies in $\pi_1(\partial V,v_0)$ if and only if $T(f)$ does, that is, $T(g)\,g_\#(T(f)) = T(g)\,T(f)$.

Now, define $\mathcal{E}_1(V,\underline{\underline{v}}',\underline{\underline{v}}'')$ to be the kernel of T, a subgroup of index at most 2. It remains to show that every element of $\mathcal{E}_1(V,\underline{\underline{v}}',\underline{\underline{v}}'')$ is admissibly homotopic to a homeomorphism.

Any element of $\mathcal{E}_1(V,\underline{\underline{v}}',\underline{\underline{v}}'')$ can be represented by an f which preserves v_0 and is homotopic to the identity by a homotopy whose trace lies in $\pi_1(\partial V,v_0)$. There is an isotopy of V (supported in a neighborhood of ∂V) starting at the identity and ending at a homeomorphism h which is the identity on A, and which has the same trace at A as does f. Therefore f is homotopic to h relative to v_0. By lemma 10.1.5, there is a deformation of this homotopy to a homotopy that preserves A, that is, which is admissible. This completes the proof in the case when $|\underline{\underline{v}}' \cup \underline{\underline{v}}''|$ is a single annulus.

The remaining case is when $|\underline{\underline{v}}' \cup \underline{\underline{v}}''|$ has at least two components A_1 and A_2 which are annuli. Let $H(\gamma,\alpha)$ be a sweep of A_1 around a boundary circle of the quotient surface $(B,\underline{b})$. Consider the annuli $\overline{\partial V - (A_1 \cup A_2)}$. The restriction of $H(\gamma,\alpha)^n$ to each of these annuli is an admissible singular annulus, and for no nonzero n are both of these singular annuli admissibly homotopic into ∂V. Therefore no nonzero power of $H(\gamma,\alpha)$ is admissibly homotopic to a homeomorphism, so these represent infinitely many distinct cosets of $\mathcal{H}(V,\underline{\underline{v}}',\underline{\underline{v}}'')$ in $\mathcal{E}(V,\underline{\underline{v}}',\underline{\underline{v}}'')$. □

The next lemma will allow us to make certain homotopies between admissible homotopy equivalences admissible, at the expense of postcomposing one of the maps by an admissible homeomorphism.

LEMMA 10.3.5. *Assume that V is not homeomorphic to a solid torus or to the I-bundle over the Klein bottle. Suppose that T is a torus boundary component of V with $T \subseteq |\underline{\underline{v}}' \cup \underline{\underline{v}}''|$. Let $f_1, f_2 \in \mathcal{E}(V,\underline{\underline{v}}',\underline{\underline{v}}'')$, and suppose f_1 is homotopic to f_2. Then there exist an admissible homeomorphism h of $(V,\underline{\underline{v}}' \cup \underline{\underline{v}}'')$ and a homotopy from f_1 to hf_2 that preserves each element of $\underline{\underline{v}}' \cup \underline{\underline{v}}''$ that lies in T.*

PROOF. Since f_1 and f_2 are in $\mathcal{E}(V,\underline{\underline{v}}',\underline{\underline{v}}'')$, they preserve each element of $\underline{\underline{v}}' \cup \underline{\underline{v}}''$. We may change f_1 and f_2 by admissible homotopy so that each preserves a basepoint v_0 in the interior of an element of $\underline{\underline{v}}' \cup \underline{\underline{v}}''$ that lies in T.

Suppose for contradiction that there is no homotopy from f_1 to f_2 that preserves T. Then the restriction of a homotopy from f_1 to f_2 to $T \times \mathrm{I}$ is an essential map from $T \times \mathrm{I}$ to $(V, \underline{\underline{\emptyset}})$. Since V is not a solid torus, $\underline{\underline{\emptyset}}$ is useful. By Waldhausen's Theorem 2.5.6, this map is admissibly homotopic to a covering map, so $\pi_1(T)$ has finite index in $\pi_1(V)$. By the Finite Index Theorem 2.1.1, V is homeomorphic to $T \times \mathrm{I}$ or to the I-bundle over the Klein bottle. The latter is excluded by hypothesis and the former is impossible since f_1 and f_2 both preserve T. Therefore there is a homotopy H from f_1 to f_2 that preserves T.

Let α be the trace of H at v_0. Since H preserves T, α lies in $\pi_1(T, v_0)$. Let J_t be an isotopy of T from id_T to id_T, whose trace at v_0 is α. Construct a Dehn twist homeomorphism h (see section 4.2) on a collar neighborhood of T using the J_t on the levels of the collar. Then there is an isotopy from id_V to h whose trace at v_0 is α^{-1}, so there is a homotopy from f_1 to hf_2 which preserves T and whose trace at v_0 is trivial. Its restriction to T admits a deformation to a homotopy that preserves each element of $\underline{\underline{v}}' \cup \underline{\underline{v}}''$ that lies in T, so (using the homotopy extension property) a homotopy from f_1 to hf_2 may be found that does the same. □

The cases (iii) with V not the I-bundle over the Klein bottle, (iv), and (v) of proposition 10.3.1 are addressed in the next lemma.

LEMMA 10.3.6. *Assume that the sum of the rank of $H_1(B)$ and the number of exceptional fibers is exactly 2, that V is not the I-bundle over the Klein bottle, and that $|\underline{\underline{v}}' \cup \underline{\underline{v}}''| \neq \partial V$. Then $\mathcal{H}(V, \underline{\underline{v}}', \underline{\underline{v}}'')$ has finite index in $\mathcal{E}(V, \underline{\underline{v}}', \underline{\underline{v}}'')$ if and only if one of the following occurs.*

1. *V is fibered over the annulus with one exceptional fiber, and no component of $|\underline{\underline{v}}' \cup \underline{\underline{v}}''|$ is an annulus.*
2. *V is fibered over a disk with two holes with no exceptional fiber, and $|\underline{\underline{v}}' \cup \underline{\underline{v}}''|$ consists of two boundary tori of V.*
3. *V is fibered over either the disk with two exceptional fibers, over the Möbius band with one exceptional fiber, or over the torus with one hole with no exceptional fiber, and $|\underline{\underline{v}}' \cup \underline{\underline{v}}''|$ is empty.*

PROOF. We will make frequent use of the presentations for the fundamental groups of Seifert-fibered 3-manifolds, which were discussed in section 2.1. For each of the manifolds V in the lemma, the fundamental group has a presentation

$$\pi_1(V) = \langle t, c_1, c_2 \mid c_1^{p_1} = t^{q_1}, c_2^{p_2} = t^{q_2}, c_1 t c_1^{-1} = t^{\epsilon_1}, c_2 t c_2^{-1} = t^{\epsilon_2} \rangle$$

where each (p_j, q_j) is either a relatively prime pair or is $(0,0)$. Each $\epsilon_j = \pm 1$, and can equal -1 only when $p_j = 0$. As explained in section 2.1, t is represented by the fiber and the c_j correspond to loops in B. For each exceptional fiber, c_j is represented by a loop whose projection to B encircles the exceptional point in B determined by the fiber, and in this case there is a relation of the form $c_j^{p_j} = t^{q_j}$ where the invariants (p_j, q_j) are associated to the exceptional fiber. The projections of the other c_j either encircle boundary components, in which case $c_j t c_j^{-1} = t$, or are orientation-reversing loops, in which case $c_j t c_j^{-1} = t^{-1}$. The projection of $c_1 c_2$ is homotopic, missing the exceptional points, to a boundary circle of B. The regular fiber represents t, which generates an infinite cyclic normal subgroup of $\pi_1(V)$, and there is an exact sequence

$$1 \to \langle t \mid \rangle \to \pi_1(V) \to \langle c_1, c_2 \mid c_1^{p_1} = c_2^{p_2} = 1\rangle \to 1\ .$$

Since the quotient group is a nontrivial free product, it has no cyclic normal subgroup. Therefore any cyclic normal subgroup is contained in the subgroup generated by t, which must then be the unique maximal cyclic normal subgroup of $\pi_1(V)$. Consequently, it must preserved by every automorphism, so any automorphism ϕ induces an automorphism $\overline{\phi}$ on

$$Q = \langle c_1, c_2 \mid c_1^{p_1} = c_2^{p_2} = 1\rangle \cong (\mathbb{Z}/p_1\mathbb{Z}) * (\mathbb{Z}/p_2\mathbb{Z}).$$

Note that when $\overline{\phi}$ is the identity on Q, we must have $\phi(c_1) = c_1 t^{m_1}$ and $\phi(c_2) = c_2 t^{m_2}$ for some m_1 and m_2.

In the parts of cases I, II, and III below for which the index is finite, the strategy will be to consider the induced automorphism $\phi \in \mathrm{Out}(\pi_1(V))$ of an element f of $\mathcal{E}(V, \underline{\underline{v}}', \underline{\underline{v}}'')$. By passing to subgroups of finite index in $\mathrm{Out}(\pi_1(V))$, we will sufficiently restrict ϕ so that its associated homotopy equivalence f is admissibly homotopic to a homeomorphism.

In cases I and II, where $\epsilon_1 = \epsilon_2 = 1$, we may pass to a subgroup of index 2 to assume that $\phi(t) = t$. In any of the three cases, when $p_j \neq 0$, the Kurosh subgroup theorem shows that $\overline{\phi}(\mathbb{Z}/p_j\mathbb{Z})$ must be conjugate to $\mathbb{Z}/p_j\mathbb{Z}$, unless $p_1 = p_2$ in which case $\overline{\phi}(\mathbb{Z}/p_j\mathbb{Z})$ must be conjugate to either $\mathbb{Z}/p_1\mathbb{Z}$ or $\mathbb{Z}/p_2\mathbb{Z}$. If the latter occurs, by passing to a smaller subgroup of finite index in $\mathrm{Out}(\pi_1(V))$, we may assume that each $\overline{\phi}(\mathbb{Z}/p_j\mathbb{Z})$ is conjugate to $\mathbb{Z}/p_j\mathbb{Z}$.

Our first case corresponds to statement 3 of the lemma when B is a disk.

Case I: $p_1 \neq 0$ and $p_2 \neq 0$.

Since there are two exceptional fibers, B is a disk, and consequently ∂V is a torus. Since V is not the I-bundle over the Klein bottle, we may assume that at least one of the p_j, say p_1, is greater than 2. In addition, we may choose notation so that $\pi_1(\partial V)$ is generated by t and c_1c_2. By lemma 9.1.2, we may assume that $\overline{\phi}(\mathbb{Z}/p_1\mathbb{Z}) = \mathbb{Z}/p_1\mathbb{Z}$ and $\overline{\phi}(\mathbb{Z}/p_2\mathbb{Z}) = \mathbb{Z}/p_2\mathbb{Z}$. So by passing to a subgroup of finite index in $\mathrm{Out}(\pi_1(V))$, we may assume that $\phi(t) = t$ and $\phi(c_j) = c_j t^{m_j}$. But the relations $c_j^{p_j} = t^{q_j}$ imply that each $m_j = 0$, so $\mathrm{Out}(\pi_1(V))$ is finite. Therefore, we may pass to a subgroup of finite index in $\mathcal{E}(V, \underline{\underline{v}}', \underline{\underline{v}}'')$ to assume that f is homotopic to the identity map.

If $|\underline{\underline{v}}' \cup \underline{\underline{v}}''|$ is empty, then such a homotopy is automatically admissible, and the proof is complete in this case.

Now suppose that $|\underline{\underline{v}}' \cup \underline{\underline{v}}''|$ is nonempty. Since $|\underline{\underline{v}}' \cup \underline{\underline{v}}''|$ is not all of ∂V, it must contain an annulus A. Let f be the admissible homotopy equivalence obtained by sweeping A around c_2c_1. Under f^k, a peripheral loop c_1c_2 based at a basepoint v_0 in A is carried to $(c_2c_1)^k c_1c_2 (c_2c_1)^{-k}$. Suppose for contradiction that some nonzero power f^k is admissibly homotopic to the identity. We may assume that f fixes v_0, and since the homotopy is admissible, the trace of homotopy at v_0 lies in the subgroup $\pi_1(A, v_0)$ of $\pi_1(V, v_0)$. Since A is a fibered annulus, this implies that the trace is a power of t. Since t is central, this means that f^k induces the identity automorphism (not just the identity outer automorphism) on Q. In particular, the image of $\phi^k(c_1c_2)$ in Q would have to equal c_1c_2. Since c_1 has order at least 3, the normal form of elements in free products shows that this can only happen when $k=0$. Therefore the f^k represent distinct cosets of $\mathcal{H}(V, \underline{\underline{v}}', \underline{\underline{v}}'')$ in $\mathcal{E}(V, \underline{\underline{v}}', \underline{\underline{v}}'')$.

The second case gives statement 1 of the lemma.

Case II: $p_1 \neq 0$, $p_2=0$, and $\epsilon_2=1$.

In this case there is one exceptional fiber, and B is an annulus since $c_2tc_2^{-1}=t$. By lemma 9.1.2, ϕ may be changed by inner automorphism so that $\overline{\phi}(\mathbb{Z}/p_1\mathbb{Z}) = \mathbb{Z}/p_1\mathbb{Z}$ and $\overline{\phi}(c_2)=c_1^{\ell}c_2^{\pm1}$. By passing to a finite index subgroup of $\operatorname{Out}(\pi_1(V))$, we may assume that $\overline{\phi}(c_1)=c_1$ and $\overline{\phi}(c_2)=c_2$, and hence (using the fact that $p_1 \neq 0$ as in case I) that $\phi(c_1)=c_1$ and $\phi(c_2)=c_2t^k$ for some k.

Assume first that no component of $|\underline{\underline{v}}' \cup \underline{\underline{v}}''|$ is an annulus; we will argue that f is admissibly homotopic to a homeomorphism. Suppose first that $\underline{\underline{v}}''$ contains a boundary component T of V. Now $\pi_1(T) \cong \mathbb{Z} \times \mathbb{Z}$ is generated either by t and c_2 or by t and c_1c_2; since the restriction of f to T is isotopic to the identity, $k=0$ and therefore f is homotopic to the identity. Suppose that $\underline{\underline{v}}''$ does not contain a torus boundary component of V. Then ϕ is induced by a homeomorphism in $\mathcal{H}(V, \underline{\underline{v}}', \underline{\underline{v}}'')$ that takes each fiber to itself, in fact by the kth power of a Dehn twist about a nonseparating essential vertical annulus that meets both boundary components of V. So again, f is homotopic to a homeomorphism. Since no component of $|\underline{\underline{v}}' \cup \underline{\underline{v}}''|$ is an annulus, lemma 10.3.5 now shows that f is admissibly homotopic to an admissible homeomorphism. So if every component of $|\underline{\underline{v}}' \cup \underline{\underline{v}}''|$ is a torus, f is admissibly homotopic to a homeomorphism.

Suppose now that some component of $|\underline{\underline{v}}' \cup \underline{\underline{v}}''|$ is an annulus A. We may choose notation so that A lies in the boundary component with fundamental group generated by t and c_2 (by replacing c_2 with c_1c_2 and c_1 with c_1^{-1}, if necessary). Let f be the sweep of A around a loop that represents c_1c_2, then an argument exactly as in case I shows that $\mathcal{H}(V, \underline{\underline{v}}', \underline{\underline{v}}'')$ has infinite index.

The next case is statement 3 of the lemma when B is a Möbius band.

Case III: $p_1 \neq 0$, $p_2 = 0$ and $\epsilon_2=-1$.

As in Case II, we may assume that $\overline{\phi}(c_1)=c_1$ and $\overline{\phi}(c_2)=c_2$. Then, $\phi(c_1)=c_1$ and $\phi(c_2) = c_2t^k$ for some k. Since $tc_2t^{-1} = c_2t^{-2}$, we may conjugate by a power of t to assume that $k = 0$ or 1, then by passing to a subgroup of finite index in $\mathcal{E}(V, \underline{\underline{v}}', \underline{\underline{v}}'')$ we may assume that each homotopy equivalence is homotopic to the identity.

If $|\underline{\underline{v}}' \cup \underline{\underline{v}}''|$ is empty, this completes the proof. Suppose now that $|\underline{\underline{v}}' \cup \underline{\underline{v}}''|$ contains an annulus A. The fundamental group of the boundary torus is generated by t and $c_1c_2^2$. An argument as in case I, sweeping A around $c_2c_1c_2$, completes the proof.

The next case gives statement 2, and statement 3 for the case when B is a torus with one hole.

Case IV: $p_1=p_2=0$ and $\epsilon_1=\epsilon_2=1$.

In this case there is no exceptional fiber and B is either a torus with one hole or a disk with two holes, so this case of the lemma follows immediately from proposition 10.2.1.

The final two cases do not appear in the three statements of the lemma; they are the remaining manifolds satisfying the hypotheses and for which the index of $\mathcal{H}(V, \underline{\underline{v}}', \underline{\underline{v}}'')$ is infinite.

Case V: $p_1=0$, $\epsilon_1=1$, and $\epsilon_2=-1$.

In this case there is no exceptional fiber, and B is a Möbius band minus a disk. Proposition 10.2.1 again applies.

Case VI: $\epsilon_2=\epsilon_1=-1$.

Again there is no exceptional fiber, and B is a Klein bottle with one hole. Once more, proposition 10.2.1 applies. □

Our final lemma of the section handles case (vi) and shows that there are no other cases for which $\mathcal{H}(V,\underline{\underline{v}}',\underline{\underline{v}}'')$ has finite index in $\mathcal{E}(V,\underline{\underline{v}}',\underline{\underline{v}}'')$. (The cases where the sum of the rank of $H_1(B)$ and the number of exceptional fibers is 0 or 1 are handled by lemmas 10.3.2, 10.3.3, and 10.3.4.)

LEMMA 10.3.7. *Suppose* $(V,\underline{\underline{v}}'\cup\underline{\underline{v}}'')$ *is Seifert fibered over* $(B,\underline{\underline{b}})$, *and* $|\underline{\underline{v}}'\cup\underline{\underline{v}}''|\neq\partial V$. *Assume that the sum of the rank of* $H_1(B)$ *and the number of exceptional fibers of* V *is at least* 3. *Then* $\mathcal{H}(V,\underline{\underline{v}}',\underline{\underline{v}}'')$ *has finite index in* $\mathcal{E}(V,\underline{\underline{v}}',\underline{\underline{v}}'')$ *if and only if* B *is the* 2*-disk,* $|\underline{\underline{v}}'\cup\underline{\underline{v}}''|$ *is empty, and there are three exceptional fibers all of type* $(2,1)$.

It is interesting to note that in the case where the index is finite, V has a 2-fold branched cover which is the product of a circle and a torus with one hole.

PROOF. We will use the presentations for the fundamental groups of Seifert-fibered 3-manifolds given in section 2.1. First we assume B is the 2-disk, $|\underline{\underline{v}}'\cup\underline{\underline{v}}''|$ is empty, and there are three exceptional fibers all of type $(2,1)$. Then $\pi_1(V)$ can be presented as

$$\langle c_1,c_2,c_3,t \mid [c_j,t]=1 \text{ and } c_j^2=t \text{ for } 1\leq j\leq 3\rangle$$

where the fundamental group of the boundary torus is generated by t and $c_1c_2c_3$. Now $\mathcal{E}(V,\underline{\underline{\emptyset}},\underline{\underline{\emptyset}})$ is just the group of all self homotopy-equivalences of V, so is isomorphic to $\mathrm{Out}(\pi_1(V))$, since V is aspherical. We will show that every automorphism of $\pi_1(V)$ preserves $\pi_1(\partial V)$ up to conjugacy, so every homotopy equivalence is homotopic to a map preserving ∂V. By Waldhausen's Theorem 2.5.6 this proves it is homotopic to a homeomorphism.

The element t generates the center Z of $\pi_1(V)$, so every automorphism carries t to either t or t^{-1}. Every automorphism of $\pi_1(V)$ induces an automorphism on $\pi_1(V)/Z$, which is isomorphic to $\mathbb{Z}/2*\mathbb{Z}/2*\mathbb{Z}/2$ generated by c_1, c_2, and c_3. From section 9.2, we have generators of the automorphism group of a free product of finitely many indecomposable groups. For the special case of $\mathbb{Z}/2*\mathbb{Z}/2*\mathbb{Z}/2$, there are no infinite cyclic free factors, so the generators $\rho_{i,j}$, $\lambda_{i,j}$, and σ_i are not present. There are no factor automorphisms φ_i, since each of the factors $\mathbb{Z}/2$ has trivial automorphism group. For a given $i\neq j$, there is only one generator of the form $\mu_{i,j}(x)$, namely $\mu_{i,j}(c_i)$. So the only generators needed are:

(i) the $\omega_{i,j}$, which interchange c_i and c_j, and

(ii) $\mu_{i,j}(c_i)$, the automorphism that sends c_j to $c_ic_jc_i$ and fixes all other c_k.

(Actually, only $\mu_{12}(c_1)$ is needed, since all other $\mu_{i,j}(c_i)$ can be conjugated to this one using the $\omega_{i,j}$.) Each of these generators takes $c_1c_2c_3$ to an element conjugate to $c_1c_2c_3$ or its inverse $c_3c_2c_1$. Therefore every automorphism of $\pi_1(V)$ takes $c_1c_2c_3$ to $\omega(c_1c_2c_3)^{\pm1}t^k\omega^{-1}$ for some $\omega\in\pi_1(V)$ and some k, and hence preserves $\pi_1(\partial V)$ up to conjugacy (since the other generator t, being central, equals $\omega t\omega^{-1}$). That is, if $(V,\underline{\underline{v}}'\cup\underline{\underline{v}}'')$ is as in case (vi), then $\mathcal{H}(V,\underline{\underline{v}}',\underline{\underline{v}}'')=\mathcal{E}(V,\underline{\underline{v}}',\underline{\underline{v}}'')$.

From now on we suppose that $(V, \underline{\underline{v}}' \cup \underline{\underline{v}}'')$ is not the exceptional case described above, and must prove that $\mathcal{H}(V, \underline{\underline{v}}', \underline{\underline{v}}'')$ has infinite index in $\mathcal{E}(V, \underline{\underline{v}}', \underline{\underline{v}}'')$. We recall from section 2.1 that $\pi_1(V)$ has one of the three presentations:

$$\langle a_1, b_1, \ldots, a_g, b_g, c_1, \ldots, c_s, t \mid [a_i, t] = [b_i, t] = 1 \text{ for } 1 \le i \le g,$$
$$[c_j, t] = 1 \text{ for } 1 \le j \le s, \text{ and } c_j^{p_j} = t^{q_j} \text{ for } 1 \le j \le r\rangle\ ,$$
$$\text{if } B \text{ is orientable and non-planar,}$$

$$\langle v_1, \ldots, v_g, c_1, \ldots, c_s, t \mid v_i t v_i^{-1} = t^{-1} \text{ for } 1 \le i \le g,$$
$$[c_j, t] = 1 \text{ for } 1 \le j \le s, \text{ and } c_j^{p_j} = t^{q_j} \text{ for } 1 \le j \le r\rangle\ ,$$
$$\text{if } B \text{ is not orientable, or}$$

$$\langle c_1, \ldots, c_s, t \mid [c_j, t] = 1 \text{ for } 1 \le j \le s, \text{ and } c_j^{p_j} = t^{q_j} \text{ for } 1 \le j \le r\rangle$$
$$\text{if } B \text{ is planar.}$$

We interpret these presentations explicitly as follows. Start with B_0, a compact surface of orientable or nonorientable genus g, and having $s+1$ boundary components C, $C_1, \ldots\ C_s$. Fix a basepoint b_0 in C. Initially, we regard the generators of $\pi_1(V)$ other than t as loops in B_0 based at b_0. If B_0 is orientable, each pair a_i, b_i corresponds to a pair of loops determined by a torus summand of B_0; a_i and b_i intersect at one point other than b_0. These are the only loops that intersect at any point other than b_0. If B_0 is nonorientable, each v_i is a loop that passes through a crosscap of B_0. For $1 \le j \le s$, c_j encircles the boundary component C_j. For the three cases, respectively, the boundary circle C represents $\prod_{i=1}^{g}[a_i, b_i]\prod_{j=1}^{s} c_j$, $\prod_{i=1}^{g} v_i^2 \prod_{j=1}^{s} c_j$, or $\prod_{j=1}^{s} c_j$. In each case, we denote this element by c.

Now, let V_0 be the S^1-bundle over B_0 with orientable total space. The fiber represents an element t that generates an infinite cyclic normal subgroup of $\pi_1(V_0)$. Finally, form V by filling in the boundary tori that are the preimages of C_j for $1 \le j \le r$ with fibered solid tori of type (p_j, q_j). This adds the remaining relations $c_j^{p_j} = t^{q_j}$ to $\pi_1(V)$. The orbit surface B is obtained by filling in the circles $C_1, \ldots,$ C_r of B_0 with disks. We may regard B_0 as a section of V_0, contained in V, obtaining explicit loops representing the generators of $\pi_1(V)$ other than t.

Observe that the quotient of $\pi_1(V)$ by the infinite cyclic normal subgroup generated by t is a group Q given in the three cases by

$$\langle a_1, b_1, \ldots, a_g, b_g, c_1, \ldots, c_s \mid c_j^{p_j} = 1 \text{ for } 1 \le j \le r\rangle\ ,$$
$$\text{if } B \text{ is orientable and non-planar,}$$

$$\langle v_1, \ldots, v_g, c_1, \ldots, c_s \mid c_j^{p_j} = 1 \text{ for } 1 \le j \le r\rangle\ ,$$
$$\text{if } B \text{ is not orientable, or}$$

$$\langle c_1, \ldots, c_s \mid c_j^{p_j} = 1 \text{ for } 1 \le j \le r\rangle \text{ if } B \text{ is planar.}$$

This is also the quotient group of $\pi_1(B_0)$ obtained by adding the relations $c_j^{p_j} = 1$.

The boundary of V consists of the $1+s-r$ tori which are the preimages of the boundary components C, $C_{r+1}, \ldots,$ C_s of B. We may choose our notation so that the preimage torus T of C is not entirely contained in $|\underline{\underline{v}}' \cup \underline{\underline{v}}''|$. The fundamental group of T is generated by t and c.

To see that $\mathcal{H}(V, \underline{\underline{v}}', \underline{\underline{v}}'')$ has infinite index in $\mathcal{E}(V, \underline{\underline{v}}', \underline{\underline{v}}'')$, we will construct an admissible homotopy equivalence H of V, no nonzero power of which is realizable by a homeomorphism of V. To construct H, we start with a certain sweep $h(\gamma, \alpha)$ of B_0, and lift it to a sweep $H(\gamma, \alpha)$ of V_0 as in the construction in section 10.2. This $H(\gamma, \alpha)$ will be the identity on all boundary components other than T, so can be extended using the identity map on the fibered solid tori of $\overline{V - V_0}$, obtaining a fiber-preserving map H. By construction, the automorphism induced by H on the quotient Q of $\pi_1(V)$ is the same as that induced on Q, regarded as a quotient of $\pi_1(B_0)$, by $h(\gamma, \alpha)$. For simplicity of terminology, we regard $h(\gamma, \alpha)$ as a sweep on B, by letting it be the identity on the disks of $\overline{B - B_0}$, and refer to H as the sweep of the preimage of γ around α.

The first case will require us to use the boundary pattern, since, as we saw above, the index is finite if $|\underline{\underline{v}}' \cup \underline{\underline{v}}''|$ is empty. In the other cases, we can make no assumption about the boundary pattern other than that it is not all of ∂V.

Case I. B is the 2-disk, and there are three exceptional fibers all of type $(2, 1)$, and $|\underline{\underline{v}}' \cup \underline{\underline{v}}''|$ is not empty.

Since ∂V is a single torus, there must be an annulus component A of $|\underline{\underline{v}}' \cup \underline{\underline{v}}''|$. Its image in B is an arc component of $|\underline{\underline{b}}|$. Let γ be the boundary of a regular neighborhood of this arc in B_0, let b_0 be a basepoint in γ, and let H be a homotopy equivalence of V obtained by sweeping the preimage of γ around a loop representing c_1c_2 in $\pi_1(B_0, b_0)$. Fix a basepoint $v_0 \in A$. Since H fixes each point of A, we have $H(v_0) = v_0$. Suppose there is an admissible homotopy from H^n to a homeomorphism. Since the homeomorphism preserves A, we may assume that it fixes v_0. Since the homotopy is admissible, its trace at v_0 must lie in the subgroup $\pi_1(A, v_0)$, which is the subgroup generated by t. Since t is central, the homeomorphism must induce the same automorphism on $\pi_1(V, v_0)$ as does H^n, and hence must induce the same automorphism on Q. Since the homeomorphism preserves ∂V and takes t to $t^{\pm 1}$, it must take c to $c^{\pm 1}t^k$ for some k. So its induced automorphism on Q must carry c to $c^{\pm 1}$. But the induced automorphism of H^n carries $c = c_1c_2c_3$ to $(c_1c_2)^{n+1}c_3(c_1c_2)^{-n}$. Using the normal form in the free product Q, we see that this equals $(c_1c_2c_3)^{\pm 1}$ only when $n = 0$. So the H^n represent distinct cosets of $\mathcal{H}(V, \underline{\underline{v}}', \underline{\underline{v}}'')$ in $\mathcal{E}(V, \underline{\underline{v}}', \underline{\underline{v}}'')$. This completes the proof in Case I.

From now on, we assume that we are not in Case I. Since t generates the unique infinite cyclic normal subgroup of $\pi_1(V)$, every automorphism must take t to $t^{\pm 1}$. In each of the remaining three cases, H will have the property that for all $n \neq 0$, the automorphism it induces on Q carries c to an element not conjugate to $c^{\pm 1}$ or to $c_j^{\pm 1}$ for any $r + 1 \leq j \leq s$. If H^n were homotopic to a homeomorphism, then a loop representing c in $\pi_1(V)$ would be freely homotopic into ∂V, so the induced automorphism on Q would carry c to an element conjugate to $c^{\pm 1}$ or to $c_j^{\pm 1}$ for some $r + 1 \leq j \leq s$. As usual, it follows that $\mathcal{H}(V, \underline{\underline{v}}', \underline{\underline{v}}'')$ has infinite index in $\mathcal{E}(V, \underline{\underline{v}}', \underline{\underline{v}}'')$.

Case II. B is orientable and not planar.

Since B is not planar, we have $g \geq 1$ and, by hypothesis, $2g + s \geq 3$. Fix an arc γ, with endpoints in $C - |\underline{\underline{v}}' \cup \underline{\underline{v}}''|$ and with $\gamma \subset B_0 \subseteq B$, that meets a_1 transversely in a single point and is disjoint from all other a_i, b_i, and c_j. Choose a loop α in B_0, compatible with γ, representing $\prod_{i=2}^{g}[a_i, b_i] \prod_{j=1}^{s} c_j$. The effect of $h(\gamma, \alpha)$ on c is to insert α at each crossing of c with γ. There are two such crossings, corresponding to the a_1 and the a_1^{-1} in the expression for c, and they occur in opposite directions.

So $h(\gamma,\alpha)^n$ carries c to

$$\left(\prod_{i=2}^{g}[a_i,b_i]\prod_{j=1}^{s}c_j\right)^n a_1b_1a_1^{-1}\left(\prod_{i=2}^{g}[a_i,b_i]\prod_{j=1}^{s}c_j\right)^{-n} b_1^{-1}\prod_{i=2}^{g}[a_i,b_i]\prod_{j=1}^{s}c_j\ .$$

For $n \neq 0$, this is not conjugate to $c^{\pm 1}$ or to any $c_j^{\pm 1}$. As explained above, the induced automorphism of the lifted sweep H^n on Q has the same effect as $h(\gamma,\alpha)$, showing that it is not admissibly homotopic to a homeomorphism.

Case III. B is nonorientable.

Since B is not planar, we have $g \geq 1$ and, by hypothesis, $g+s \geq 3$. Select γ in B_0 having one transverse crossing with v_1 and disjoint from all other v_i and c_j. Choose α representing $\prod_{i=2}^{g} v_i^2 \prod_{j=1}^{s} c_j$. There are two crossings of c with γ, corresponding to the v_1 in the expression for c, and as in Case II this implies that $h(\gamma,\alpha)^n$ carries c to $\left((\prod_{i=2}^{g} v_i^2 \prod_{j=1}^{s} c_j)^n v_1\right)^2 \prod_{i=2}^{g} v_i^2 \prod_{j=1}^{s} c_j$, which is not conjugate to $c^{\pm 1}$ or to any $c_j^{\pm 1}$ when $n \neq 0$. Again, the lifted sweep H^n is not homotopic to a homeomorphism.

Case IV. B is planar.

By hypothesis $s \geq 3$. Select γ in B_0 crossing c_1 twice and disjoint from the other c_j, and not parallel into ∂B_0 (i. e. γ cuts off the exceptional point or boundary component encircled by c_1 from the rest of the exceptional points and boundary circles). Let α represent c_3c_2. Then $H^n(\gamma,\alpha)^n$ carries c to $(c_3c_2)^n c_1 (c_3c_2)^{-n} c_2c_3c_4\cdots c_s$. If $s \geq 4$, this is not conjugate in Q to c or any c_j, when $n \neq 0$. Suppose $s=3$. Since we are not in case I, we may assume that c_3 does not have order 2 in Q. With this condition, $(c_3c_2)^n c_1 (c_3c_2)^{-n} c_2c_3$ is not conjugate to $c^{\pm 1}$ or any $c_j^{\pm 1}$, when $n \neq 0$. As before, this produces the desired H.

We saw at the outset of the proof of lemma 10.3.7 that case IV fails if $s = 3$ and all c_j have order 2 in Q. Indeed, we then have $(c_3c_2)^n c_1 (c_3c_2)^{-n} c_2c_3 = (c_3c_2)^n c_1c_2c_3 (c_3c_2)^{-n}$. □

This completes the proof of proposition 10.3.1. □

10.4. Proof of Main Topological Theorem 2

In this section we deduce Main Topological Theorem 2, which was stated in section 8.1. For convenience, we will restate it here.

Main Topological Theorem 2: *Let M be a compact orientable irreducible 3-manifold with nonempty boundary and a (possibly empty) boundary pattern $\underline{\underline{m}}$ whose completion is useful. Assume that the elements of $\underline{\underline{m}}$ are disjoint. Let $\underline{\underline{m}}'$ be the set of elements of $\underline{\underline{m}}$ that are not annuli. Then $\mathcal{R}(M,\underline{\underline{m}})$ has finite index in* $\mathrm{Out}(\pi_1(M),\pi_1(\underline{\underline{m}}))$ *if and only if every Seifert-fibered component V of the characteristic submanifold of $(M,\underline{\underline{m}})$ that meets $\partial M - |\underline{\underline{m}}|$ satisfies one of the following:*

(1) *V is a solid torus, or*

(2) *V is either $S^1\times S^1\times I$ or the I-bundle over the Klein bottle, and no boundary component of V contains more than one component of $\overline{V\cap\partial M - |\underline{\underline{m}}'|}$, or*

(3) *V is fibered over the annulus with one exceptional fiber, and no component of $V\cap\partial M - |\underline{\underline{m}}'|$ is an annulus, or*

(4) *V is fibered over the disk with two holes with no exceptional fibers, and $V \cap \overline{\partial M - |\underline{\underline{m}}'|}$ is one of the boundary tori of V, or*
(5) *$V = M$ and either V is fibered either over the disk with two exceptional fibers, or over the Möbius band with one exceptional fiber, or over the torus with one hole with no exceptional fiber, or*
(6) *$V = M$ and V is fibered over the disk with three exceptional fibers, each of type $(2,1)$,*

and every I-bundle component V of the characteristic submanifold of $(M, \overline{\underline{\underline{m}}})$ which has all of its lids contained in $|\underline{\underline{m}}|$ and meets $\partial M - |\underline{\underline{m}}|$ satisfies one of the following:

(7) *V is a 3-ball, or*
(8) *V is I-fibered over a topological annulus or Möbius band and no component of $V \cap \overline{\partial M - |\underline{\underline{m}}|}$ is a square which meets two different components of the frontier of V, or*
(9) *V is I-fibered over the disk with two holes, and $V \cap \overline{\partial M - |\underline{\underline{m}}|}$ is an annulus, or*
(10) *$V = M$ and V is I-fibered over the torus with one hole.*

Proof of Main Topological Theorem 2. Theorem 10.1.1 showed that $\mathcal{R}(M, \underline{\underline{m}})$ has finite index in $\mathrm{Out}(\pi_1(M), \pi_1(\underline{\underline{m}}))$ if and only if for each component $(V, \widehat{\underline{\underline{v}}})$ of the characteristic submanifold of $(M, \overline{\underline{\underline{m}}})$, the subgroup $\mathcal{H}(V, \underline{\underline{v}}', \underline{\underline{v}}'')$ of elements of $\mathcal{E}(V, \underline{\underline{v}}', \underline{\underline{v}}'')$ realizable by orientation-preserving homeomorphisms has finite index in $\mathcal{E}(V, \underline{\underline{v}}', \underline{\underline{v}}'')$ (the notations $\underline{\underline{v}}'$, $\underline{\underline{v}}''$, $\mathcal{H}(V, \underline{\underline{v}}', \underline{\underline{v}}'')$, and $\mathcal{E}(V, \underline{\underline{v}}', \underline{\underline{v}}'')$ are defined at the beginning of chapter 10). We will examine the possibilities for each component $(V, \widehat{\underline{\underline{v}}})$ of the characteristic submanifold of $(M, \overline{\underline{\underline{m}}})$, and show that the index of $\mathcal{H}(V, \underline{\underline{v}}', \underline{\underline{v}}'')$ in $\mathcal{E}(V, \underline{\underline{v}}', \underline{\underline{v}}'')$ is as given in the following assertions:

I. If $(V, \widehat{\underline{\underline{v}}})$ is Seifert-fibered and does not meet $\partial M - |\underline{\underline{m}}|$, then the index is finite.
II. If $(V, \widehat{\underline{\underline{v}}})$ is Seifert-fibered and meets $\partial M - |\underline{\underline{m}}|$, then the index is finite if and only if V is one of the cases listed in items (1)-(6) of the theorem.
III. If $(V, \widehat{\underline{\underline{v}}})$ is an I-bundle with a lid that meets $\partial M - |\underline{\underline{m}}|$, then the index is finite.
IV. If $(V, \widehat{\underline{\underline{v}}})$ is an I-bundle which does not meet $\partial M - |\underline{\underline{m}}|$, then the index is finite.
V. If $(V, \widehat{\underline{\underline{v}}})$ is an I-bundle whose lid or lids are contained in $|\underline{\underline{m}}|$, but which does meet $\partial M - |\underline{\underline{m}}|$, then the index is finite if and only if V is one of the cases listed in items (7)-(10) of the theorem.

We begin with a Seifert-fibered component $(V, \widehat{\underline{\underline{v}}})$ of the characteristic submanifold of $(M, \overline{\underline{\underline{m}}})$. By definition, $\underline{\underline{v}}'$ consists of the components of the intersections of V with elements of $\underline{\underline{m}}'$, so we have the following observation, which will be used repeatedly:

$$(*) \qquad V \cap (\partial M - |\underline{\underline{m}}'|) = \partial V - |\underline{\underline{v}}' \cup \underline{\underline{v}}''| \,.$$

If V meets an annulus of $\underline{\underline{m}}$, then by lemma 2.10.8 V contains a regular neighborhood of that annulus, so V meets $\partial M - |\underline{\underline{m}}|$. So if V does not meet $\partial M - |\underline{\underline{m}}|$, it also does not meet $\partial M - |\underline{\underline{m}}'|$. In particular, if V does not meet $\partial M - |\underline{\underline{m}}|$, then $|\underline{\underline{v}}' \cup \underline{\underline{v}}''| = \partial V$ and proposition 10.3.1 shows that $\mathcal{H}(V, \underline{\underline{v}}', \underline{\underline{v}}'') = \mathcal{E}(V, \underline{\underline{v}}', \underline{\underline{v}}'')$, giving assertion I. Therefore, we may assume that $|\underline{\underline{v}}' \cup \underline{\underline{v}}''| \neq \partial V$.

If $\underline{\underline{v}}' \cup \underline{\underline{v}}''$ does not have useful completion, then lemma 2.6.1 implies that V is a solid torus. Lemma 10.3.2 then applies to show that $\mathcal{E}(V, \underline{\underline{v}}', \underline{\underline{v}}'')$ is finite. (This case is covered by item (1).)

Proposition 10.3.1 shows that if $\underline{\underline{v}}' \cup \underline{\underline{v}}''$ has useful completion, then $\mathcal{H}(V, \underline{\underline{v}}', \underline{\underline{v}}'')$ has finite index in $\mathcal{E}(V, \underline{\underline{v}}', \underline{\underline{v}}'')$ if and only if $(V, \widehat{\underline{\underline{v}}})$ is as described in items (1)-(6) of the theorem. Items (1)-(6) correspond to items (i)-(vi) in proposition 10.3.1, but we have altered the description in order to avoid mentioning $\underline{\underline{v}}'$ and $\underline{\underline{v}}''$ in the statement of the Theorem.

In item (2), we may use (∗) to see that the hypothesis in item (ii) of proposition 10.3.1 (that no boundary component of V contains more than one annulus of $|\underline{\underline{v}}' \cup \underline{\underline{v}}''|$) says exactly that no boundary component of V contains more than one component of $V \cap (\partial M - |\underline{\underline{m}}'|)$.

In item (3), we may again use (∗) to see that the hypothesis in item (iii) of proposition 10.3.1 (that no component $|\underline{\underline{v}}' \cup \underline{\underline{v}}''|$ is an annulus) corresponds to the condition that no component of $V \cap (\partial M - |\underline{\underline{m}}'|)$ is an annulus.

In item (4), we use (∗) to see that the hypothesis in item (iv) of proposition 10.3.1 (that $|\underline{\underline{v}}' \cup \underline{\underline{v}}''|$ consists of two boundary tori) says exactly that $V \cap (\partial M - |\underline{\underline{m}}'|)$ is a single boundary torus.

In items (5) and (6), the condition that $V = M$ corresponds exactly to the condition that $\underline{\underline{v}}''$ is empty. Similarly, the hypothesis that $V \cap (\partial M - |\underline{\underline{m}}'|)$ is nonempty is equivalent to $\underline{\underline{m}}$ consisting of annuli and hence to the condition that $\underline{\underline{v}}'$ is empty. This establishes assertion II.

Consider now an I-bundle component $(V, \widehat{\underline{\underline{v}}})$ of the characteristic submanifold of $(M, \overline{\underline{\underline{m}}})$, I-fibered over $(B, \widehat{\underline{\underline{b}}})$. Recall that $\underline{\underline{v}}' = \underline{\underline{v}}$, since V is an I-bundle, and $|\underline{\underline{v}}| = V \cap |\underline{\underline{m}}|$. Let $\underline{\underline{b}}'$ denote the images in B of the elements of $\underline{\underline{v}}'$ that are not lids, and let $\underline{\underline{b}}''$ denote the images of the elements of $\underline{\underline{v}}''$.

If $\underline{\underline{v}}' \cup \underline{\underline{v}}''$ does not have useful completion, then lemma 2.6.1 implies that V is a $(V, \overline{\underline{\underline{v}}' \cup \underline{\underline{v}}''})$ is an I-bundle over a small-faced disk. It is easy to check that $\mathcal{E}(V, \underline{\underline{v}}', \underline{\underline{v}}'')$ is finite in this case, so we may assume that $\underline{\underline{v}}' \cup \underline{\underline{v}}''$ has useful completion.

Suppose first that not all lids of $(V, \widehat{\underline{\underline{v}}})$ are contained in $|\underline{\underline{m}}|$. Then some lid lies in $\overline{\partial M - |\underline{\underline{m}}|}$, so every side of the I-bundle $(V, \overline{\underline{\underline{v}}})$ lies either in $|\underline{\underline{m}}|$ or in the frontier of V. That is, all sides of $(V, \overline{\underline{\underline{v}}' \cup \underline{\underline{v}}''})$ lie in $\underline{\underline{v}}' \cup \underline{\underline{v}}''$, so $|\underline{\underline{b}}' \cup \underline{\underline{b}}''| = \partial B$. By proposition 10.2.1, $\mathcal{H}(V, \underline{\underline{v}}', \underline{\underline{v}}'')$ has finite index in $\mathcal{E}(V, \underline{\underline{v}}', \underline{\underline{v}}'')$, establishing assertion III.

Assume now that all lids of $(V, \widehat{\underline{\underline{v}}})$ lie in $|\underline{\underline{m}}|$. This implies that $\underline{\underline{v}}'$ must equal the set of lids of $(V, \widehat{\underline{\underline{v}}})$. For if a side of $(V, \widehat{\underline{\underline{v}}})$ were contained in $|\underline{\underline{m}}|$, then since the elements of $\underline{\underline{m}}$ are disjoint, there would be a lid in $\overline{\partial M - |\underline{\underline{m}}|}$. So $\underline{\underline{v}}'$ equals the set of lids of $(V, \widehat{\underline{\underline{v}}})$, and $\underline{\underline{b}}'$ is empty.

If V does not meet $\overline{\partial M - |\underline{\underline{m}}|}$, then $|\underline{\underline{b}}''| = \partial B$ and by proposition 10.2.1, $\mathcal{H}(V, \underline{\underline{v}}', \underline{\underline{v}}'')$ has finite index in $\mathcal{E}(V, \underline{\underline{v}}', \underline{\underline{v}}'')$, giving assertion IV. Suppose that V does meet $\overline{\partial M - |\underline{\underline{m}}|}$. Since $|\underline{\underline{b}}' \cup \underline{\underline{b}}''| \neq \partial B$, proposition 10.2.1 then shows that $\mathcal{H}(V, \underline{\underline{v}}', \underline{\underline{v}}'')$ has finite index in $\mathcal{E}(V, \underline{\underline{v}}', \underline{\underline{v}}'')$ if and only if V is as described in items (7)-(10) of the theorem, giving assertion V and completing the proof. □

CHAPTER 11

Dehn Flips

Dehn flips are certain kinds of homotopy equivalences between 2-manifolds or 3-manifolds, which are homeomorphisms outside a square (in the 2-dimensional case) or a solid torus (in the 3-dimensional case). Precise definitions are given below. Johannson used Dehn flips to give a description of the homotopy type of a Haken 3-manifold (see pp. 3, 227, and 243-249 of [**58**]). According to theorem 29.1 of [**58**], each Haken 3-manifold M with incompressible boundary contains a union W of disjoint essential solid tori so that any Haken 3-manifold homotopy equivalent to M is obtained by Dehn flips along components of W. That is, Dehn flips "generate" the homotopy type of a boundary-incompressible Haken manifold, which consequently contains only finitely many homeomorphism types. Moreover, the proof is constructive in the sense that one can obtain W in a finite number of steps from a triangulation of M. This compares with our results in section 4.2, where the finiteness of the admissible homotopy type is obtained whenever the boundary pattern has useful completion (theorem 4.2.1), or even is only usable (theorem 4.2.3), but the proof is nonconstructive.

When the solid torus used to construct a Dehn flip is a component of the characteristic submanifold of M, and the core circles of its frontier annuli generate the fundamental group of the solid torus, the Dehn flip is called a primitive shuffle homotopy equivalence. These play a central role in [**9**] (see section 13.2 below).

In this section we will show explicitly how to write sweeps as compositions of Dehn flips. While not essential to our main program, it seems of interest to clarify this relationship.

In dimension 2, a *Dehn flip* is a homotopy equivalence $f\colon F_1 \to F_2$ between 2-manifolds such that for some square $W_1 \subset F_1$, meeting ∂F_1 in two opposite sides, and a corresponding square $W_2 \subset F_2$, we have $f^{-1}(W_2) = W_1$ and $f|_{\overline{F_1 - W_1}}\colon \overline{F_1 - W_1} \to \overline{F_2 - W_2}$ a homeomorphism. A good example to keep in mind is a homotopy equivalence from an annulus to a Möbius band that is a homeomorphism outside a regular neighborhood of an arc that connects the two boundary circles. Note that the composition of any Dehn flip with a homeomorphism is a Dehn flip. The definition is analogous in dimension 3, with an essential solid torus (or possibly, in the context of manifolds with boundary pattern, an I-bundle over the disk) in the role of W_1. The frontier of the torus consists of two or more essential annuli.

A sweep and a Dehn flip are similar in that their restriction to the complement of a set of squares (or solid tori and 3-balls, in the 3-dimensional case) is a homeomorphism, but a sweep will generally move points in the square into its complement, while Dehn flips will preserve the squares. We will show explicitly that every sweep around a simple closed curve can be written as a composition of three Dehn flips. Since every sweep is a product of sweeps around simple closed

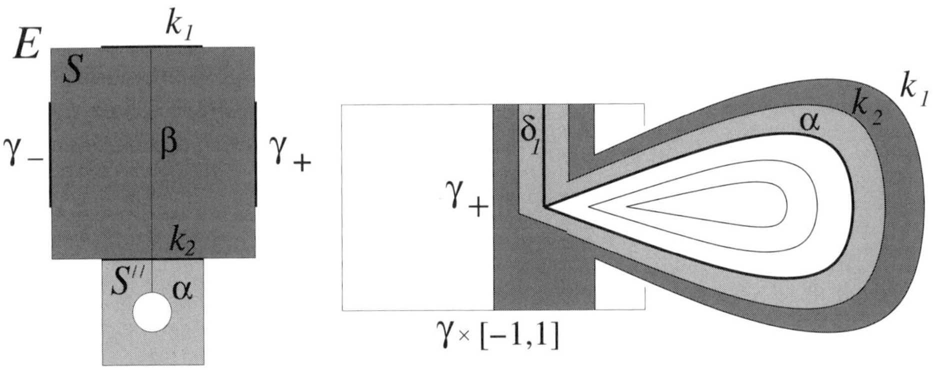

FIGURE 11.1. The surface E

curves (because $\pi_1(B - \gamma)$ is generated by imbedded loops), this gives an explicit way to write any sweep (up to admissible isotopy) as a product of Dehn flips.

We first focus on the 2-dimensional case. The idea of the proof is first to isolate the key part of the construction. Any sweep $h(\gamma, \alpha)$ with α a simple loop is supported on a regular neighborhood of the union of α with the arc $\gamma \times \{1/2\}$ (for sweeps we will use the notation given in the definition of $h(\gamma, \alpha)$ in section 10.1). This neighborhood is an annulus whose frontier consists of one boundary circle parallel to α and two arcs γ_+ and k_1 in its other boundary circle. We first coordinatize an annulus E and describe in detail how a sweep of γ about α on a surface containing E is homotopic to a composition of three Dehn flips. To finish the proof, we tell how this particular coordinatization can placed on any sweep of an arc about a simple closed loop.

In dimension 2, the only kind of Dehn flip we will need is when F_2 is a surface $S(F_1, k)$ obtained by cutting F_1 along k and regluing using an orientation-reversing homeomorphism of k. The Dehn flip is the identity homeomorphism outside a regular neighborhood W_1 of k in F_1, and is a half twist from W_1 to a regular neighborhood of k in F_2. In the remainder of our discussion, all Dehn flips will be of this type.

We will use a surface E which will support the sweep and the Dehn flips. Figure 11.1 shows E, abstractly and also as it will appear later as s submanifold of a more complicated surface. To begin the construction of E, let $S = [-2, 2] \times [-2, 2] \subset \mathbb{R}^2$. Let $k_1 = [-1, 1] \times \{2\}$, $k_2 = [-1, 1] \times \{-2\}$, $\gamma_+ = \{2\} \times [-1, 1]$, and $\gamma_- = \{-2\} \times [-1, 1]$, these are four disjoint arcs in ∂S. Let S' be the square $[-1, 1] \times [-4, -2] \subset \mathbb{R}^2$, which intersects S in k_2, and let S'' be obtained from S' by removing a small open disk centered at $(0, -3)$, leaving a boundary circle α. Let $E = S \cup S''$. Finally, let β be the arc in E which starts at $(0, 2)$ and travels straight down $\{0\} \times [-4, 2]$ until it meets α. We regard α as having the clockwise orientation.

Now let X' be a compact surface, connected or not, let α' be a boundary circle of X', and let k_1' and γ' be two disjoint copies of $[-1, 1]$ in $\partial X' - \alpha'$. Form compact surfaces X_+ (respectively, X_-) from the disjoint union $X' \sqcup E$ by identifying k_1 with k_1', α with α', and γ_+ (respectively, γ_-) with γ'. The identifications are made in such a way that the $[-1, 1]$-coordinates in the two identified arcs agree.

We now describe a homeomorphism h from X_- to $S(X_+, \gamma')$. On E, it is the final homeomorphism of an isotopy of E that starts at the identity. The isotopy "slides" γ_- down the left side of S, along $S \cap \partial E$ to S'', along $S'' \cap \partial E$ to the other side of S'', then up the right side of S and onto γ_+. Thus h carries γ_- to γ_+. More precisely, it carries $(-2, t) \in \gamma_-$ to $(2, -t)$ in γ_+. That is, it carries the point identified with a given t in γ' in X_- to the point identified with t in $S(X_+, \gamma')$, and so it extends using the identity map of X' to a homeomorphism h from X_- to $S(X_+, \gamma')$. We can easily arrange that the isotopy fix $\alpha \cup \beta \cup k_1$, so that h will be the identity map on $\alpha \cup \beta \cup k_1$.

We also need a homeomorphism k from $S(S(X_+, k_2), k_1)$ to X_-. It is the identity on $X' \cup S''$, and on S it sends (x, y) to $(-x, y)$. Note that it is the identity map on β.

Now, let f_1 and f_2 be Dehn flips along k_1 and k_2 respectively, so $f_1 f_2$ carries X_+ to $S(S(X_+, k_2), k_1)$. We may select f_1 and f_2 to be the identity on $X' \cup S'' \cup \beta$. The composition hkf_1f_2 carries X_+ to $S(X_+, \gamma')$ and is the identity on $X' \cup \alpha \cup \beta$. Let f_0 be a Dehn flip of $S(X_+, \gamma')$ along γ', it may also be selected to be the identity on $X' \cup S'' \cup \beta$. Thus, the composition $f_0hkf_1f_2$ carries X_+ to X_+ and is the identity on $X' \cup \alpha \cup \beta$. We will identify $f_0hkf_1f_2$ as a sweep.

Let ω be the component of $E \cap \partial X_+$ that contains $(2, 2)$, that is, the arc in $S \cap \partial E$ that connects γ_+ to k_1. There is a deformation retraction from E to $k_1 \cup \alpha \cup \beta \cup \omega \cup \gamma_+$, and it follows that a map of X_+ that is the identity on $X' \cup \alpha \cup \beta$ is determined up to homotopy by its effect on ω. Now f_2 fixes ω, and f_1 moves it to an arc in S near $k_1 \cup \omega$ that connects $(2, 1)$ to $(-1, 2)$. Then, k moves this arc to an arc in S that connects $(-2, 1)$ to $(1, 2)$, and h move this to an arc in E that connects $(2, -1)$ to $(1, 2)$ and which circles clockwise around α (i. e. α and $(2, 2)$ lie in the same component of the complement of the arc in E). Finally, f_0 moves it to an arc in E that connects $(2, 1)$ to $(1, 2)$ and circles clockwise around α. This agrees with the effect on ω of a sweep that carries γ' clockwise around α (and which is also the identity on $X' \cup \alpha \cup \beta$), so $f_0hkf_1f_2$ is homotopic to this sweep. In fact, the homotopy can be supported on E, so will be admissible provided that E is disjoint from the boundary pattern of X_+.

Now we can explain how to write a sweep $h(\gamma, \alpha)$ of a surface B as a product of three Dehn flips. Let $\gamma_1 = \gamma \times \{1/2\}$, and let δ_1 be the closure of one of the components of $\gamma_1 - (\{b_0\} \times \{1/2\})$. Let N be a regular neighborhood of $\alpha \cup \gamma_1$, disjoint from γ (and from all elements of the boundary pattern on the surface). The frontier of N has three components: a loop parallel to α, an arc parallel to γ, and an arc k_1. Let N_1 be a regular neighborhood of $\alpha \cup \delta_1$, contained in the interior of N, and let k_2 be the component of the frontier of N_1 that is an arc. We now have exactly the configuration described in the previous paragraphs; the square S corresponds to the submanifold of B bounded by γ, k_1, and k_2, and the region S'' corresponds to the submanifold bounded by k_2 and α. The sweep can now be factored into a product of three Dehn flips exactly as described in the previous paragraph. The direction of the sweep around α is determined by which component of $\gamma_1 - (\{b_0\} \times \{1/2\})$ is used to define δ_1.

By taking the product of this 2-dimensional construction with I or S^1, we can write a 3-dimensional sweep as a product of Dehn flips on cubes or solid tori.

CHAPTER 12

Finite Index Realization For Reducible 3-Manifolds

In our work on the Finite Index Realization Problem, we have restricted attention to Haken 3-manifolds, and it is natural to wonder what happens for larger classes of compact 3-manifolds. In this chapter, we will prove some results for the case of reducible 3-manifolds. They apply when the 3-manifolds are nonorientable, as well. In chapter 13, we will give a conjectural view of the Finite Index Realization Problem for all compact 3-manifolds.

As in the irreducible case, the Finite Index Realization Problem breaks into the cases when ∂M is compressible and when it is incompressible. But first, there is a technical reduction. As we will review in section 12.1, any compact 3-manifold can be factored in an (almost) unique way as a connected sum of prime 3-manifolds. There are at most two ways that a simply-connected summand can occur. First, there can be summands that are 3-balls. These appear as product neighborhoods of 2-sphere boundary components. Second, if the Poincaré Conjecture turns out to be false, then there could be simply-connected prime summands that are homotopy 3-spheres. These would appear in M as "fake 3-cells". To make our main results apply even when there are simply-connected summands, we use the Poincaré associate $P(M)$, which is the manifold obtained from M by replacing each simply-connected prime summand with a 3-sphere (see section 12.1 below for a more precise definition, also see Appendix I of [**69**] or p. 88 of [**51**]). We prove in proposition 12.1.4 that $\mathcal{R}(M)$ has finite index in $\mathrm{Out}(\pi_1(M))$ if and only if $\mathcal{R}(P(M))$ has finite index in $\mathrm{Out}(\pi_1(P(M)))$. By virtue of this reduction, we need only consider the Poincaré associate of M.

The main results of this section apply when $P(M)$ is reducible, and make no assumption of orientability. When M has compressible boundary, we prove in theorem 12.2.1 that $\mathcal{R}(M)$ has finite index in $\mathrm{Out}(\pi_1(M))$ only when M is a connected sum of a 3-manifold with finite fundamental group and a 3-manifold that is a solid torus or solid Klein bottle. This is similar in flavor to the results of section 9.3, indeed a significant portion of the proof of theorem 12.2.1 is identical to a portion of the proof of proposition 9.3.1. When M has incompressible boundary, we prove in theorem 12.3.1 that $\mathcal{R}(M)$ has finite index in $\mathrm{Out}(\pi_1(M))$ if and only if $\mathcal{R}(M_i)$ has finite index in $\mathrm{Out}(\pi_1(M_i))$ for each irreducible prime summand M_i of M.

The proofs of the reduction to $P(M)$ and the main result for the case when ∂M is incompressible use the concept of uniform mapping classes developed in [**86**]. In section 12.1, we review the necessary ideas and results from [**86**], after setting up notation and reviewing the general theory of reducible 3-manifolds. The reduction to $P(M)$ is the final result in that section. Sections 12.2 and 12.3 contain the main results for the cases when ∂M is compressible and incompressible, respectively.

12.1. Homeomorphisms of connected sums

Recall that the *connected sum* $P\#Q$ of two (connected) 3-manifolds P and Q is constructed by removing the interior of a closed 3-cell from each of P and Q, obtaining manifolds that we shall denote by P' and Q', and identifying the resulting 2-sphere boundary components by a homeomorphism. In particular, $P\#S^3$ is homeomorphic to P.

Any two 3-balls in P are ambiently isotopic, and there are two isotopy classes of homeomorphisms that one can use to identify two 2-spheres, so at most two manifolds can result from a given P and Q. If P and Q are orientable and have fixed orientations, then we identify the 2-sphere boundary components using an orientation-reversing homeomorphism so that $P\#Q$ will have an orientation that restricts to the given ones on each of P' and Q'. It is possible that $P\#Q$ and $P\#(-Q)$ are not homeomorphic, where $-Q$ denotes Q with the other orientation. When dealing with orientable manifolds, one usually assumes that they carry fixed orientations, so that the connected sum operation is well-defined. If one of the summands is nonorientable, then the homeomorphism type of the sum is uniquely determined. For if Q is nonorientable, there is a homeomorphism of Q' that reverses orientation on the 2-sphere boundary component, and this allows a homeomorphism to be defined between the two manifolds that result by gluing the 2-spheres using the two isotopy classes of gluing homeomorphism. The homeomorphism of Q' comes from sliding the 3-ball $\overline{Q - Q'}$ around an orientation-reversing loop in Q. Homeomorphisms based on this type of sliding construction play a key role in the results we develop in this chapter, and they will be discussed in detail below.

A 3-manifold is *prime* if it is not the 3-sphere, and whenever it is written as a connected sum, one of the summands must be the 3-sphere. We will tacitly assume throughout the remainder of this chapter that our 3-manifolds are not the 3-sphere. Since our main results are trivially true for the 3-sphere, it does not need to be excluded from their statements.

The standard references on 3-manifolds, such as [**51**], detail the Kneser-Milnor factorization of a compact 3-manifold $M \neq S^3$ as a connected sum $(\#_{i=1}^{r} M_i)\#(\#_{j=1}^{s} N_j)$ with each M_i and N_j prime. The M_i are irreducible, and each of the N_j is one of the two S^2-bundles over S^1, either $S^1 \times S^2$ or the nontrivial S^2-bundle over S^1, which has nonorientable total space. The irreducible prime summands M_i are unique up to homeomorphism. More precisely, if there is a homeomorphism h from $(\#_{i=1}^{r} M_i)\#(\#_{j=1}^{s} N_j)$ to $(\#_{i=1}^{r'} W_i)\#(\#_{j=1}^{s'} T_j)$, then $r = r'$, and after permutation of the indices, there are homeomorphisms h_i from M_i to W_i for $1 \leq i \leq r$. If M is orientable and h is orientation-preserving, then all h_i may be chosen to be orientation-preserving. For the summands that are S^2-bundles over S^1, the statement is a bit complicated. One always has $s = s'$. If M and N are orientable, then all N_j and T_j are orientable so must be $S^2 \times S^1$. If some M_i is nonorientable, then all N_j and T_j may be chosen to be $S^2 \times S^1$. If all M_i are orientable, but M is nonorientable, then one may choose N_1 and T_1 to be the nontrivial S^2-bundle over S^1 and all other N_j and T_j to be $S^2 \times S^1$. All that is going on is that if N is the nontrivial S^2-bundle over S^1, and P is any nonorientable

3-manifold, then $P\#N$ is homeomorphic to $P\#(S^2 \times S^1)$, and this accounts for all nonuniqueness of the summands.

We will now fix some more precise notation. Since we will be drawing on ideas and results from [**86**], we will stay close to the notation used there. Let Σ be the result of removing from a 3-sphere the interiors of $r + 2s$ disjoint (smoothly imbedded) 3-balls B_1, $B_2, \ldots, B_r$, D_1, E_1, $D_2, E_2, \ldots, D_s, E_s$. For $1 \leq i \leq r$, let M_i' result from removing from M_i the interior of a 3-ball C_i. Construct M from the disjoint union of Σ and the M_i' and s copies $S_j \times \mathrm{I}$ of $S^2 \times \mathrm{I}$ by identifying each ∂B_i with ∂C_i, each $S_j \times \{0\}$ with ∂D_j, and each $S_j \times \{1\}$ with ∂E_j.

We will use some of the generating automorphisms (and later, some of the relations) of $\mathrm{Aut}(\pi_1(M))$ given in Fouxe-Rabinovitch [**41**]. These were detailed at the start of section 9.2, and we will continue to use the notation and terminology established there. To set notation in the present situation, assume that the fundamental group $\pi_1(M)$ is based at a point in the interior of Σ, and regard it as a free product $G_1 * \cdots * G_r * G_{r+1} * \cdots * G_{r+s}$, where $G_i = \pi_1(M_i)$ for $1 \leq i \leq r$ and $G_{r+j} = \pi_1(N_j)$ for $1 \leq j \leq s$. Thus all G_{r+j} are infinite cyclic, and a G_i with $i \leq r$ may be infinite cyclic as well, when M_i is a solid torus or solid Klein bottle.

For now, assume that M is orientable. In [**86**], five kinds of homeomorphisms of M are defined: Dehn twists about 2-spheres (called "rotations" in [**86**]), factor homeomorphisms (supported in one of the M_i'), basic slide homeomorphisms (that slide an M_j' or one of the two ends of an $S_j \times \mathrm{I}$ around a loop α in M that passes through Σ and exactly one M_i or $S_i \times \mathrm{I}$), interchanges (of a pair of homeomorphic M_i and M_j, or of an $S_i \times \mathrm{I}$ and an $S_j \times \mathrm{I}$), and spins (that interchange the ends ∂D_j and ∂E_j of an $S_j \times \mathrm{I}$). Their induced automorphisms on $\pi_1(M)$ are as follows:

(1) Dehn twists about 2-spheres induce the identity automorphism, since they are supported on simply-connected subsets.
(2) Factor homeomorphisms induce factor automorphisms ϕ_i.
(3) Basic slide homeomorphisms induce $\mu_{i,j}(x)$, when sliding ∂M_j, $\lambda_{i,r+j}(x)$, when sliding the ∂D_j-end of $S_j \times \mathrm{I}$, and $\rho_{i,r+j}(x)$, when sliding the ∂E_j-end of $S_j \times \mathrm{I}$. In all cases, the element x represented by α lies in G_i, that is, in $\pi_1(M_i)$ if $i \leq r$ and in $\pi_1(N_{i-r})$ if $i > r$.
(4) Interchanges of homeomorphic M_i and M_j induce interchange automorphisms $\omega_{i,j}$ of the free factors $\pi_1(M_i)$ and $\pi_1(M_j)$ of $\pi_1(M)$. Interchanges of $S_i \times \mathrm{I}$ and $S_j \times \mathrm{I}$ induce interchange automorphisms $\omega_{r+i,r+j}$ of the infinite cyclic factors $\pi_1(N_i)$ and $\pi_1(N_j)$ of $\pi_1(M)$.
(5) Spins induce σ_{r+j}.

All of these automorphisms were defined in section 9.2, with the exception of the interchange automorphisms of isomorphic $\pi_1(M_i)$ and $\pi_1(M_j)$. Since these will be involved in the proof of theorem 12.3.1 below, we will take a moment to define them carefully here. For notational simplicity, suppose that $\pi_1(M_1)$ is isomorphic to $\pi_1(M_i)$ for $2 \leq i \leq \ell$. Fix isomorphisms $\alpha_i \colon \pi_1(M_1) \to \pi_1(M_i)$, and let α_1 be the identity on $\pi_1(M_1)$. Put $\alpha_{i,j} = \alpha_j \circ \alpha_i^{-1}$. Then, for $1 \leq i, j \leq \ell$, $\omega_{i,j}$ is defined to be $\alpha_{i,j}(x)$ for $x \in \pi_1(M_i)$ and $\alpha_{j,i}(x)$ for $x \in \pi_1(M_j)$, and to fix elements in all other factors. Sending each $\omega_{i,j}$ to the permutation it induces on the set $\{\pi_1(M_1), \ldots, \pi_1(M_\ell)\}$ defines an isomorphism from the subgroup of $\mathrm{Aut}(\pi_1(M))$ generated by these $\omega_{i,j}$ to the symmetric group on ℓ letters. This follows since the orbit of each element x of $\pi_1(M_1)$ is exactly $\{x, \alpha_2(x), \alpha_3(x), \ldots, \alpha_\ell(x)\}$, so if any composition of the $\omega_{i,j}$ preserves $\pi_1(M_k)$, it must actually restrict to the identity automorphism on $\pi_1(M_k)$. In particular, any composition that induces the identity permutation on $\{\pi_1(M_1), \ldots, \pi_1(M_\ell)\}$ must actually be the identity automorphism of $\pi_1(M)$.

As in chapter 9, we will always assume that the subscripts of the Fouxe-Rabinovitch generators lie in the appropriate ranges for which they are defined. That is, for $\mu_{j,k}(x)$, we have $1 \le j \le r+s$ and $1 \le k \le r$, while for $\rho_{j,k}(x)$ and $\lambda_{j,k}(x)$, we have $1 \le j \le r+s$ and $r+1 \le k \le r+s$. For ϕ_i, $1 \le i \le m$, and for σ_j, $r+1 \le j \le r+s$. For $\omega_{i,j}$, either $1 \le i,j \le r$ and $\pi_1(M_i) \cong \pi_1(M_j)$, or $r+1 \le i,j \le r+s$.

The slide homeomorphisms we will use are analogous to the slide homeomorphisms of compression bodies that were described in section 9.2, except that 2-spheres are in the role of the 2-disks used there. Since they play a pivotal role in our arguments, we describe them more precisely here. Let α be an arc imbedded in $\overline{M - M_j'}$, meeting M_j' only in its endpoints. Let M_0 be the manifold resulting from M by replacing M_j' by the 3-ball B_j. Since M is orientable, a regular neighborhood of $B_j \cup \alpha$ must be a solid torus. Choose an isotopy J_t of M_0 such that

(i) J_0 is the identity map,
(ii) each J_t is the identity outside a regular neighborhood of $B_j \cup \alpha$,
(iii) J_1 is the identity on B_j, and
(iv) during the isotopy J_t, B_j travels once around α.

Define a homeomorphism h of M by taking J_1 on $\overline{M - M_j'}$ and the identity on M_j'. We call h a *slide homeomorphism* which *slides* M_j *around* α. Although it will not have any direct effect on our work, we mention that the slide homeomorphism is not well-defined up to isotopy by the description we have given here. Different choices of the isotopy J_t can result in homeomorphisms that differ by a Dehn twist about the 2-sphere ∂B_j, and such Dehn twists may not be isotopically trivial in M. In all of our constructions, when we say to slide M_j around α, we really mean to select one of the possible slide homeomorphisms determined by the choice of M_j and α.

In order to slide the ∂D_j-end or the ∂E_j-end of $S_j \times I$ around a loop, a similar sliding construction can be performed using the manifold obtained from M by replacing $S_j \times I$ by the balls D_j and E_j for some fixed j. In this case, the isotopy J_t slides one of the 3-balls while keeping the other fixed. For interchanges of homeomorphic prime summands and spins of an $S_j \times I$ (that interchange ∂D_j and ∂E_j), one uses a sliding construction in which J_t moves both of the filled-in balls, and J_1 interchanges them.

When an automorphism of $\pi_1(M_i)$ can be realized by a homeomorphism, we can try to realize the corresponding factor automorphism of $\pi_1(M)$ using a factor homeomorphism. If the homeomorphism of M_i is orientation-preserving, we may change it by isotopy to restrict to the identity on C_i. Then, we restrict it to M_i' and use the identity map to extend over the rest of M. However, if the homeomorphism is orientation-reversing on M_i, this is not possible. For nonorientable M, the same kind of problem arises in the definition of a slide homeomorphism of M_j around an arc α which reverses the local orientation. Then, a regular neighborhood of $B_j \cup \alpha$ is a solid Klein bottle, and the isotopy cannot be chosen so that its final map J_1 is the identity on B_j. To overcome these difficulties, we will use the idea of a uniform homeomorphism. This appeared in [**86**], but we will give a complete and self-contained treatment here.

From now on, we will allow M to be nonorientable. Form a collection $\{M(\mu_1, \ldots, \mu_r, \tau_1, \ldots, \tau_s) \mid \mu_i, \tau_j \in \{1, -1\}\}$, where $M(\mu_1, \ldots, \mu_r, \tau_1, \ldots, \tau_s)$ is the manifold obtained from Σ, the M_i', and the $S_j \times \mathrm{I}$ as follows. Identify each ∂C_i to ∂B_i using the same homeomorphism as was used for M, if $\mu_i = 1$, but using the composition of this homeomorphism with a standard orientation-reversing reflection of ∂C_i if $\mu_i = -1$. Identify $S_j \times \{0\}$ to ∂D_j as it was identified to form M, but identify $S_j \times \{1\}$ to ∂E_j using either the same identification or its composition with a standard orientation-reversing reflection, as governed by the values of the additional parameters $\tau_1, \ldots, \tau_s$. That is, if $\tau_i = 1$, we identify them as they are identified to form M, while if $\tau_i = -1$, we change the identification by composing it with a standard orientation-reversing reflection of ∂E_j. Notice that M may be identified with $M(1, \ldots, 1, 1, \ldots, 1)$.

The disjoint union of all the $M(\mu_1, \ldots, \mu_r, \tau_1, \ldots, \tau_s)$ is a 3-manifold with 2^{r+s} components, and will be denoted by $\mathcal{M}$. We use μ as an abbreviation for $\mu_1, \ldots, \mu_r$ and similarly for τ. Thus, a typical component of $\mathcal{M}$ may be written as $M(\mu, \tau)$. Also, we shorten r 1's or s 1's to a single 1, writing $M(1, 1)$ for $M(1, \ldots, 1, 1, \ldots, 1)$.

A homeomorphism of M_i is isotopic to one which is either the identity or the standard orientation-reversing reflection on C_i, and its restriction to M_i' can then be extended on each component of $\mathcal{M}$, provided that we regard it as sending $M(\mu_1, \ldots, \mu_r, \tau)$ to $M(\mu_1, \ldots, \epsilon\mu_i, \ldots, \mu_r, \tau)$, where ϵ is 1 if the homeomorphism is the identity on ∂C_i and is -1 if it is the reflection. The resulting homeomorphism of $\mathcal{M}$ is called a uniform factor homeomorphism. A technical point arises here in realizing factor automorphisms of $\pi_1(M)$. A factor outer automorphism ϕ_i of $\pi_1(M)$ corresponds to a uniquely determined automorphism ϕ (not just outer automorphism) of $\pi_1(M_i)$. So in trying to realize ϕ by a uniform factor homeomorphism, one must work with homeomorphisms of M_i that fix a basepoint in C_i, and isotopies that preserve this basepoint. Even when M_i is nonorientable, a realizable automorphism of $\pi_1(M_i)$ is not necessarily realizable by a homeomorphism of M_i that is the identity on C_i.

We will now discuss uniform slide homeomorphisms of $\mathcal{M}$, beginning with the case of a uniform slide of an M_j, which realizes $\mu_{i,j}(x)$. We start with a sliding arc α in one of the components of $\mathcal{M}$, that meets M_j' only in its endpoints. It defines corresponding arcs in the other components, provided that we are careful enough to choose α so that it intersects all the 2-spheres ∂B_k, ∂D_k, and ∂E_k in points that lie in the circles fixed by the standard orientation-reversing reflections used to construct the $M(\mu, \tau)$. This can always be achieved by isotopy of α. When this condition holds, the portions of α in the copies of the M_k', the $S_k \times \mathrm{I}$, and Σ in every $M(\mu, \tau)$ fit together to form a sliding path.

Even though α is not a loop, it makes sense to say that α is orientation-preserving or orientation-reversing, since it starts and ends in the orientable submanifold Σ. Notice that α may be orientation-preserving in some components of $\mathcal{M}$ and orientation-reversing in others. For example, suppose that an orientation-preserving sliding arc α in $M(1, 1)$ travels once over $S_i \times \mathrm{I}$. In a component $M(\mu, \tau)$ with $\tau_i = -1$, the corresponding sliding arc will be orientation-reversing.

In a component $M(\mu, \tau)$ in which α is orientation-preserving, the slide homeomorphism is constructed as in the orientable case. That is, one replaces M_j' with a ball and chooses an isotopy J_t that moves this ball around α in such a way that

J_1 restricts to the identity on the ball. Then, the slide homeomorphism is J_1 on $M(\mu, \tau) - M_j'$ and the identity on M_j'.

If α is orientation-reversing in $M(\mu, \tau)$, we cannot choose J_t so that J_1 restricts to the identity on the filled-in ball, instead we choose it to restrict to the standard orientation-reversing reflection. The slide homeomorphism is defined on $M(\mu, \tau)$ exactly as before, but this time $M(\mu, \tau)$ is carried to $M(\mu_1, \ldots, -\mu_j, \ldots, \mu_r, \tau)$.

The uniform slide homeomorphisms of the ∂D_j-end or ∂E_j-end of $S_j \times \mathrm{I}$ are defined similarly. One starts with a sliding arc α in a component of $\mathcal{M}$, with both ends in ∂D_j or in ∂E_j. It meets $S_j \times \mathrm{I}$ only in its endpoints, and as in the case of sliding M_j we may choose α so that it defines sliding arcs in all other components of $\mathcal{M}$. In a component $M(\mu, \tau)$ in which α is orientation-preserving, the construction is exactly as in the orientable case, and preserves the component $M(\mu, \tau)$. Suppose instead that α is orientation-reversing in $M(\mu, \tau)$. If J_t slides the ∂E_j-end of $S_j \times \mathrm{I}$, we define the slide homeomorphism to be the identity on $S_j \times \mathrm{I}$, but when J_t slides the ∂D_j-end, we define it to be the standard orientation-reversing reflection on each $S_j \times \{t\}$. This is in accordance with the fact that the $M(\mu, \tau)$ are always constructed with $S_j \times \{0\}$ identified with ∂D_j as it is in M, while the identification of $S_j \times \{1\}$ with ∂E_j is either the same as in M or is composed with the standard orientation-reversing reflection. In either case, $M(\mu, \tau)$ is carried to $M(\mu, \tau_1, \ldots, -\tau_j, \ldots, \tau_s)$.

We use similar constructions for uniform interchanges of homeomorphic M_i and M_j, uniform interchanges of $S_i \times \mathrm{I}$ and $S_j \times \mathrm{I}$, and uniform spins of the $S_j \times \mathrm{I}$. Interchanges slide M_i and M_j along arcs in Σ connecting ∂B_i and ∂B_j. (A technical point here is that one should choose each isomorphism α_i in the definition of interchange isomorphism to be realizable by a homeomorphism $h_i \colon M_1 \to M_i$ that carries C_1 to C_i, and then define the interchange homeomorphism realizing $\omega_{i,j}$ to be $h_j h_i^{-1}$ on M_i and $h_i h_j^{-1}$ on M_j. The interchange homeomorphism then sends $M(\mu_1, \ldots, \mu_i, \ldots, \mu_j, \ldots, \mu_r, \tau)$ to $M(\mu_1, \ldots, \mu_j, \ldots, \mu_i, \ldots, \mu_r, \tau)$ if $h_i h_j^{-1}$ is consistent with the identifications of ∂B_i and ∂B_j with ∂C_i and ∂C_j, but sends $M(\mu_1, \ldots, \mu_i, \ldots, \mu_j, \ldots, \mu_r, \tau)$ to $M(\mu_1, \ldots, -\mu_j, \ldots, -\mu_i, \ldots, \mu_r, \tau)$ if not.) A uniform interchange of $S_i \times \mathrm{I}$ and $S_j \times \mathrm{I}$ interchanges the left ends and interchanges the right ends by sliding all four along arcs in Σ, and interchanges $S_i \times \mathrm{I}$ and $S_j \times \mathrm{I}$ by a homeomorphism which is the identity with respect to the coordinates on each of these copies of $S^2 \times \mathrm{I}$ used in the original construction of M. This sends $M(\mu, \tau_1, \ldots, \tau_i, \ldots, \tau_j, \ldots, \tau_s)$ to $M(\mu, \tau_1, \ldots, \tau_j, \ldots, \tau_i, \ldots, \tau_s)$. Uniform spins slide the two ends of $S_j \times \mathrm{I}$ along arcs in Σ connecting ∂D_j and ∂E_j, and on $S_j \times \mathrm{I}$ the homeomorphism is the product of a reflection in the I-factor and the standard orientation-reversing reflection in each $S_j \times \{t\}$. Therefore, uniform spins preserve each component of $\mathcal{M}$.

There are two other kinds of uniform homeomorphism. First, we may fix an orientation-reversing homeomorphism of Σ which restricts to the standard orientation-reversing reflection on each of its boundary spheres, and define R to be this homeomorphism on Σ and the identity on each M_i and $S_j \times \mathrm{I}$. This R sends $M(\mu_1, \ldots, \tau_s)$ to $M(-\mu_1, \ldots, -\tau_s)$. Second, we may perform a Dehn twist about a 2-sphere in the copy of Σ in one of the $M(\mu, \tau)$ (while being the identity map in all other components of $\mathcal{M}$). Any such Dehn twist is considered a uniform homeomorphism of $\mathcal{M}$.

The *uniform mapping class group* is defined to be the group of mapping classes $\mathcal{U}(\mathcal{M})$ generated by the uniform homeomorphisms we have described: uniform factor homeomorphisms, uniform slide homeomorphisms, uniform interchanges and spins, R, and Dehn twists about 2-spheres in the copies of Σ.

Since all the manifolds in $\mathcal{M}$ differ only by cutting and reattaching along simply-connected subsets, there is a natural way to identify their fundamental groups with $\pi_1(M)$. Using these identifications, we may regard a uniform homeomorphism as inducing an (outer) automorphism on $\pi_1(M)$.

Observe that any slide homeomorphism is isotopic to a composition of slide homeomorphisms which use sliding paths α which travel around a loop in a single M_i, or once over $S_i \times \mathrm{I}$, and do not meet any of the other M_k or $S_k \times \mathrm{I}$. This just corresponds to writing α as a word in the elements of $\pi_1(M_i)$ and $\pi_1(N_j)$ in the free product decomposition $\pi_1(M) = (*_{i=1}^r \pi_1(M_i)) * (*_{j=1}^s \pi(N_j))$. Slide homeomorphisms using these restricted sliding paths are called "basic" slide homeomorphisms, and they induce the standard generating automorphisms $\rho_{i,j}(x)$, $\lambda_{i,j}(x)$, and $\mu_{i,j}(x)$ for $\mathrm{Aut}_p(\pi_1(M))$ that were defined in section 9.2 (where the p subscript means the automorphisms that take each $\pi_1(M_i)$ to a conjugate of itself). Similarly, uniform spins and uniform interchanges induce the spin automorphisms σ_j and the interchange automorphisms $\omega_{i,j}$ respectively. The uniform homeomorphism R and the Dehn twists about 2-spheres induce the identity automorphism.

When all ∂M_i are incompressible, each $\pi_1(M_i)$ is indecomposable (see for example theorem 7.1 of [**51**]), and is not infinite cyclic. So in this case, we have given uniform homeomorphisms that realize all Fouxe-Rabinovitch generators of $\mathrm{Out}(\pi_1(M))$, except for factor automorphisms coming from automorphisms of the $\pi_1(M_i)$ that cannot be realized by homeomorphisms of M_i, and interchange automorphisms of $\pi_1(M_i)$ and $\pi_1(M_j)$ that are isomorphic but for which M_i and M_j are not homeomorphic. When ∂M_i is compressible, two other phenomena lead to nonrealizable Fouxe-Rabinovitch generators. First, when M_i is either a solid torus or a solid Klein bottle, $\pi_1(M_i)$ is infinite cyclic, but uniform homeomorphisms inducing $\lambda_{j,i}(x)$ or $\rho_{j,i}(x)$ cannot exist. This can be shown by methods along the lines of the proof of proposition 9.3.1 adapted to the reducible case as in the proof of theorem 12.2.1 below. Second, when M_i other than solid tori or solid Klein bottles have compressible boundary, their fundamental groups will be decomposable as nontrivial free products. In this case, the Fouxe-Rabinovitch generators defined with respect to the free product decomposition $\pi_1(M) = (*_{i=1}^r \pi_1(M_i)) * (*_{j=1}^s \pi(N_j))$ are far short of a full generating set.

We will now prove a result that relates the uniform homeomorphisms of $\mathcal{M}$ to the homeomorphisms of M. It will furnish one of the main steps in the proof of theorem 12.3.1 below. Let $\mathrm{St}(M(\mu,\tau))$ denote the elements of the uniform mapping class group that preserve the component $M(\mu,\tau)$. Restricting a homeomorphism of $\mathcal{M}$ to a homeomorphism of $M(\mu,\tau)$ induces a homomorphism from $\mathrm{St}(M(\mu,\tau))$ to $\mathcal{H}(M)$. The following result appears as Theorem 2.1 in [**86**].

THEOREM 12.1.1. *The homomorphisms* $\mathrm{St}(M(\mu,\tau)) \to \mathcal{H}(M(\mu,\tau))$ *induced by restriction are surjective for all* (μ,τ).

Before beginning the proof, we give the key topological argument in the form of a lemma.

LEMMA 12.1.2. *Let T be a 2-sphere in a component $M(\mu,\tau)$ of $\mathcal{M}$. Then there is a composition f of uniform slide homeomorphisms and isotopies such that $f(T)$ is contained in Σ (that is, in the copy of Σ in $f(M(\mu,\tau))$). If T does not separate*

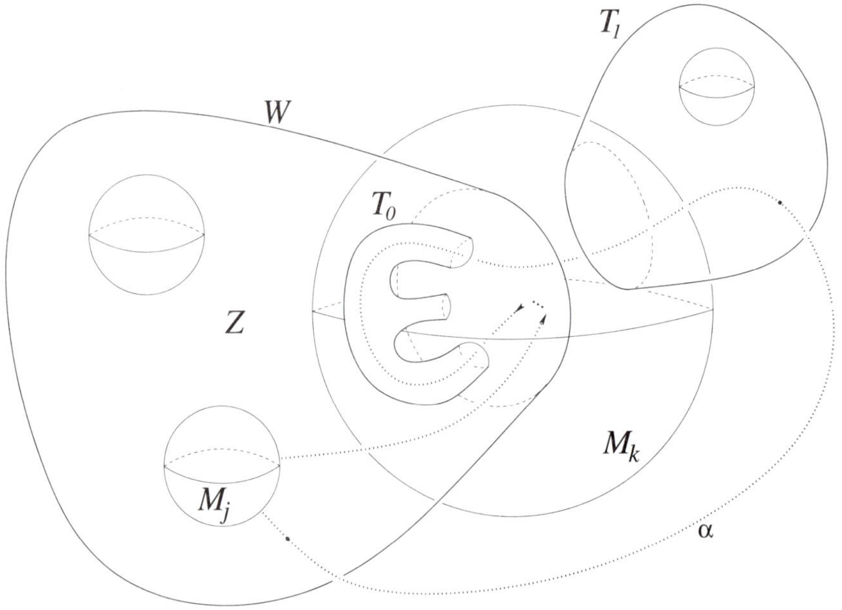

FIGURE 12.1. A sliding path α for clearing out the region Z

$M(\mu, \tau)$, then f may be chosen so that $f(T) = \partial D_k$ for some k. Moreover, if X is the union of all the ∂B_i, ∂D_j, and ∂E_j in all components of $\mathcal{M}$, then f may be selected so that any 2-sphere components of $X \cap M(\mu, \tau)$ that are disjoint from T are carried by f into the corresponding spheres of $X \cap f(M(\mu, \tau))$.

PROOF. We may assume that T meets X transversely. We will apply a sequence of uniform slide homeomorphisms and isotopies to reduce their number of intersection circles.

Consider an intersection circle innermost on T, bounding a disk W on T for which $W \cap X = \partial W$. If W lies in an M_i' or in one of the $S_j \times \mathrm{I}$, then there is an ambient isotopy of $M(\mu, \tau)$ that pulls W across X, reducing the intersection of T with X. This isotopy may be selected to fix each component of X in $M(\mu, \tau)$ that is disjoint from T; from now on we tacitly assume that all adjustments will be carried out so as to disturb components of X only when necessary; this will achieve the condition in the last sentence of the statement of the lemma. We may assume there are no isotopies of this first kind, otherwise the intersection can be reduced.

Suppose now that W lies in Σ. There is a disk W' on X, bounded by ∂W, and $W \cup W'$ is a 2-sphere S in $\mathcal{M}$. Now S together with some of the 2-spheres of X bounds a punctured cell Z in Σ. By reselecting W if necessary, we may choose Z to be an innermost such punctured cell in Σ.

Applying a sequence of uniform slide homeomorphisms and isotopies, we may "slide the 2-spheres of X that meet Z out of Z." Here are the details. Suppose, for example, that ∂B_j lies in Z. As shown in figure 12.1, construct a sliding path α that starts in ∂B_j, travels through Z until it meets W' near T, and exits Z through W' into some M_k' or $S_k \times \mathrm{I}$. Since there are no isotopies of the first kind, no component of $T \cap M_k'$ or $T \cap (S_k \times \mathrm{I})$ is a disk. Therefore α can continue, near T but not crossing it, until it crosses X and reenters Σ. It may be alongside a planar portion of T that lies inside Z, such as the T_0 shown in figure 12.1, but we just continue

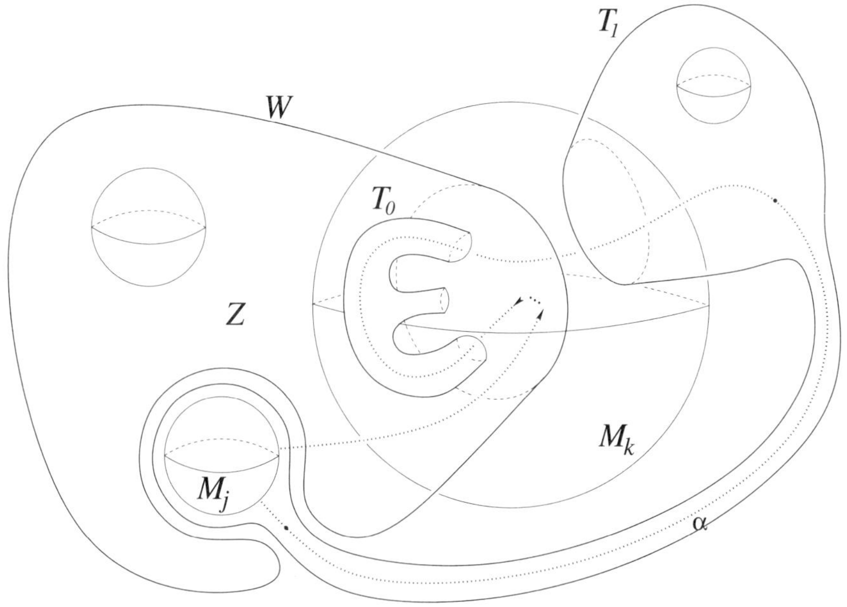

FIGURE 12.2. Result of a slide that clears M_j out of Z

along T until we emerge into Σ alongside a disk component T_1 of $T \cap \Sigma$ distinct from W. Since Z was innermost, T_1 cannot lie in Z. Continue α through $\Sigma - Z$, crossing T_1 and crossing W into Z, and on to terminate in ∂B_j. Now, slide M_j around this α (which may, of course, move us to a different component of $\mathcal{M}$). One must choose the isotopy J_t in the definition of the slide homeomorphism with some care so that it has the correct effect on W. Initially, the J_t should move ∂B_j while keeping W fixed, until ∂B_j approaches the point on α where α passes through W. The remaining J_t move W, but keep it in Σ so that no additional intersection of W with $\partial\Sigma$ is created. Figure 12.2 shows the resulting configuration after this slide. The disk W of T is repositioned in Σ so that the sphere S no longer encloses ∂B_j. The new Z has fewer boundary spheres. Continuing this process with the other spheres of X that lie in Z, we reach the situation where Z is a ball in Σ bounded by S, and W can be pushed across Z to eliminate $\partial W \cap X$ from $T \cap X$, as well as any other intersection circles that lie in W'. Repeating, we find a composition g of uniform slide homeomorphisms (and isotopies) such that $g(T)$ does not meet X (in $g(M(\mu, \tau))$). If $g(T)$ does not already lie in Σ, then it lies in one of the M_i' or $S_j \times \mathrm{I}$. In this case it is either compressible (i. e. bounds a 3-ball) or parallel to a 2-sphere of X in the frontier of M_i' or $S_j \times \mathrm{I}$, so it can be moved by isotopy into Σ.

Suppose that T was nonseparating. Since Σ is simply-connected, $g(T)$ must separate Σ, so $g(T)$ together with some 2-spheres of X bounds a punctured cell Z in Σ that meets some ∂D_j but not ∂E_j. We may apply uniform slide homeomorphisms to clear out the region Z until it is a product between $g(T)$ and ∂D_j, then $g(T)$ is isotopic to ∂D_j. □

We isolate the next step, since it will be used again in the last part of the proof of theorem 12.3.1 below.

LEMMA 12.1.3. *Let h be a homeomorphism from $M(1,1)$ to some $M(\mu,\tau)$, and let X be the union of all the ∂B_i, ∂D_j, and ∂E_j in all components of $\mathcal{M}$. Then there is a composition f of uniform slide homeomorphisms and isotopies such that fh carries $X \cap M(1,1)$ to $X \cap fh(M(1,1))$.*

PROOF. Start with the copy of ∂B_1 in $M(1,1)$ and consider $h(\partial B_1)$. By lemma 12.1.2, we may apply a composition f of uniform slide homeomorphisms (and isotopies) so that $fh(\partial B_1)$ lies in the copy of Σ in $fh(M(1,1))$. Since fh is a homeomorphism, $fh(\partial B_1)$ bounds a submanifold Y homeomorphic to M_1'. Since M_1' is irreducible, Y can contain only one 2-sphere of X, which bounds one of the M_k' in $fh(M(1,1))$. So $Y \cap \Sigma$ consists of a product region between $fh(\partial B_1)$ and ∂B_k. Changing f by an isotopy that moves $fh(\partial B_1)$ to ∂B_k, we may assume that $fh(\partial B_1)$ equals ∂B_k in $fh(M(1,1))$, and $fh(M_1') = M_k'$. From now on, we will simply call the composition h again. Repeating this process with the other ∂B_i, we may assume that h carries the union of the ∂B_i into X.

Now consider $h(\partial D_1)$. Since ∂D_1 is nonseparating, lemma 12.1.2 shows that we may apply uniform slide homeomorphisms until $h(\partial D_1)$ is isotopic to ∂D_k. By isotopy, we may assume that either $h(D_1) = D_k$ and $h(E_1) = E_k$, or that $h(D_1) = E_k$ and $h(E_1) = D_k$. Continuing this process completes the proof. □

PROOF OF THEOREM 12.1.1. Without loss of generality we may choose notation so that $(\mu,\tau) = (1,1)$. Let h represent an element of $\mathcal{H}(M(1,1))$. Applying lemma 12.1.3, we find a product f of uniform slide homeomorphisms and isotopies such that fh carries $X \cap M(1,1)$ to $X \cap f(M(1,1))$. Composing with uniform interchanges and spins, if necessary, we may achieve that fh carries each 2-sphere of X to the corresponding 2-sphere of X in $f(M(1,1))$ (i. e. it carries ∂B_1 to ∂B_1, and so on). In particular, fh carries each M_i' to M_i' and preserves each end of an $S_j \times \mathrm{I}$. Applying factor homeomorphisms in the M_i, we may assume that fh is the identity on $\cup_{i=1}^{r} M_i'$. Applying R, we may assume that the restriction of fh to Σ is orientation-preserving. Since each $S_j \times \mathrm{I}$ is attached to ∂D_j by the standard homeomorphism, we may further assume that fh is the identity on all the $S_j \times \mathrm{I}$. We then have $f(M(1,1)) = M(1,1)$, and fh is the identity except on Σ, where it restricts to the identity on $\partial\Sigma$. It is then isotopic, relative to $\partial\Sigma$, to a composition of Dehn twists about 2-spheres in Σ (see for example Lemma 5.4 of [**68**]). Composing with their inverses (in fact, they are mapping classes of order 2), we make fh the identity on Σ as well. So h equals the restriction of f^{-1} to $M(1,1)$. □

In the introduction we mentioned the *Poincaré associate* of M, which is the 3-manifold $P(M)$ obtained from M by replacing each M_i' that is simply-connected with a 3-ball K_i. Note that none of the prime summands of $P(M)$ is simply-connected, so $P(M)$ has no 2-sphere boundary components. In $P(M)$, any simply-connected submanifold bounded by a 2-sphere is a 3-ball.

PROPOSITION 12.1.4. *$\mathcal{R}(M)$ has finite index in* $\mathrm{Out}(\pi_1(M))$ *if and only if $\mathcal{R}(P(M))$ has finite index in* $\mathrm{Out}(\pi_1(P(M)))$.

PROOF. First we note that $\pi_1(M) \cong \pi_1(P(M))$. In fact, we may assume that $M_1, \ldots, M_q$ are not simply connected, and $M_{q+1}, \ldots, M_r$ are simply connected, so that $P(M)$ is constructed as $(\#_{i=1}^{q} M_i)\#(\#_{j=1}^{s} N_j)$ and M as $(\#_{i=1}^{r} M_i)\#(\#_{j=1}^{s} N_j)$. Then, M is obtained from $P(M)$ by replacing 3-balls $B_{q+1}, \ldots, B_r$ with the simply-connected manifolds $M_{q+1}', \ldots, M_r'$, and there is a natural identification of $\pi_1(P(M))$ with $\pi_1(M)$. Moreover, provided that we choose sliding paths and their regular neighborhoods so as to avoid the 3-balls $B_{q+1}, \ldots, B_r$, a uniform homeomorphism of $P(M)$ that does not involve the summands $M_{q+1}, \ldots, M_r$ (or a path that passes through one of them) can be restricted to the complement of the B_i with $i > q$ and then extended to a uniform homeomorphism of M by taking it to be the identity on each M_i' with $i > q$. With respect to the natural identification of the fundamental groups, this uniform homeomorphism will induce the same automorphism.

The uniform homeomorphisms for $P(M)$ take place on a disjoint union $\mathcal{P}$ of copies of $P(M)$. It may be regarded as obtained from the submanifold of $\mathcal{M}$ consisting of all $M(\mu, \tau)$ having $\mu_{q+1} = \cdots = \mu_r = 1$, by replacing every M_i' with $q+1 \leq i \leq r$ with a ball. As usual, $P(M)(1,1) = P(M)$.

Suppose now that ϕ is an outer automorphism of $\pi_1(M)$ that can be realized by a homeomorphism h of M. By theorem 12.1.1, there is a composition of uniform homeomorphisms of $\mathcal{M}$ that stabilizes $M = M(1,1)$ and restricts to h on $M(1,1)$. We choose each slide homeomorphisms to be "basic", i. e. so that its sliding path passes through only one of the M_i' or $S_j \times \mathrm{I}$. Now, from this composition of uniform homeomorphisms, omit those that are slides or interchanges of simply-connected prime summands, or factor homeomorphisms supported on simply-connected summands, or slide homeomorphisms around loops in M_i with $i > q$. Since all omitted homeomorphisms induce the identity automorphism, the composition h_0 of the remaining uniform homeomorphisms of $\mathcal{M}$ still induces ϕ. It might not stabilize $M(1,1)$, since a slide homeomorphism of a simply-connected summand around a nonorientable loop, or an interchange of simply-connected summands, or a factor homeomorphism which reverses orientation on a simply-connected summand could move $M(1,1)$ to some other $M(\mu, \tau)$. Since the uniform homeomorphisms of $\mathcal{M}$ other than these three kinds have corresponding uniform homeomorphisms on $\mathcal{P}$, we obtain a composition of uniform homeomorphisms $P(h_0)$ of $\mathcal{P}$ that corresponds to h_0, and in particular induces ϕ. The action of the omitted uniform homeomorphisms on the components of $\mathcal{M}$ could only change μ_i that correspond to simply-connected summands, that is, μ_i with $q+1 \leq i \leq r$. This might occur, for example, if one of the uniform homeomorphisms were a factor homeomorphism which was orientation-reversing on its M_i'. But the simply-connected M_i are all orientable, so none of the slides of other summands around loops in those M_i can change a μ_j or a τ_j. Therefore $P(h_0)$ will stabilize the component $P(M)(1,1)$, and the restriction of $P(h_0)$ to $P(M)(1,1)$ induces ϕ.

For the converse direction, this subtlety does not arise. One can take any composition of uniform homeomorphisms in $\mathcal{P}$ that stabilizes $P(M)(1,1)$ and apply the corresponding uniform homeomorphisms to $\mathcal{M}$, obtaining an element of $\mathrm{St}(M(1,1))$ that induces the same automorphism on $\pi_1(M)$. □

By virtue of proposition 12.1.4, there is no loss of generality in assuming that no prime summand of M is simply-connected, when considering finite-index realization questions. In particular, our previous results apply to many manifolds for which $P(M)$ is irreducible. That is, when there is one irreducible summand M_1 and all other summands are simply-connected, $P(M)$ will equal M_1. If M_1 is Haken, then we can use the Main Topological Theorems to solve the Finite Index Realization Problem for M_1 and hence for M.

12.2. Reducible 3-manifolds with compressible boundary

Let M be a 3-manifold for which $P(M)$ is reducible. We say that M is *small* if $P(M)$ is a connected sum $P\#Q$, where P is either a solid torus or a solid Klein bottle, and Q is a 3-manifold with finite fundamental group.

THEOREM 12.2.1. *Let M be a compact* 3*-manifold with compressible boundary, with $P(M)$ reducible. Then $\mathcal{R}(M)$ has finite index in* $\mathrm{Out}(\pi_1(M))$ *if and only if M is small.*

We will prove the two directions of theorem 12.2.1 as separate propositions.

PROPOSITION 12.2.2. *Let M be a compact* 3*-manifold with compressible boundary, with $P(M)$ reducible. If M is small, then* $\mathrm{Out}(\pi_1(M))$ *is finite, and consequently $\mathcal{R}(M)$ has finite index in* $\mathrm{Out}(\pi_1(M))$.

PROOF. By proposition 12.1.4, we may assume that $M = P(M)$, so that no prime summand is simply-connected. Write $M = P\#Q$ as in the definition of reducible small manifold. Then $\pi_1(M)=\pi_1(P)*\pi_1(Q)$ where $\pi_1(P)$ is infinite cyclic. Let ϕ represent an element of $\mathrm{Out}(\pi_1(M))$. Since $\pi_1(Q)$ is indecomposable and not infinite cyclic, the Kurosh subgroup theorem shows that $\phi(\pi_1(Q))$ is conjugate to $\pi_1(Q)$. So by lemma 9.1.2, we may assume that $\phi(\pi_1(Q))=\pi_1(Q)$ and $\phi(\omega)=\omega^\epsilon\gamma$ where ω is a generator of $\pi_1(P)$, $\epsilon \in \{1,-1\}$, and $\gamma \in \pi_1(Q)$. Moreover, the element γ, the power ϵ, and the restriction ϕ_1 of ϕ to $\pi_1(Q)$ are uniquely determined by the element of $\mathrm{Out}(\pi_1(M))$ represented by ϕ. Sending ϕ to ϕ_1 determines a homomorphism $\mathrm{Out}(\pi_1(M)) \to \mathrm{Aut}(\pi_1(Q))$ whose kernel has finite index. Sending an element of the kernel to ϵ determines a homomorphism to $\mathbb{Z}/2$, and sending an element of its kernel to γ determines an injective homomorphism to Q. So $\mathrm{Out}(\pi_1(M))$ is finite. □

PROPOSITION 12.2.3. *Let M be a compact* 3*-manifold with compressible boundary, with $P(M)$ reducible. If M is not small, then $\mathcal{R}(M)$ has infinite index in* $\mathrm{Out}(\pi_1(M))$.

PROOF. Again by proposition 12.1.4, we may assume that $M = P(M)$. In particular, no component of ∂M is a 2-sphere.

Let F be a compressible boundary component. Now F cannot be a 2-sphere, and cannot be a projective plane since two-sided projective planes always inject on fundamental groups (if not, the Loop Theorem would yield a properly imbedded disk D with ∂D a one-sided loop in F, but the normal bundle of D must be a product since D is simply-connected).

In this proof, we allow compression bodies to be nonorientable, and to have constituents that are 2-spheres or projective planes. If we carry out the construction of a (minimally imbedded) compression body neighborhood F as in section 3.2, we obtain a compression body neighborhood V of F with incompressible frontier. Since there is no boundary pattern involved, the constituents of V will be closed surfaces. (Unlike the case when M is irreducible, V need not be unique up to isotopy, or even up to ambient homeomorphism. For example, suppose M is the connected sum of a compression body W with three irreducible manifolds P_1, P_2, and P_3, where W has two constituents F_1 and F_2. One choice for V would be a copy of W whose complementary components are homeomorphic to $(F_1 \times [0, 1/2))\#P_1$ and $(F_2 \times [0, 1/2))\#P_2\#P_3$, another choice would have complementary components homeomorphic to $(F_1 \times [0, 1/2))\#P_2$ and $(F_2 \times [0, 1/2))\#P_1\#P_3$.)

Let $F_1, \ldots, F_m$ be the constituents of V. Since M is reducible, $m \geq 1$. Let W_i be the component of $\overline{M - V}$ that contains F_i, of course we might have $W_i = W_j$ for some $i \neq j$. As usual, let k be the number of 1-handles of V.

If at least one constituent of V is not a 2-sphere, then we may select V so that no constituent is a 2-sphere. For suppose that F_m is a 2-sphere and F_1 is not. Choose the compression body structure on V so that only one 1-handle meets $F_m \times \{1\}$, and that the other end of this 1-handle lies in $F_1 \times \{1\}$. Take an arc α that runs from $F_1 \times \{0\}$ straight up to $F_1 \times \{1\}$, then through the 1-handle (as the core 1-disk) to $F_m \times \{1\}$, and straight down to $F_m \times \{0\}$. Let N be a regular neighborhood of α in V, and remove from V the topological interior of N in V. This produces a new compression body neighborhood of F with both k and m decreased by 1. Repeating this process, all 2-sphere constituents may be eliminated.

If all constituents are 2-spheres, then the same construction can be used to eliminate them, until only one remains. Suppose for now that this occurs, so $m = 1$ and $F_1 = S^2$. Since F is not a 2-sphere, we have $k \geq 1$. Let $D \times I$ be a 1-handle of V; since $m = 1$, $D \times \{0\}$ does not separate V. If $\pi_1(W_1)$ is infinite, then the argument of Case Ib of the proof of proposition 9.3.1 applies to show that $\mathcal{R}(M)$ has infinite index in $\mathrm{Out}(\pi_1(M))$. If $\pi_1(W_1)$ is finite, then we must have $k \geq 2$, since M is not small. So then, the argument of Case Ic of the proof of proposition 9.3.1 applies. So from now on, we may assume that no constituent of V is a 2-sphere.

One of the W_i, say W_1, must be reducible, since otherwise M would be irreducible. So $\pi_1(W_1)$ is of the form $A * B$ with A and B nontrivial and $\pi_1(F_1) \subset A$. By the normal form for elements of free products, the normalizer of $\pi_1(F_1)$ must be contained in A, so it has infinite index in $\pi_1(W_1)$. We will now apply the arguments from various cases of the proof of proposition 9.3.1.

Suppose first that $k \geq m$, so that there is a 1-handle in V whose cocore disk does not separate V. Then one of Cases Ia, Ib, or Id applies. If $m \geq 3$, then one of Cases IIa or IIb applies. So we may assume that $k < m$ and $m \leq 2$. We cannot have $m = 1$, since then we would have $k = 0$ and F would be incompressible, so $k = 1$ and $m = 2$. Since the normalizer of $\pi_1(F_1)$ in $\pi_1(W_1)$ has infinite index in $\pi_1(W_1)$, either Case IIIa or IIIb applies to complete the proof. □

12.3. Reducible 3-manifolds with incompressible boundary

When the boundary of M is incompressible, the Finite Index Realization Problem simply reduces to the irreducible prime summands.

THEOREM 12.3.1. *Let M be a compact 3-manifold with incompressible boundary. Write the prime factorization of $P(M)$ as $(\#_{i=1}^{r} M_i)\#(\#_{j=1}^{s} N_j)$ with the M_i irreducible. Then $\mathcal{R}(M)$ has finite index in $\mathrm{Out}(\pi_1(M))$ if and only if $\mathcal{R}(M_i)$ has finite index in $\mathrm{Out}(\pi_1(M_i))$ for all $1 \leq i \leq r$.*

PROOF. Using proposition 12.1.4, we may assume that $M = P(M)$. If $i + j \leq 1$, the theorem is immediate, so we assume that $P(M)$ is reducible.

We will use the manifold $\mathcal{M}$ and its uniform homeomorphisms, discussed in section 12.1. As usual, we regard M as the component $M(1,1)$ of $\mathcal{M}$. Let $\Phi_{\mathcal{U}} \colon \mathcal{U}(\mathcal{M}) \to \mathrm{Out}(\pi_1(M))$ carry each uniform homeomorphism to the outer automorphism it induces on $\pi_1(M)$, and let $\Phi_0 \colon \mathrm{St}(M(1,1)) \to \mathrm{Out}(\pi_1(M))$ be the restriction of $\Phi_{\mathcal{U}}$ to $\mathrm{St}(M(1,1))$.

Notice that Φ_0 is the composition $\mathrm{St}(M(1,1)) \to \mathcal{H}(M) \to \mathrm{Out}(\pi_1(M))$. Since by theorem 12.1.1 the restriction $\mathrm{St}(M(1,1)) \to \mathcal{H}(M)$ is surjective, it follows that $\mathcal{R}(M)$ has finite index in $\mathrm{Out}(\pi_1(M))$ if and only if the image of Φ_0 has finite index. Since $\mathrm{St}(M(1,1))$ has finite index in $\mathcal{U}(\mathcal{M})$, the latter occurs if and only if the image of $\Phi_{\mathcal{U}}$ has finite index. So the theorem is reduced to showing that the image of $\Phi_{\mathcal{U}}$ has finite index if and only if $\mathcal{R}(M_i)$ has finite index in $\mathrm{Out}(\pi_1(M_i))$ for $1 \leq i \leq r$.

The next part of the proof will use algebraic information about $\mathrm{Aut}(\pi_1(M))$ to show that the image of $\Phi_{\mathcal{U}}$ has finite index if and only if the image of a certain composite homomorphism $\Theta \circ \Phi_{\mathcal{U}}$ carrying $\mathrm{Out}(\pi_1(M))$ to a quotient group of $\mathrm{Out}(\pi_1(M))$ has finite index. Roughly speaking, this is the quotient by the subgroup generated by all slide homeomorphisms.

As in section 12.1, we will use the Fouxe-Rabinovitch generators of $\mathrm{Aut}(\pi_1(M))$ as detailed in section 9.2. That is, we regard $\pi_1(M)$ as a free product $G_1 * \cdots * G_r * G_{r+1} * \cdots * G_{r+s}$ where $G_i = \pi_1(M_i)$ for $1 \leq i \leq r$ and $G_{r+j} = \pi_1(N_j)$ for $1 \leq j \leq s$. Since each M_i has incompressible boundary, each $\pi_1(M_i)$ is an indecomposable group (see for example theorem 7.1 of [**51**]), and is not infinite cyclic. So the Fouxe-Rabinovitch generators defined with respect to the free factors $\pi_1(M_i)$ and $\pi_1(N_j)$ generate all of $\mathrm{Aut}(\pi_1(M))$.

Recall the interchange automorphisms $\omega_{i,j}$ for $1 \leq i, j \leq r$ that were defined in section 12.1. As noted there, they generate a subgroup Ω of $\mathrm{Aut}(\pi_1(M))$ isomorphic to a direct product of finite permutation groups. The interchange automorphisms interact in simple ways with the other generating automorphisms. In fact, the following relations are easily checked, where i, j, k, and ℓ are assumed distinct, with $i, j \leq r$.

(1) $\omega_{i,j}\rho_{k,\ell}(x)\omega_{i,j} = \rho_{k,\ell}(x)$, $\omega_{i,j}\rho_{k,j}(x)\omega_{i,j} = \rho_{k,i}(x)$,
$\omega_{i,j}\rho_{j,k}(x)\omega_{i,j} = \rho_{i,k}(\omega_{i,j}(x))$, $\omega_{i,j}\rho_{i,j}(x)\omega_{i,j} = \rho_{j,i}(\omega_{i,j}(x))$,
(2) $\omega_{i,j}\lambda_{k,\ell}(x)\omega_{i,j} = \lambda_{k,\ell}(x)$, $\omega_{i,j}\lambda_{k,j}(x)\omega_{i,j} = \lambda_{k,i}(x)$,
$\omega_{i,j}\lambda_{j,k}(x)\omega_{i,j} = \lambda_{i,k}(\omega_{i,j}(x))$, $\omega_{i,j}\lambda_{i,j}(x)\omega_{i,j} = \lambda_{j,i}(\omega_{i,j}(x))$,
(3) $\omega_{i,j}\mu_{k,\ell}(x)\omega_{i,j} = \mu_{k,\ell}(x)$, $\omega_{i,j}\mu_{k,j}(x)\omega_{i,j} = \mu_{k,i}(x)$,
$\omega_{i,j}\mu_{j,k}(x)\omega_{i,j} = \mu_{i,k}(\omega_{i,j}(x))$, $\omega_{i,j}\mu_{i,j}(x)\omega_{i,j} = \mu_{j,i}(\omega_{i,j}(x))$,
(4) $\omega_{i,j}\omega_{k,\ell}\omega_{i,j} = \omega_{k,\ell}$,
(5) $\omega_{i,j}\sigma_k\omega_{i,j} = \sigma_k$,
(6) if $\phi \in \mathrm{Aut}(\pi_1(M_k))$, then $\omega_{i,j}\phi_k = \phi_k\omega_{i,j}$,
(7) if $\phi \in \mathrm{Aut}(\pi_1(M_i))$, then $\omega_{i,j}\phi_i\omega_{i,j}^{-1} = (\alpha_{i,j}\phi\alpha_{j,i})_j$, where $\alpha_{i,j}$ and $\alpha_{j,i}$ are as in the definition of $\omega_{i,j}$.

As a consequence of the last two relations, Ω acts on the direct products $\prod_{i=1}^r \mathrm{Aut}(\pi_1(M_i))$ and $\prod_{i=1}^r \mathrm{Out}(\pi_1(M_i))$. In particular, there is a semidirect product $(\prod_{i=1}^r \mathrm{Out}(\pi_1(M_i))) \circ \Omega$.

Let N be the subgroup of $\mathrm{Aut}(\pi_1(M))$ generated by all $\mu_{i,j}(x)$, $\lambda_{i,j}(x)$, $\rho_{i,j}(x)$, all $\omega_{r+i,r+j}$ (that is, the $\omega_{k,\ell}$ that interchange infinite cyclic factors), all σ_j, and all ϕ_i for which ϕ is an inner automorphism of $\pi_1(M_i)$. Using the Fouxe-Rabinovitch relations, in fact just the easily-verified ones listed in section 9.2, together with those listed in the previous paragraph, one can check that N is a normal subgroup of $\mathrm{Aut}(\pi_1(M))$. Notice that N contains $\mathrm{Inn}(G)$. For example, conjugation of $\pi_1(M)$ by an element x from $\pi_1(M_1)$ can be written as

$(\mu(x))_1 \prod_{i=2}^{r} \mu_{1,i}(x) \prod_{j=1}^{s} \rho_{1,r+j}(x)\lambda_{1,r+j}(x)$, where $\mu(x) \in \operatorname{Inn}(\pi_1(M_1))$ is the inner automorphism defined by $\mu(x)(g) = x^{-1}gx$ for all $g \in \pi_1(M_1)$. It follows that $\operatorname{Aut}(\pi_1(M))/N$ is isomorphic to $\operatorname{Out}(\pi_1(M))/(N/\operatorname{Inn}(\pi_1(M)))$.

We define a surjective homomorphism from $\operatorname{Aut}(\pi_1(M))/N$ to the semidirect product $(\prod_{i=1}^{r} \operatorname{Out}(\pi_1(M_i))) \circ \Omega$, by sending each factor automorphism to the outer automorphism it induces on its $\pi_1(M_i)$, and each interchange automorphism of non-infinite-cyclic factors $\omega_{i,j}$ to the element of Ω that it determines. Since the factor automorphisms defined using inner automorphisms of the $\pi_1(M_i)$ lie in N, one can define an inverse for this homomorphism by sending an outer automorphism ϕ of $\pi_1(M_i)$ to the factor automorphism defined using one of its representatives, and sending $\omega_{i,j}$ back to the corresponding interchange automorphism. Let $\Theta\colon \operatorname{Out}(\pi_1(M)) \to (\prod_{i=1}^{r} \operatorname{Out}(\pi_1(M_i))) \circ \Omega$ be the composition of the quotient map from $\operatorname{Out}(\pi_1(M))$ to $\operatorname{Out}(\pi_1(M))/(N/\operatorname{Inn}(\pi_1(M)))$, followed by the natural isomorphism from $\operatorname{Out}(\pi_1(M))/(N/\operatorname{Inn}(\pi_1(M)))$ to $\operatorname{Aut}(\pi_1(M))/N$, followed by the isomorphism identifying the latter with $(\prod_{i=1}^{r} \operatorname{Out}(\pi_1(M_i))) \circ \Omega$.

The composition $\Theta \circ \Phi_{\mathcal{U}}$ carries $\mathcal{U}(\mathcal{M})$ to $(\prod_{i=1}^{r} \operatorname{Out}(\pi_1(M_i))) \circ \Omega$. Since each generator of N is induced by a uniform homeomorphism, $N/\operatorname{Inn}(\pi_1(M))$ lies in the image of $\Phi_{\mathcal{U}}$. Therefore the image of $\Phi_{\mathcal{U}}$ has finite index if and only if the image of $\Theta \circ \Phi_{\mathcal{U}}$ has finite index. So the theorem is reduced to showing that the image of $\Theta \circ \Phi_{\mathcal{U}}$ has finite index if and only if $\mathcal{R}(M_i)$ has finite index in $\operatorname{Out}(\pi_1(M_i))$ for $1 \leq i \leq r$.

Since any automorphism in $\mathcal{R}(M_i)$ can be induced by a uniform factor homeomorphism, the subgroup $\prod_{i=1}^{r} \mathcal{R}(M_i)$ of $(\prod_{i=1}^{r} \operatorname{Out}(\pi_1(M_i))) \circ \Omega$ lies in the image of $\Theta \circ \Phi_{\mathcal{U}}$. So if each $\mathcal{R}(M_i)$ has finite index in $\operatorname{Out}(\pi_1(M_i))$, the image of $\Theta \circ \Phi_{\mathcal{U}}$ has finite index.

For the converse, suppose that the image of $\Theta \circ \Phi_{\mathcal{U}}$ has finite index. Then it meets each of the subgroups $\operatorname{Out}(\pi_1(M_k))$ of $(\prod_{i=1}^{r} \operatorname{Out}(\pi_1(M_i))) \circ \Omega$ in a subgroup of finite index. We must show that for each k, $\mathcal{R}(M_k)$ has finite index in $\operatorname{Out}(\pi_1(M_k))$. For this, it is sufficient to show that if ϕ is an automorphism of a $\pi_1(M_k)$, and the factor automorphism ϕ_k lies in the image of $\Theta \circ \Phi_{\mathcal{U}}$, then ϕ lies in $\mathcal{R}(M_k)$.

Let h be a uniform homeomorphism of $\mathcal{M}$ with $\Theta \circ \Phi_{\mathcal{U}}(h) = \phi_k$. By lemma 12.1.3, there is a composition of slide homeomorphisms f so that fh preserves X (that is, it carries the copy of X in $M(1,1)$ to the copy of X in $fh(M(1,1))$). Since the induced automorphism of f lies in N, $\Theta \circ \Phi_{\mathcal{U}}(fh) = \phi_k$. Consequently, fh must preserve M_k' (for any permutation of irreducible summands would show up in Ω). In particular, its restriction to M_k' induces ϕ. Adding a 3-ball to M_k', we may extend the restriction of fh to a homeomorphism of M_k inducing ϕ. □

CHAPTER 13

Epilogue

In this section, we summarize some related work and discuss the future direction of research in the topics that have been examined in this book. In section 13.1, we examine the Finite Index Realization Problem, and in section 13.2 we discuss recent advances in understanding $\mathrm{AH}(\pi_1(M))$.

13.1. More topology

Main Topological Theorem 1 and theorem 12.2.1 combine to give a rather general solution of the Finite Index Realization Problem for compact 3-manifolds with compressible boundary. For those with incompressible boundary, theorem 12.3.1 shows that one need only solve the Finite Index Realization Problem for irreducible 3-manifolds.

In this section we will examine what is known and what is conjectured for this case. We will say that M "has finite index realization" when $\mathcal{R}(M)$ has finite index in $\mathrm{Out}(\pi_1(M))$.

From now on we assume that M is irreducible with incompressible boundary, and for the time being we suppose that M is orientable. Then M satisfies exactly one of the following:

(1) $\pi_1(M)$ is finite.
(2) $\pi_1(M)$ is infinite, but M is not Haken.
(3) M is Haken.

In particular, case (3) occurs whenever M has nonempty boundary (see for example Theorem III.10 of [**54**]).

In case 1, $\mathrm{Out}(\pi_1(M))$ is finite so M automatically has finite index realization. For case 2, it is conjectured that $\mathrm{Out}(\pi_1(M))$ is finite. Two classes of non-Haken 3-manifolds with infinite fundamental group are known. There is a class of Seifert-fibered 3-manifolds, determined by Heil [**50**]. For these, $\mathrm{Out}(\pi_1(M))$ is known to be finite (p. 21 of [**87**]). Conjecturally, all non-Haken 3-manifolds which are not Seifert-fibered admit hyperbolic structures, in which case $\mathrm{Out}(\pi_1(M))$ would be finite for them as well. Recent work of Gabai and Kazez shows that large classes of non-Haken 3-manifolds have finite $\mathrm{Out}(\pi_1(M))$. In [**42**] they show that the fundamental groups of closed, atoroidal, genuinely laminar 3-manifolds are negatively curved in the sense of Gromov. By Paulin [**109**], such a group has finite outer automorphism group unless it admits an isometric action on some $\mathbb{R}$-tree with almost cyclic edge stabilizers and without a global fixed point. But the main corollary of [**97**] shows that such actions do not exist. Gabai and Kazez have also shown that the mapping class groups of closed atoroidal genuinely laminar 3-manifolds are finite [**43**].

As for case 3, Waldhausen's Theorem 2.5.6 shows that closed Haken 3-manifolds have finite index realization. For Haken 3-manifolds with incompressible boundary, our Main Topological Theorem 2 give a complete solution. Simplified to the case of Haken manifolds with empty boundary pattern, and combined with Waldhausen's theorem 2.5.6, it becomes the following:

THEOREM 13.1.1. *Let M be a Haken 3-manifold with incompressible boundary, possibly empty. Then $\mathcal{R}(M)$ has finite index in* $\operatorname{Out}(\pi_1(M))$ *if and only if every Seifert-fibered component V of the characteristic submanifold Σ of $(M, \underline{\underline{\emptyset}})$ that meets ∂M satisfies one of the following:*

(1) *V is a solid torus, or*
(2) *V is an S^1-bundle over the Möbius band or annulus, and no component of ∂V contains more than one component of $V \cap \partial M$, or*
(3) *V is fibered over the annulus with one exceptional fiber, and no component of $V \cap \partial M$ is an annulus, or*
(4) *V is fibered over the disk with two holes with no exceptional fibers, and $V \cap \partial M$ is one of the boundary tori of V, or*
(5) *$V = M$, and either V is fibered over the disk with two exceptional fibers, or V is fibered over the Möbius band with one exceptional fiber, or V is fibered over the torus with one hole with no exceptional fibers, or*
(6) *$V = M$, and V is fibered over the disk with three exceptional fibers each of type* $(2, 1)$.

In summary, we expect that a compact orientable 3-manifold M has finite index realization if and only if every prime summand of M that is a Haken 3-manifold satisfies the conditions in theorem 13.1.1.

We turn now to the nonorientable case. A one-sided projective plane $\mathbb{RP}^2$ in a 3-manifold can only appear in a prime summand which is a real projective 3-space $\mathbb{RP}^3$, since it must have a tubular neighborhood which is the I-bundle over $\mathbb{RP}^2$ with orientable total space. As the I-bundle over $\mathbb{RP}^2$ with orientable total space is homeomorphic to the complement of a 3-ball in $\mathbb{RP}^3$, the summand must be homeomorphic to $\mathbb{RP}^3$.

However, two-sided projective planes can appear in more complicated ways inside nonorientable irreducible 3-manifolds. A simple family of examples is provided by the *Jaco manifolds* [**55**]. These are constructed by starting with a surface F and an involution τ of F whose fixed-point set consists of a nonempty finite collection of points $\{p_1, \ldots, p_r\}$. An involution of $F \times [-1, 1]$ is defined by sending (x, t) to $(\tau(x), -t)$. Its fixed point set is the collection of points $\{p_1 \times \{0\}, \ldots, p_r \times \{0\}\}$. One may remove invariant open 3-ball neighborhoods of each point in the fixed point set, to obtain a 3-manifold W. If we restrict the involution to an involution ρ on W, the quotient $J = W/\rho$ is a 3-manifold having one boundary component which is a projective plane for each fixed point of the original involution τ. It will be irreducible provided that F was not S^2, and will have incompressible boundary if F is closed (or $F = D^2$, in which case J is simply $\mathbb{RP}^2 \times \mathrm{I}$).

A 3-manifold is called $\mathbb{RP}^2$-irreducible (or $\mathbb{P}^2$-irreducible, in the original terminology of [**49**]) if it is irreducible and contains no two-sided projective planes. A nonorientable irreducible 3-manifold M must satisfy exactly one of the following:

(1) $\pi_1(M)$ is finite. In this case, a theorem of Epstein [**34**] shows that M is homotopy equivalent to $\mathbb{RP}^2 \times \mathrm{I}$.
(2) M is $\mathbb{RP}^2$-irreducible, but does not contain a two-sided incompressible surface.

(3) M is $\mathbb{RP}^2$-irreducible, and does contain a two-sided incompressible surface.
(4) M contains two-sided projective planes.

With respect to finite-index realization, one expects the first three cases to be analogous to the three orientable cases discussed above. The homeomorphisms for manifolds as in case 4 are not very well understood, although progress has been made by G. A. Swarup [**119**] and by J. Kalliongis [**60**].

A conjectural description of 3-manifold mapping class groups is given in [**88**].

13.2. More geometry

In chapter 7 we reviewed the classical deformation theory of Kleinian groups, which provides a very satisfying parameterization of the space $\mathrm{GF}(\pi_1(M), \pi_1(P))$ of (marked) geometrically finite uniformizations of pared 3-manifolds homotopy equivalent to (M, P). The work of Marden and Sullivan tells us that $\mathrm{GF}(\pi_1(M), \pi_1(P))$ is the interior of the space $\mathrm{AH}(\pi_1(M), \pi_1(P))$ of all (marked) hyperbolic 3-manifolds homotopy equivalent to M (with cusps associated to every component of P). The structure of the complement of $\mathrm{GF}(\pi_1(M), \pi_1(P))$ in $\mathrm{AH}(\pi_1(M), \pi_1(P))$ is still quite mysterious. However, in recent years, there has been some progress in our understanding of this space. In many cases, that progress revealed that $\mathrm{AH}(\pi_1(M), \pi_1(P))$ is a much more complicated object than had originally been suspected. We will review some of the major conjectures and the progress that has been made on them.[1]

13.2.1. Bers' Density Conjecture. The most basic conjecture concerning the structure of $\mathrm{AH}(\pi_1(M), \pi_1(P))$ is Bers' Density Conjecture which predicts that $\mathrm{AH}(\pi_1(M), \pi_1(P))$ is the closure of $\mathrm{GF}(\pi_1(M), \pi_1(P))$. Bers [**12**] originally posed a related conjecture in the context of the Bers Slice. The full conjecture was later formulated by Sullivan [**116**] and Thurston [**121**].

Bers' Density Conjecture: *For a pared* 3*-manifold* (M, P)*,* $\mathrm{AH}(\pi_1(M), \pi_1(P))$ *is the closure of* $\mathrm{GF}(\pi_1(M), \pi_1(P))$.

Bers' Density Conjecture is an easy consequence of Mostow's Rigidity Theorem [**98**] in the case that every component of $\partial M - P$ is a sphere with three holes; in this case there is a unique hyperbolic structure on (M, P). (For an approach to this fact using Marden's Isomorphism Theorem see Theorem III in Keen-Maskit-Series [**62**].)

The only nontrivial case in which Bers' Density Conjecture is known to be true is the case when M is an I-bundle over the torus with one hole and $M - P$ is the associated ∂I-bundle. This deep result was recently established by Minsky [**94**]. In this case, $\mathrm{GF}(\pi_1(M), \pi_1(P))$ is topologically a 4-ball, but McMullen [**91**] established that $\mathrm{AH}(\pi_1(M), \pi_1(P))$ is not a manifold.[2]

---

[1] Since the initial submission of this manuscript substantial progress has been made in our understanding of $\mathrm{AH}(\pi_1(M))$. Surveys of this more recent work are currently being written and will likely appear contemporaneously with this paper. We will simply note some of the highlights of this recent progress in a series of footnotes.

[2] Bromberg [**25**] and Brock-Bromberg [**22**] proved that if M has incompressible boundary, $\rho \in \mathrm{AH}(\pi_1(M))$, and $\rho(\pi_1(M))$ contains no maximal cyclic parabolic subgroups, then ρ lies in the closure of $\mathrm{GF}(\pi_1(M))$. The Bers Density Conjecture for pared 3-manifolds (M, P) with $\partial M - P$ incompressible is a corollary of The Ending Lamination Theorem of Brock-Canary-Minsky [**24**].)

13.2.2. Bumping. In chapter 7, we studied the map

$$\Theta\colon \mathrm{GF}(\pi_1(M), \pi_1(P)) \to \mathcal{A}(M, P)$$

which records the (marked) pared homeomorphism type of a geometrically finite hyperbolic 3-manifold. One may extend Θ to a map defined on all of $\mathrm{AH}(\pi_1(M), \pi_1(P))$.

In order to describe this decomposition, we briefly recall the thick-thin decomposition of a hyperbolic 3-manifold. If $\epsilon > 0$, we let

$$N_{thin(\epsilon)} = \{x \in N \mid \mathrm{inj}_N(x) < \epsilon\}$$

be the ϵ-*thin-part* of the hyperbolic 3-manifold, where the injectivity radius $\mathrm{inj}_N(x)$ at N at a point x is defined to be half the length of the shortest homotopically nontrivial closed curve through x. It is a consequence of the Margulis Lemma, see Chapter D in [**10**], that there exists a universal constant μ_3, such that if $\epsilon < \mu_3$, then every component of the ϵ-thin-part of a hyperbolic 3-manifold N is either (a) a solid torus neighborhood of a closed geodesic in N, or (b) a "cusp", i. e. a quotient of a horoball in $\mathbb{H}^3$ by a group of parabolic isometries. The submanifold $N^0(\rho)$ is obtained from $N(\rho)$ by removing the cuspidal components of $N(\rho)_{thin(\epsilon)}$. (Here we are assuming that one has fixed a choice of $\epsilon < \mu_3$.) A result of McCullough [**85**] (see also Kulkarni-Shalen [**67**]) guarantees that there exists a compact core $M(\rho)$ for $N^0(\rho)$ such that $M(\rho)$ intersects each noncompact component of the boundary of $N^0(\rho)$ in a single incompressible annulus and each compact component of the boundary of $N^0(\rho)$ in a torus. The manifold pair $(M(\rho), P(\rho))$ is known as the *relative compact core* of $N(\rho)$ where $P(\rho) = M(\rho) \cap \partial N^0(\rho)$.

We recall that if ρ is an element of an equivalence class of representations in $\mathrm{AH}(\pi_1(M))$, then there exists a homotopy equivalence $r_\rho\colon M \to N(\rho)$ in the homotopy class determined by ρ. In order to extend Θ, we consider the subset $P(\rho)$ consisting of the components of $P(\rho)$ which are homotopic to elements of $r_\rho(P)$. Then, by proposition 5.2.3, r_ρ is homotopic to a pared homotopy equivalence $h_\rho\colon (M, P) \to (M(\rho), \widehat{P}(\rho))$. We then set $\Theta(\rho) = ((M(\rho), \widehat{P}(\rho)), h_\rho)$. One may use (a generalization of the) work of McCullough, Miller and Swarup [**90**] to show that Θ is well-defined and agrees with Θ on $\mathrm{GF}(\pi_1(M), \pi_1(P))$.

One of the recently discovered surprises was that the homeomorphism type of a 3-manifold does not vary continuously over $\mathrm{AH}(\pi_1(M), \pi_1(P))$. Anderson and Canary [**8**] gave the first examples. In their situation, M_n was a book of I-bundles (see example 2.10.4), and P_n was empty. Explicitly, M_n was obtained from a collection of n I-bundles over surfaces of genus $1, \ldots, n$, each with one boundary component, by attaching the sides of the I-bundles to n disjoint parallel longitudinal annuli in the boundary of a solid torus. In this case, $\mathcal{A}(M_n, P_n) = \mathcal{A}(M_n)$ contains $(n-1)!$ elements which correspond to the different cyclic orderings of the annuli to which the I-bundles are attached. It is shown that every two components of $\mathrm{GF}(\pi_1(M_n))$ have intersecting closures in $\mathrm{AH}(\pi_1(M_n))$. The proof explicitly constructs a sequence of geometrically finite uniformizations of M_n which converge, in $\mathrm{AH}(\pi_1(M_n))$, to a hyperbolic 3-manifold which is homotopy equivalent, but not homeomorphic, to M_n.

Holt [**52**] has further analyzed the collection of examples in [**8**]. He proves that, for all n, there is a point in $\mathrm{AH}(\pi_1(M_n))$ which lies in the closure of every component of $\mathrm{GF}(\pi_1(M_n))$. In later work, he has observed that the intersection locus of any two components of $\mathrm{GF}(\pi_1(M_n))$ is disconnected.

Anderson, Canary and McCullough [**9**] have completely analyzed this phenomenon in the case when M has incompressible boundary and P consists of the toroidal boundary components of M. They show that marked homeomorphism type *is* locally constant modulo primitive shuffles. Roughly, primitive shuffles are homotopy equivalences obtained by "shuffling" or "rearranging" the way in which the manifold is glued together along primitive solid torus components of its characteristic submanifold. Moreover, if the marked homeomorphism types associated to any two components of $\mathrm{GF}(\pi_1(M))$ differ by a primitive shuffle, then the two components have intersecting closures. As a result, one obtains a complete enumeration of the components of the closure of $\mathrm{GF}(\pi_1(M))$ and hence a conjectural enumeration of the components of $\mathrm{AH}(\pi_1(M))$.

We now state these results more precisely. For $i = 1, 2$, let M_i be a compact, orientable, irreducible 3-manifold and let $\mathcal{V}_i$ be a codimension-zero submanifold of M_i. Denote by $\mathrm{Fr}(\mathcal{V}_i)$ the frontier of $\mathcal{V}_i$ in M_i. We always assume that $\mathrm{Fr}(\mathcal{V}_i)$ is incompressible in M_i; in particular, no component of $\mathrm{Fr}(\mathcal{V}_i)$ is a 2-sphere or a boundary-parallel 2-disk. To avoid trivial cases, we assume that $\mathcal{V}_i$ is a nonempty proper subset of M_i. A homotopy equivalence $h\colon M_1 \to M_2$ is a *shuffle*, with respect to $\mathcal{V}_1$ and $\mathcal{V}_2$, if $h^{-1}(\mathcal{V}_2) = \mathcal{V}_1$ and h restricts to a homeomorphism from $\overline{M_1 - \mathcal{V}_1}$ to $\overline{M_2 - \mathcal{V}_2}$.

Assume now that M_1 and M_2 have nonempty incompressible boundary. Let $(\Sigma_i, \underline{\underline{\sigma_i}})$ denote the characteristic submanifold of $(M_i, \underline{\underline{\emptyset}})$. A solid torus component V of Σ_i is said to be *primitive* when for each frontier annulus A of V, $\pi_1(A)$ surjects onto $\pi_1(V)$, i. e. its core curve is longitudinal. A shuffle $s\colon M_1 \to M_2$ with respect to $\mathcal{V}_1$ and $\mathcal{V}_2$ is called a *primitive shuffle* if

(1) each $\mathcal{V}_i$ is a collection of primitive solid torus components of $\Sigma(M_i)$, and
(2) s restricts to an *orientation-preserving* homeomorphism carrying $\overline{M_1 - \mathcal{V}_1}$ to $\overline{M_2 - \mathcal{V}_2}$.

We say that two elements (M_1, P_1) and (M_2, P_2) of $\mathcal{A}(M)$ are *primitive shuffle equivalent* if there exists a primitive shuffle $s\colon M_1 \to M_2$ such that $[(M_2, h_2)] = [(M_2, s \circ h_1)]$. It is established in [**9**] that primitive shuffle equivalence is a finite-to-one equivalence relation on $\mathcal{A}(M)$. Define $\widehat{\mathcal{A}}(M)$ to be the collection of equivalence classes, and let $q\colon \mathcal{A}(M) \to \widehat{\mathcal{A}}(M)$ be the quotient map. One can then define $\widehat{\Theta}\colon \mathrm{AH}(\pi_1(M)) \to \widehat{\mathcal{A}}(M)$ by setting $\widehat{\Theta} = q \circ \Theta$. The map $\widehat{\Theta}$ records the marked homeomorphism type up to primitive shuffle equivalence. One of the key results in [**9**] is that $\widehat{\Theta}$ is continuous.

THEOREM 13.2.1. *If M is a compact, hyperbolizable 3-manifold with nonempty incompressible boundary, then $\widehat{\Theta}\colon \mathrm{AH}(\pi_1(M)) \to \widehat{\mathcal{A}}(M)$ is continuous.*

Conversely, it is also shown in [**9**] that if two elements of $\mathcal{A}(M)$ are primitive shuffle equivalent, then the corresponding components of $\mathrm{GF}(\pi_1(M))$ have intersecting closures.

THEOREM 13.2.2. *Let M be a compact, hyperbolizable 3-manifold with nonempty incompressible boundary, and let $[(M_1, h_1)]$ and $[(M_2, h_2)]$ be two elements of $\mathcal{A}(M)$. If $[(M_2, h_2)]$ is primitive shuffle equivalent to $[(M_1, h_1)]$, then the associated components of $\mathrm{GF}(\pi_1(M))$ have intersecting closures.*

As an immediate consequence of these two results one sees that two components of $\mathrm{GF}(\pi_1(M))$ have intersecting closures if and only if their associated marked homeomorphism types are primitive shuffle equivalent.

COROLLARY 13.2.3. *Let M be a compact, hyperbolizable 3-manifold with nonempty incompressible boundary, and let $[(M_1, h_1)]$ and $[(M_2, h_2)]$ be two elements of $\mathcal{A}(M)$. The associated components of $\mathrm{GF}(\pi_1(M))$ have intersecting closures if and only if $[(M_2, h_2)]$ is primitive shuffle equivalent to $[(M_1, h_1)]$*

It follows that the components of the closure of $\mathrm{GF}(\pi_1(M))$ are enumerated by the elements of $\widehat{\mathcal{A}}(M)$.

COROLLARY 13.2.4. *Let M be a compact, hyperbolizable 3-manifold with nonempty incompressible boundary. Then, the components of the closure $\overline{\mathrm{GF}(\pi_1(M))}$ of $\mathrm{GF}(\pi_1(M))$ are in a one-to-one correspondence with the elements of $\widehat{\mathcal{A}}(M)$.*

One may then combine corollary 13.2.4 with the Main Hyperbolic Corollary to see that the closure of $\mathrm{GF}(\pi_1(M))$ has finitely many components if and only if M does not have double trouble.

COROLLARY 13.2.5. *Let M be a compact, hyperbolizable 3-manifold with nonempty incompressible boundary. Then, $\overline{\mathrm{GF}(\pi_1(M))}$ has infinitely many components if and only if M has double trouble. Moreover, if M has double trouble, then $\mathrm{AH}(\pi_1(M))$ itself has infinitely many components.*

The Bers' Density Conjecture predicts that $\overline{\mathrm{GF}(\pi_1(M))} = AH(\pi_1(M))$, so it is natural to expect that $\mathrm{AH}(\pi_1(M))$ itself has infinitely many components if and only if M has double trouble.

Conjecture: *Let M be a compact, hyperbolizable 3-manifold with nonempty incompressible boundary. Then $\mathrm{AH}(\pi_1(M))$ has infinitely many components if and only if M has double trouble.*[3]

Holt [**53**] has recently generalized the work in [**52**] to show that if M has incompressible boundary and $\{[(M_1, h_1)], \ldots, [(M_k, h_k)]\}$ are a collection of primitive shuffle equivalent elements of $\mathcal{A}(M)$, then there exists a point $\rho \in \mathrm{AH}(\pi_1(M))$ which lies in the closure of the component of $\mathrm{GF}(\pi_1(M))$ associated to $[(M_i, h_i)]$ for all $i = 1, \ldots, k$.

McMullen [**91**] used the construction in [**8**] to show that if S is a closed hyperbolic surface then $\mathrm{AH}(\pi_1(S))$ is not a manifold. One can further use theorem 13.2.1 to show that $\mathrm{AH}(\pi_1(M))$ is not a manifold whenever M is the domain of a nontrivial primitive shuffle equivalence, see [**9**] for details.

COROLLARY 13.2.6. *Let M be a compact, hyperbolizable 3-manifold with nonempty incompressible boundary. If $q\colon \mathcal{A}(M) \to \widehat{\mathcal{A}}(M)$ is not injective, then $\mathrm{AH}(\pi_1(M))$ is not a manifold.*

Bromberg and Holt [**26**] recently generalized both corollary 13.2.6 and McMullen's result to prove that $\mathrm{AH}(\pi_1(M))$ is not a manifold whenever M contains an essential annulus A such that $\pi_1(A)$ is a maximal cylic subgroup of $\pi_1(M)$.

[3] This conjecture is also a corollary of the Ending Lamination Theorem of Brock-Canary-Minsky [**24**].

THEOREM 13.2.7 (Bromberg-Holt). *Let M be a compact, hyperbolizable 3-manifold with nonempty incompressible boundary. If M contains an essential annulus A such that $\pi_1(A)$ is a maximal cyclic subgroup of $\pi_1(M)$), and C is a component of* $\mathrm{GF}(\pi_1(M))$, *then the closure of C is not a manifold. Moreover, $AH(\pi_1(M))$ is not a manifold.*

13.2.3. Marden's Tameness Conjecture. One of the major conjectures in Kleinian groups is that every hyperbolic 3-manifold with finitely generated fundamental group is *topologically tame*, i. e. homeomorphic to the interior of a compact 3-manifold. Notice that geometrically finite hyperbolic 3-manifolds are topologically tame by definition.

Marden's Tameness Conjecture: *If a hyperbolic 3-manifold has finitely generated fundamental group, then it is homeomorphic to the interior of a compact 3-manifold.*

Bonahon [**17**] established Marden's Tameness Conjecture for hyperbolic 3-manifolds with freely indecomposable fundamental group. More generally, Bonahon's result holds for any hyperbolic 3-manifold N such that each component of the frontier, in N^0, of its relative compact core is incompressible. Marden's conjecture predicts that N is homeomorphic to the interior of its relative compact core.

Canary [**28**] showed that topologically tame hyperbolic 3-manifolds are also geometrically well-behaved. In particular, Marden's Tameness Conjecture implies Ahlfors' Measure Conjecture and a variety of other conjectured geometric properties of hyperbolic 3-manifolds. We recall that Ahlfors' Measure Conjecture predicts that if Γ is a finitely generated discrete subgroup of $\mathrm{PSL}(2,\mathbb{C})$, then either its limit set $\Lambda(\Gamma)$ has measure zero or $\Lambda(\Gamma) = \overline{\mathbb{C}}$ and Γ acts ergodically on $\overline{\mathbb{C}}$.

Marden's conjecture has also been verified for various limits of tame hyperbolic 3-manifolds, see [**30**], [**102**] or [**37**]. The best of these results, due to Evans [**37, 38**], shows that if $\{\rho_i\}$ is a sequence in $\mathrm{AH}(\pi_1(M))$ which converges to $\rho \in \mathrm{AH}(\pi_1(M))$, ρ_i is topologically tame for all i, $\rho_i(g)$ is parabolic for all i whenever $\rho(g)$ is parabolic, and $\{\rho_i(\pi_1(M))\}$ converges geometrically to $\rho(\pi_1(M))$, then ρ is also topologically tame. He also shows that if M is not homotopy equivalent to a compression body then one need not assume that $\{\rho_i(\pi_1(M))\}$ converges geometrically to $\rho(\pi_1(M))$. In particular, this result implies that a "typical" representation in the boundary of any component of $\mathrm{GF}(\pi_1(M))$ is topologically tame.[4]

Although this conjecture is not directly about the deformation theory of Kleinian groups, we will see in section 13.2.5 that topological tameness also plays a role in Thurston's conjectural classification of hyperbolic 3-manifolds.

13.2.4. Laminations and Thurston's Compactification of Teichmüller space. In this section we will recall some basic facts about geodesic laminations and Thurston's compactification of Teichmüller space by the space of projective measured laminations. This material will underlie our upcoming discussion of Thurston's Ending Lamination Conjecture and Thurston's Masur Domain Conjecuture.

[4]Brock, Bromberg, Evans and Souto [**23**] have recently shown that if ρ lies in the boundary of $\mathrm{GF}(\pi_1(M))$ and either a) $\Omega(\rho)$ is non-empty or b) M is not homotopy equivalent to a compression body, then $N(\rho)$ is topologically tame.

A *geodesic lamination* on a hyperbolic surface S is a closed subset L of S which is a disjoint union of simple geodesics. A *measured lamination* is a geodesic lamination L together with an invariant transverse measure with support L. (An invariant transverse measure assigns positive numbers to arcs transverse to L, is additive, and is invariant with respect to projection along L.) By $\mathrm{ML}(S)$ we denote the space of all measured laminations on S, and the projective lamination space $\mathrm{PL}(S)$ is $(\mathrm{ML}(S) - \{\emptyset\}) / \mathbb{R}^+$ (i. e. two measured laminations are considered projectively equivalent if they have the same support and their transverse measures are linear multiples of one another).

Measured laminations are natural generalizations of simple closed curves. A simple closed geodesic may be thought of as a measured lamination, where we simply give transverse arcs the counting measure. In fact, (weighted) simple closed geodesics are dense in $\mathrm{ML}(S)$, so we may think of $\mathrm{ML}(S)$ as the closure of the set of simple closed geodesics. We may also speak of the length of a measured lamination and the intersection number of two measured laminations, both of which are continuous on $\mathrm{ML}(S)$ and agree with the corresponding notions for simple closed geodesics. (For a more detailed treatment of laminations see Bonahon [**18**], Hatcher [**48**] or Thurston [**120**] and for the parallel theory of measured foliations see Fathi-Laudenbach-Poénaru [**39**].)

Thurston showed that one can use $PL(S)$ to compactify Teichmüller space in a natural geometric manner (see Thurston [**120**]):

THEOREM 13.2.8 (Thurston). *If S is a closed hyperbolic surface of genus g, $\mathrm{PL}(S)$ is a sphere of dimension $6g-7$ which may be identified as the compactification of the Teichmüller space of S, i. e. $\mathcal{T}(S) \cup \mathrm{PL}(S) \cong B^{6g-6}$. This compactification is natural in the sense that the action of the mapping class group $\mathrm{Mod}(S)$ of S on $\mathcal{T}(S)$ extends continuously to an action on $\mathcal{T}(S) \cup \mathrm{PL}(S)$.*

The main property of this compactification is that it keeps track of which curves are getting stretched the most as a sequence goes to infinity in $\mathcal{T}(S)$. (This property is stated in Thurston [**124**] and may be proven using the techniques of [**123**]; versions of this result are established in Fathi-Laudenbach-Poenaru [**39**] and Wolf [**129**].)

THEOREM 13.2.9 (Thurston). *Let S be a closed hyperbolic surface of genus g, and $\tau_n \in \mathcal{T}(S)$ a sequence of hyperbolic structures converging to $\mu \in \mathrm{PL}(S)$ in the Thurston compactification of Teichmüller space. Then there exist a sequence of measured laminations $\mu_n \in \mathrm{ML}(S)$ and a constant $K > 0$, such that for any measured lamination λ*

$$i(\mu_n, \lambda) + K\, l_{\tau_0}(\lambda) \geq l_{\tau_n}(\lambda) \geq i(\mu_n, \lambda).$$

Also, $l_{\tau_n}(\mu_n)$ remains bounded and $[\mu_n]$ converges to μ in $\mathrm{PL}(S)$.

13.2.5. The Ending Lamination Conjecture. Thurston's Ending Lamination Conjecture is a conjectural classification of the elements of $\mathrm{AH}(\pi_1(M))$. A full discussion of this conjecture is beyond the scope of this epilogue, but we will give an outline of the conjecture in our language. (See [**93**] or [**94**] for a more complete discussion.) Roughly, Thurston proposes that a hyperbolic 3-manifold is determined up to isometry by its (relative) marked homeomorphism type, captured by its relative compact core, and geometric invariants, called *ending invariants*, which capture the asymptotic geometry of its ends. We will assume for the bulk of the section that M has incompressible boundary.

Given an element $N(\rho)$ of $\mathrm{AH}(\pi_1(M))$, let $(M(\rho), P(\rho))$ be its relative compact core. Each component G of the conformal boundary $\partial\widehat{N}(\rho)$ lies on the boundary of a component X of $N^0(\rho) - M(\rho)$ such that $X \cong G \times (0,1)$ and $\partial X = F \cup G$ where F is a free side of $(M(\rho), P(\rho))$. In this case, F is called a geometrically finite free side and the ending invariant is the element of $\mathcal{T}(F)$ determined by the Riemann surface G. If F is a free side of $(M(\rho), P(\rho))$ which is not geometrically finite, Bonahon [**17**] showed that there is a sequence $\{\alpha_i\}$ of simple closed curves on F whose geodesic representatives in $N(\rho)$ lie in the component of $N^0(\rho) - M(\rho)$ bounded by F and leave every compact subset of $N(\rho)$. The geodesic representatives of α_i, in some fixed hyperbolic structure on F, converge to a geodesic lamination μ (at least up to subsequence). The maximal sublamination λ_F of μ which is the support of a measured lamination is the *ending invariant* associated to F. (Bonahon also shows that λ_F is independent of the sequence $\{\alpha_i\}$ chosen and that λ_F is filling, i.e. intersects every simple closed geodesic in F transversely.) If M has incompressible boundary Thurston's Ending Lamination Conjecture asserts that the element of $\mathrm{AH}(\pi_1(M))$ is determined by the marked pared homeomorphism type of $(M(\rho), P(\rho))$ and its collection of ending invariants.

It follows from the classical theory, see The Parameterization Theorem in chapter 7, that the Ending Lamination Conjecture is valid for geometrically finite hyperbolic manifolds. The Ending Lamination Conjecture has only been verified for very special classes of geometrically infinite manifolds. Minsky verified it for punctured torus groups in [**94**] and for hyperbolic 3-manifolds with freely indecomposable fundamental group and a lower bound on the injectivity radius in [**92**].[5] However, one may use Thurston's double limit theorem and relative compactness theorems to determine exactly which collections of ending invariants can arise, see Ohshika [**101**]. Moreover, Thurston's Ending Lamination Conjecture implies Bers' Density Conjecture as one can produce representations with every allowable collection of ending invariants on the boundary of $\mathrm{GF}(\pi_1(M))$.

One can generalize Thurston's Ending Lamination Conjecture to the case where M has compressible boundary. However, one does not know that the ending invariants are defined unless one first knows that the hyperbolic manifold is topologically tame. Moreover, the ending invariants are naturally only well-defined modulo the group $\mathrm{Mod}_0(M(\rho), P(\rho))$, see section 10 of [**28**] for more details.

13.2.6. Thurston's Masur Domain Conjecture. It is also interesting to study the issue of determining when a sequence in $\mathrm{GF}(M)$ converges in $\mathrm{AH}(\pi_1(M))$. Since each component of $\mathrm{GF}(M)$ is parameterized by data lying in $\mathcal{T}(M)$ it is very natural to attempt to determine the convergence from this data.

Thurston's double limit theorem gives a fairly complete answer to this question in the situation where $M = S \times \mathrm{I}$. It should be remarked that Thurston's double limit theorem is the key tool in the proof that atoroidal 3-manifolds which fiber over the circle are hyperbolizable, see Thurston [**124**] or Otal [**107**] for more details.

[5] Minsky [**95**] and Brock-Canary-Minsky [**24**] have established the Ending Lamination Conjecture in the case that M has incompressible boundary. More generally, they prove the Ending Lamination Conjecture for $\mathrm{AH}(\pi_1(M), \pi_1(P))$ when each component of $\partial M - P$ is incompressible. The proof makes extensive use of the geometric theory of the curve complex developed by Masur and Minsky [**80, 81**].

Two measured laminations λ and μ are said to *bind* the surface S, if

$$i(\lambda, \nu) + i(\mu, \nu) > 0$$

for every measured lamination $\nu \in \mathrm{ML}(S)$.

THEOREM 13.2.10 (Thurston [**124**]). *Let λ and μ be a pair of measured laminations which bind S. Any sequence $\{\rho_i\} = \{(g_i, h_i)\}$ in*

$$\mathrm{GF}(S \times I, \partial S \times I) \cong \mathcal{T}(S) \times \mathcal{T}(S)$$

which converges to (λ, μ) in

$$(\mathcal{T}(S) \cup \mathrm{PL}(S)) \times (\mathcal{T}(S) \cup \mathrm{PL}(S))$$

has a convergent subsequence in $\mathrm{AH}(S \times I, \partial S \times I)$.

One may combine the double limit theorem with Thurston's relative compactness theorem to obtain a very satisfying generalization of this to the setting of 3-manifolds with incompressible boundary, see Ohshika [**100**]. The slightly different version given here is from [**27**].

Thurston defines the *window* W of M to consist of the I-bundle components of the characteristic submanifold Σ of M together with regular neighborhoods of the annular components of the frontiers of Seifert fibred components of Σ. (We eliminate any redundancies which develop by absorbing any components of W which are homotopic into other components into those components.) W is an I-bundle whose lids lie in ∂M, let w denote the base of this I-bundle.

If μ is any measured lamination on w, then it is double covered by a measured lamination $\widetilde{\mu}$ in the lids of W. Let $\{S_1, \ldots, S_k\}$ denote the nontoroidal boundary components of M and let

$$\tau = (\tau_1, \ldots, \tau_k) \in (\mathcal{T}(S_1) \cup \mathrm{PL}(S_1)) \times \cdots (\mathcal{T}(S_k) \cup \mathrm{PL}(S_k))$$

We will say that τ *binds the window base* w if whenever μ is a connected measured lamination in $\mathrm{ML}(w)$, then there is a component ν of $\widetilde{\mu}$ such that $\nu \subset S_i$ and either $\tau_i \in \mathcal{T}(S_i)$ or $i(\nu, \tau_k) > 0$.

In the theorem below we assume that C is a component of $\mathrm{GF}(M)$ which has been parameterized by $\mathcal{T}(S_1) \times \cdots \times \mathcal{T}(S_k)$ as in chapter 7.

THEOREM 13.2.11. *Suppose that M is a hyperbolizable 3-manifold with incompressible boundary. Let $\{\rho_i\}$ be a sequence in a component C of $\mathrm{GF}(M)$ which converges to $\tau \in (\mathcal{T}(S_1) \cup \mathrm{PL}(S_1)) \times \cdots \times (\mathcal{T}(S_k) \cup \mathrm{PL}(S_k))$. If τ binds the window base, then $\{\rho_i\}$ has a convergent subsequence in* $\mathrm{AH}(\pi_1(M))$.

However, very little of this theory generalizes directly to the setting of manifolds with compressible boundary. We will describe a conjectural generalization, due to Thurston, in the case when M is a compression body. The general conjecture is somewhat more complicated but is in the same spirit.

If M is a compression body then each component C of $\mathrm{GF}(M)$ is parameterized by $\mathcal{T}(\partial M)/\mathrm{Mod}_0(M)$. Let F be the compressible boundary component of M and let $\{S_1, \ldots, S_k\}$ denote the other nontoroidal boundary components of M, then

$$\mathcal{T}(\partial M) = \mathcal{T}(F) \times \mathcal{T}(S_1) \times \cdots \times \mathcal{T}(S_k).$$

Since $\mathrm{Mod}_0(M)$ is generated by Dehn twists about compressing disks, $\mathrm{Mod}_0(M)$ acts trivially on the $\mathcal{T}(S_1) \times \cdots \times \mathcal{T}(S_k)$ factors, so

$$C \cong (\mathcal{T}(F)/\mathrm{Mod}_0(M)) \times \mathcal{T}(S_1) \times \cdots \times \mathcal{T}(S_k).$$

Masur [**79**] and Otal [**105**] have extensively studied the action of $\mathrm{Mod}_0(M)$ on the boundary $\mathrm{PL}(F)$ of $\mathcal{T}(F)$. Masur identified an open subset $\mathcal{M}$ of $\mathrm{PL}(F)$, called the *Masur domain*, on which $\mathrm{Mod}_0(M)$ acts properly discontinuously. Kerckhoff [**63**] established that the Masur domain has full measure in $\mathrm{PL}(F)$. Let $\mathcal{L} \subset \mathrm{PL}(F)$ denote the closure of the set of compressible curves in F. Further let $\mathcal{L}'$ denote the set of laminations which have zero intersection number with some element of $\mathcal{L}$; in particular, $\mathcal{L} \subset \mathcal{L}'$. If M is not a boundary connected sum of two I-bundles, then the Masur domain $\mathcal{M} = \mathrm{PL}(F) - \mathcal{L}'$. If M is a boundary connected sum of two I-bundles, then let $\mathcal{L}''$ denote the set of laminations which have zero intersection number with some element of $\mathcal{L}'$, and define $\mathcal{M} = PL(F) - \mathcal{L}''$.

We may append $\mathcal{M}/\mathrm{Mod}_0(M)$ to $\mathcal{T}(F)/\mathrm{Mod}_0(M)$ as a sort of "boundary at infinity." It is not, however, a compactification. If $\{\rho_i\}$ is a sequence in C, let $\{\tau_i\}$ be the associated sequence in $\mathcal{T}(F)/\mathrm{Mod}_0(M)$. We say that $\{\rho_i\}$ *converges into the Masur domain* when $\{\tau_i\}$ converges to a point in $\mathcal{M}/\mathrm{Mod}_0(M)$ within $(\mathcal{T}(F) \cup \mathcal{M})/\mathrm{Mod}_0(M)$. We are now ready to state Thurston's Masur Domain Conjecture.

Thurston's Masur Domain Conjecture: *Suppose that M is a compression body and C is a component of* $\mathrm{GF}(M)$. *If a sequence $\{\rho_i\}$ in C converges into the Masur domain, then $\{\rho_i\}$ has a subsequence which converges in* $\mathrm{AH}(\pi_1(M))$.

Until recently very little progress had been made on this conjecture, although see [**29**] and [**105**]. However, Kleineidam and Souto [**65**] recently proved Thurston's Masur Domain Conjecture in the case that $\{\tau_i\}$ converges to a minimal arational measured lamination in the Masur domain (which is the generic case).

Bibliography

[1] W. Abikoff, *The Real Analytic Theory of Teichmüller Space*, Lecture Notes in Math. 820, Springer-Verlag, 1980.

[2] W. Abikoff, "The Euler characteristic and inequalities for Kleinian groups," *Proc. Amer. Math. Soc.* **97** (1986), 593–601.

[3] W. Abikoff and B. Maskit, "Geometric decompositions of Kleinian groups," *Amer. J. Math.* **99** (1977), 687–697.

[4] L. Ahlfors, "Finitely generated Kleinian groups," *Amer. J. Math.* **86** (1964), 413–429.

[5] L. Ahlfors, "Fundamental polyhedrons and limit point sets of Kleinian groups," *Proc. Nat. Acad. Sci. USA* **55** (1966), 251–254.

[6] L. Ahlfors and L. Bers, "Riemann's mapping theorem for variable metrics," *Annals of Math.* **72** (1960), 385–404.

[7] J. W. Anderson and R. D. Canary, "Cores of hyperbolic 3-manifolds and limits of Kleinian groups," *Amer. J. Math.* **118** (1996), 745–779.

[8] J. W. Anderson and R. D. Canary, "Algebraic limits of Kleinian groups which rearrange the pages of a book, " *Invent. Math.* **126** (1996), 205–214.

[9] J. W. Anderson, R. D. Canary and D. McCullough, "On the topology of deformation spaces of Kleinian groups, *Annals of Math.* **152** (2000), 693-741.

[10] R. Benedetti and C. Petronio, *Lectures on Hyperbolic Geometry*, Springer-Verlag Universitext, 1992.

[11] L. Bers, "Quasiconformal mappings and Teichmüller's theorem," in *Analytic functions*, Princeton University Press, 1960, 89–119.

[12] L. Bers, "On boundaries of Teichmüller spaces and on Kleinian groups," *Annals of Math.* **91** (1970), 570–600.

[13] L. Bers, "Spaces of Kleinian groups," in *Maryland Conference in Several Complex Variables I*. Springer-Verlag Lecture Notes in Math, No. 155 (1970), 9–34.

[14] L. Bers, "On moduli of Kleinian groups," *Russian Math. Surveys* **29** (1974), 88–102.

[15] L. Bers, "On Sullivan's proof of the finiteness theorem and the eventual periodicity theorem," *Amer. J. Math.* **109** (1987), 833–852.

[16] F. Bonahon, "Cobordism of automorphisms of surfaces" *Ann. Sci. Ecole Norm. Sup.* **(4) 16** (1983), 237–270.

[17] F. Bonahon, "Bouts des variétés hyperboliques de dimension 3," *Annals of Math.* **124** (1986), 71–158.

[18] F. Bonahon, "The geometry of Teichmüller space via geodesic currents," *Invent. Math.* **92** (1988), 139–62.

[19] F. Bonahon and L. Siebenmann, "The characteristic toric splitting of irreducible compact 3-orbifolds," *Math. Ann.* **278** (1987), 441–479.

[20] B. Bowditch, "Geometrical finiteness for hyperbolic groups," *J. Funct. Anal.* **113** (1993), 245–317.

[21] M. Brin, K. Johansson and P. Scott, "Totally peripheral 3-manifolds," *Pacific J. Math.* **118** (1985), 37–51.

[22] J. Brock and K. Bromberg, "On the density of geometrically finite Kleinian groups," to appear in *Acta. Math.*

[23] J. Brock, K. Bromberg, R. Evans and J. Souto, "Tameness on the boundary and Ahlfors' measure conjecture," to appear in *Publ. Math. I. H. E. S.*

[24] J. Brock, R.Canary and Y. Minsky, "The Classification of Kleinian Surface Groups II: The Ending Lamination Conjecture," in preparation.

[25] K. Bromberg, "Projective structures with degenerate holonomy and the Bers density conjecture," preprint.

[26] K. Bromberg and J. Holt, "Self-bumping of deformation spaces of hyperbolic 3-manifolds," *J. Differential Geom.* **57** (2001), 47-65.

[27] R.D. Canary, "Hyperbolic structures on 3-manifolds with compressible boundary," Ph. D. thesis, Princeton University, June 1989.

[28] R. D. Canary, "Ends of hyperbolic 3-manifolds," *J. Amer. Math. Soc.* **6** (1993), 1–35.

[29] R. D. Canary, "Algebraic convergence of Schottky groups," *Trans. Amer. Math. Soc.* **337** (1993), 235–258.

[30] R. D. Canary and Y. N. Minsky, "On limits of tame hyperbolic 3-manifolds," *J. Differential Geom.* **43** (1996), 1–41.

[31] M. Culler and P. B. Shalen, "Volumes of hyperbolic Haken manifolds, I" *Invent. Math.* **118** (1994), 285-329.

[32] A. Douady and C. J. Earle, "Conformally natural extensions of homeomorphisms of the circle," *Acta Math.* **157** (1986) 23–48.

[33] C. Earle and C. McMullen, "Quasiconformal isotopies,"in *Holomorphic Functions and Moduli I*, M. S. R. I. Publication 10, Springer-Verlag, 1988, 143–154.

[34] D. B. A. Epstein, "Projective planes in 3-manifolds," *Proc. London Math. Soc.* **(3) 11** (1961), 469–484.

[35] D. B. A. Epstein, "Curves on 2-manifolds and isotopies," *Acta Math.* **115** (1966) 83–107.

[36] B. Evans, "Boundary respecting maps of 3-manifolds," *Pacific J. Math.* **42** (1972), 639-655.

[37] R. Evans, "Tameness persists in type-preserving strong limits," preprint.

[38] R. Evans, "Deformation spaces of hyperbolic 3-manifolds: strong convergence and tameness," Ph. D. thesis, University of Michigan, 2000.

[39] A. Fathi, F. Laudenbach and V. Poénaru, Travaux de Thurston sur les surfaces, *Astérisque* **66-67** (1979).

[40] D. I. Fouxe-Rabinovitch, "On the automorphism group of free products I," *Math. Sb.* **8** (1940), 265-276.

[41] D. I. Fouxe-Rabinovitch, "On the automorphism group of free products II," *Math. Sb.* **51** (1941), 183–220.

[42] D. Gabai and W. Kazez, "Group negative curvature for 3-manifolds with genuine laminations," *Geom. Top.* **2** (1998), 65–77.

[43] D. Gabai and W. Kazez, "The finiteness of the mapping class group for atoroidal 3-manifolds with genuine laminations," *J. Differential Geom.* **50** (1998), 123–127.

[44] F. Gardiner, *Teichmüller Theory and Quadratic Differentials*, John Wiley and Sons, 1987.

[45] N. Gilbert, "Presentations of the automorphism group of a free product, *Proc. London Math. Soc.* **(3) 54** (1987), 115-140

[46] M. Handel and W. P. Thurston, "New proofs of some results of Nielsen," *Adv. in Math.* **56** (1985), 173–191.

[47] W.J. Harvey, "Spaces of discrete groups," in *Discrete Groups and Automorphic Functions*, Academic Press, 1977, 295–348.

[48] A.E. Hatcher, "Measured lamination spaces for surfaces, from the topological viewpoint," *Top. and its Appl.* **30**(1988), 63–88.

[49] W. Heil, "On $\mathbb{P}^2$-irreducible 3-manifolds," *Bull. Amer. Math. Soc.* **75** (1969), 772–775.

[50] W. Heil, "Almost sufficiently large Seifert fiber spaces," *Michigan Math. J.* **20** (1973), 217–223.

[51] J. Hempel, *3-manifolds*, Princeton University Press, 1976.

[52] J. Holt, "Some new behaviour in the deformation theory of Kleinian groups," *Comm. Anal. Geom* **9** (2001), 757-775.

[53] J. Holt, "Multiple bumping of components of deformation spaces of hyperbolic 3-manifolds," Amer. J. Math. **125** (2003), 691-736.

[54] W. Jaco, *Lectures on Three-manifold Topology,* CBMS Regional Conference Series No. 43 (1977).

[55] W. Jaco, "Finitely presented subgroups of three-manifold groups," *Invent. Math.* **13** (1971), 335-346.

[56] W. Jaco and P. Shalen, "Seifert fibered spaces in 3-manifolds," *Mem. Amer. Math. Soc.* **220** (1979).

[57] B. Jiang, S. Wang, and Y.-Q. Wu, "Homeomorphisms of 3-manifolds and the realization of Nielsen number," *Comm. Anal. Geom.* **9** (2001), 825-877.

[58] K. Johannson, *Homotopy Equivalences of 3-manifolds with Boundary*, Lecture Notes in Mathematics, vol. 761, Springer-Verlag, 1979.

[59] K. Johannson, "On exotic homotopy equivalences of 3-manifolds," *Geometric Topology*, ed. by J.C. Cantrell, Academic Press (1979), 101–112.

[60] J. Kalliongis, "Realizing automorphisms of the fundamental group of irreducible 3-manifolds containing two-sided projective planes," *Math. Ann.* **293** (1992), 707–727.

[61] M. Kapovich, *Hyperbolic Manifolds and Discrete groups: Lectures on Thurston's Hyperbolization*, Birkhauser, 2000.

[62] L. Keen, B. Maskit and C. Series, "Geometric finiteness and uniqueness of groups with circle packing limit sets," *J. Reine Angew. Math.* **436** (1993), 209–219.

[63] S. P. Kerckhoff, "The measure of the limit set of the handlebody group," *Topology* **29** (1990), 27–40.

[64] S. P. Kerckhoff and W. P. Thurston, "Non-continuity of the action of the mapping class group at Bers' boundary of Teichmüller space," *Invent. Math.* **100** (1990), 25–47.

[65] G. Kleineidam and J. Souto, "Algebraic convergence of function groups," *Comment. Math. Helv.* **77** (2002), 244-269.

[66] I. Kra, "On spaces of Kleinian groups," *Comm. Math. Helv.* **47** (1972), 53–69.

[67] R. Kulkarni and P. Shalen, "On Ahlfors' finiteness theorem," *Adv. in Math.* **76** (1989), 155–169.

[68] F. Laudenbach, "Sur les 2-sphères d'une variété de dimension 3," *Annals of Math.* **97** (1973), 57–81.

[69] F. Laudenbach, "Topologie de dimension trois. Homotopie et isotopie," *Astérisque* **12** (1974) 11–152.

[70] O. Lehto, *Univalent Functions and Teichmüller Spaces*, Springer-Verlag, New York, 1987.

[71] O. Lehto and K. I. Virtanen, *Quasiconformal Mappings in the Plane*, Springer-Verlag, 1973.

[72] R. Lickorish, "Homeomorphisms of nonorientable two-manifolds," *Proc. Cambridge Phil. Soc.* **59** (1963), 307-317.

[73] R. C. Lyndon and P. E. Schupp, *Combinatorial group theory,* Springer-Verlag, 1977.

[74] A. M. MacBeath and D. Singerman, "Spaces of subgroups and Teichmüller space," *Proc. London Math. Soc.* **31** (1975), 211–256.

[75] A. Marden, "The geometry of finitely generated Kleinian groups," *Annals of Math.* **99** (1974), 383–462.

[76] B. Maskit,"Self-maps of Kleinian groups," *Amer. J. Math.* **93** (1971), 840–856.

[77] B. Maskit, "Isomorphisms of function groups," *Journal d'Analyse Math.* **32** (1977), 63–82.

[78] B. Maskit, *Kleinian Groups,* Springer-Verlag, 1988.

[79] H. Masur, "Measured foliations and handlebodies," *Erg. Th. Dyn. Sys.* **6** (1986), 99–116.

[80] H. Masur and Y. Minsky, "Geometry of the complex of curves I: hyperbolicity," *Invent. Math.* **138**(1999), 103–149.

[81] H. Masur and Y. Minsky, "Geometry of the complex of curves II: hierarchical structure," *G.A.F.A.* **10**(2000), 902–974.

[82] K. Matsuzaki and M. Taniguchi, *Hyperbolic manifolds and Kleinian groups,* Oxford Mathematical Monographs, Clarendon Press, 1998.

[83] J. McCool, "A presentation for the automorphism group of a free group of finite rank," *J. London Math. Soc.* **(2) 8** (1974), 259–266.

[84] J. McCool, "On Nielsen's presentation of the automorphism group of a free group," *J. London Math. Soc.* **(2) 10** (1975), 265–270.

[85] D. McCullough, "Compact submanifolds of 3-manifolds with boundary," *Quart. J. Math. Oxford* **37** (1986), 299–307.

[86] D. McCullough, "Topological and algebraic automorphisms of 3-manifolds," in *Groups of Self-Equivalences and Related Topics,* Lecture Notes in Mathematics, vol. 1425, Springer-Verlag (1990), 102–113.

[87] D. McCullough, "Virtually geometrically finite mapping class groups of 3-manifolds," *J. Differential Geom.* **33** (1991), 1–65.

[88] D. McCullough, "3-manifolds and their mappings," Lecture Notes Series, 26. Seoul National University, Research Institute of Mathematics, Global Analysis Research Center, Seoul (1995).

[89] D. McCullough and A. Miller, "Homeomorphisms of 3-manifolds with compressible boundary," *Mem. Amer. Math. Soc.* vol. 61 no. 344 (1986).

[90] D. McCullough, A. Miller and G. A. Swarup, "Uniqueness of cores of noncompact 3-manifolds," *J. London Math. Soc* **32** (1985), 548–556.

[91] C. T. McMullen, "Complex earthquakes and Teichmüller theory," *J. Amer. Math. Soc.* **11** (1998), 283–320.

[92] Y. N. Minsky, "On rigidity, limit sets and end invariants of hyperbolic 3-manifolds," *J. Amer. Math. Soc.* **7** (1994), 539–588.

[93] Y. N. Minsky, "On Thurston's ending lamination conjecture," Low-dimensional topology (Knoxville, TN, 1992), ed. by Klaus Johannson, International Press, 1994, 109–122.

[94] Y. N. Minsky, "The classification of punctured torus groups," *Annals of Math.*, **149** (1999), 559–626.

[95] Y.N. Minsky, "The classification of Kleinian surface groups I: models and bounds," preprint.

[96] J. W. Morgan, "On Thurston's uniformization theorem for three-dimensional manifolds," in *The Smith Conjecture*, ed. by J. Morgan and H. Bass, Academic Press, 1984, 37–125.

[97] J. W. Morgan and P. B. Shalen, "Degeneration of hyperbolic structures, III: Actions of 3-manifold groups on trees and Thurston's compactness theorem," *Annals of Math.* **127** (1988), 457–519.

[98] G. D. Mostow, "Quasiconformal mappings in n-space and the rigidity of hyperbolic space forms," *Publ. I. H. E. S.* **34** (1968), 53–104.

[99] J. Nielsen, "Die Isomorphismengruppe der freien gruppe," *Math. Ann.* **91** (1924), 169-209.

[100] K. Ohshika, "On limits of quasiconformal deformations of Kleinian groups," *Math. Z.* **201** (1989), 167–176.

[101] K. Ohshika, "Ending laminations and boundaries for deformation spaces of Kleinian groups," *J. London Math. Soc.* **42** (1990), 111–121.

[102] K. Ohshika, "Kleinian groups which are limits of geometrically finite groups," preprint.

[103] P. Orlik, *Seifert Manifolds,* Springer-Verlag Lecture Notes in Mathematics **291** (1972).

[104] P. Orlik, E. Vogt, and H. Zieschang, Zur Topologie gefaserter dreidimensionaler Mannigfaltigkeiten, *Topology* **6** (1967), 49–64.

[105] J. P. Otal, "Courants geodesiques et produits libres", preprint.

[106] J. P. Otal, Sur la dégénérescence des groupes de Schottky," *Duke Math. J.* **74** (1994), 777–792.

[107] J. P. Otal, "Le théorème d'hyperbolisation pour les variétes fibrées de dimension 3," *Astérisque* **235** (1996).

[108] R. S. Palais, "Local triviality of the restriction map for embeddings," *Comment. Math. Helv.* **34** (1960), 305–312.

[109] F. Paulin, "Outer automorphisms of hyperbolic groups and small actions on R-trees," in *Arboreal group theory (Berkeley, CA, 1988),* M. S. R. I. Publ. 19, Springer, New York (1991), 331–343.

[110] G. Prasad, "Strong rigidity of $\mathbb{Q}$-rank 1 lattices," *Invent. Math.* **21** (1973), 255–286.

[111] H. M. Reimann, "Invariant extensions of quasiconformal deformations," *Ann. Acad. Sci. Fenn.* **10** (1985), 477–492.

[112] P. Scott, "Compact submanifolds of 3-manifolds," *J. London Math. Soc.* **7** (1974), 246–250.

[113] P. Scott, "The geometries of 3-manifolds," *Bull. London Math. Soc.* **15** (1983), 401–487.

[114] H. Seifert, "Topologie dreidimensionaler gefaseter Raume," *Acta Math.* **60** (1933), 147–238.

[115] D. Sullivan, "On the ergodic theory at infinity of an arbitrary discrete group of hyperbolic motions," in *Riemann Surfaces and Related Topics, Proceedings of the 1978 Stony Brook Conference*, Annals of Math. Studies no. 97 (1980), 465–496.

[116] D. Sullivan, "Quasiconformal homeomorphisms and dynamics. II. Structural stability implies hyperbolicity for Kleinian groups," *Acta Mathematica* **155** (1985), pp. 243–260.

[117] G. A. Swarup, "On incompressible surfaces in the complements of knots," *J. Indian Math. Soc.* **37** (1973), 9-24.

[118] G. A. Swarup, "Two finiteness properties in 3-manifolds," *Bull. London Math. Soc.* **12** (1980), 296–302.

[119] G. A. Swarup, "Homeomorphisms of compact 3-manifolds," *Topology* **16** (1977),119–130.

[120] W. P. Thurston, *The Geometry and Topology of 3-manifolds*, lecture notes.

[121] W. P. Thurston, "Three dimensional manifolds, Kleinian groups, and hyperbolic geometry," *Bull. A.M.S.* **6** (1982), 357–381.

[122] W. P. Thurston, "On the geometry and dynamics of diffeomorphisms of surfaces," *Bull. A.M.S.* **19** (1988), 417–431.

[123] W. P. Thurston, "Minimal stretch maps between hyperbolic surfaces," preprint, at `http://front.math.ucdavis.edu/math.GT/9801039`.

[124] W. P. Thurston, "Hyperbolic structures on 3-manifolds, II: Surface groups and 3-manifolds which fiber over the circle," preprint, at `http://front.math.ucdavis.edu`.

[125] P. Tukia, "Quasiconformal extensions of quasisymmetric mappings compatible with a Möbius group," *Acta Math.* **154** (1985), 153–193.

[126] F. Waldhausen, Eine Klasse von 3-dimensionalen Mannigfaltigkeiten I, II, *Invent. Math.* **3** (1967), 308–333, **4** (1967), 87–117.

[127] F. Waldhausen, Gruppen mit Zentrum und 3-dimensionale Mannigfaltigkeiten, *Topology* **6** (1967), 505-517.

[128] F. Waldhausen, "On irreducible 3-manifolds which are sufficiently large," *Annals of Math.* **87** (1968), 56–88.

[129] M. Wolf, "The Teichmüller theory of harmonic maps," *J. Differential Geom.* **29** (1989), 449–479

Index

Mathematical symbols appear separate-ly at the beginning of the index, and are repeated in their lexicographical position later. Page numbers given in **boldface** indicate the location of the main definition of the item, and those in *italics* indicate statements of theorems.

Editorial Information

To be published in the *Memoirs*, a paper must be correct, new, nontrivial, and significant. Further, it must be well written and of interest to a substantial number of mathematicians. Piecemeal results, such as an inconclusive step toward an unproved major theorem or a minor variation on a known result, are in general not acceptable for publication. Papers appearing in *Memoirs* are generally longer than those appearing in *Transactions*, which shares the same editorial committee.

As of August 1, 2004, the backlog for this journal was approximately 5 volumes. This estimate is the result of dividing the number of manuscripts for this journal in the Providence office that have not yet gone to the printer on the above date by the average number of monographs per volume over the previous twelve months, reduced by the number of volumes published in four months (the time necessary for preparing a volume for the printer). (There are 6 volumes per year, each containing at least 4 numbers.)

A Consent to Publish and Copyright Agreement is required before a paper will be published in the *Memoirs*. After a paper is accepted for publication, the Providence office will send a Consent to Publish and Copyright Agreement to all authors of the paper. By submitting a paper to the *Memoirs*, authors certify that the results have not been submitted to nor are they under consideration for publication by another journal, conference proceedings, or similar publication.

Information for Authors

Memoirs are printed from camera copy fully prepared by the author. This means that the finished book will look exactly like the copy submitted.

The paper must contain a *descriptive title* and an *abstract* that summarizes the article in language suitable for workers in the general field (algebra, analysis, etc.). The *descriptive title* should be short, but informative; useless or vague phrases such as "some remarks about" or "concerning" should be avoided. The *abstract* should be at least one complete sentence, and at most 300 words. Included with the footnotes to the paper should be the 2000 *Mathematics Subject Classification* representing the primary and secondary subjects of the article. The classifications are accessible from `www.ams.org/msc/`. The list of classifications is also available in print starting with the 1999 annual index of *Mathematical Reviews*. The Mathematics Subject Classification footnote may be followed by a list of *key words and phrases* describing the subject matter of the article and taken from it. Journal abbreviations used in bibliographies are listed in the latest *Mathematical Reviews* annual index. The series abbreviations are also accessible from `www.ams.org/publications/`. To help in preparing and verifying references, the AMS offers MR Lookup, a Reference Tool for Linking, at `www.ams.org/mrlookup/`. When the manuscript is submitted, authors should supply the editor with electronic addresses if available. These will be printed after the postal address at the end of the article.

Electronically prepared manuscripts. The AMS encourages electronically prepared manuscripts, with a strong preference for $\mathcal{A}\mathcal{M}\mathcal{S}$-LaTeX. To this end, the Society has prepared $\mathcal{A}\mathcal{M}\mathcal{S}$-LaTeX author packages for each AMS publication. Author packages include instructions for preparing electronic manuscripts, the *AMS Author Handbook*, samples, and a style file that generates the particular design specifications of that publication series. Though $\mathcal{A}\mathcal{M}\mathcal{S}$-LaTeX is the highly preferred format of TeX, author packages are also available in $\mathcal{A}\mathcal{M}\mathcal{S}$-TeX.

Authors may retrieve an author package from e-MATH starting from `www.ams.org/tex/` or via FTP to `ftp.ams.org` (login as `anonymous`, enter username as password, and type `cd pub/author-info`). The *AMS Author Handbook* and the *Instruction Manual* are available in PDF format following the author packages link from `www.ams.org/tex/`. The author package can be obtained free of charge by sending email to `pub@ams.org` (Internet) or from the Publication Division, American Mathematical Society, 201 Charles St., Providence, RI 02904, USA. When requesting an author package, please specify $\mathcal{A}\mathcal{M}\mathcal{S}$-LaTeX or $\mathcal{A}\mathcal{M}\mathcal{S}$-TeX, Macintosh or IBM (3.5) format, and the publication in which your paper will appear. Please be sure to include your complete mailing address.

Sending electronic files. After acceptance, the source file(s) should be sent to the Providence office (this includes any TeX source file, any graphics files, and the DVI or PostScript file).

Before sending the source file, be sure you have proofread your paper carefully. The files you send must be the EXACT files used to generate the proof copy that was accepted for publication. For all publications, authors are required to send a printed copy of their paper, which exactly matches the copy approved for publication, along with any graphics that will appear in the paper.

TeX files may be submitted by email, FTP, or on diskette. The DVI file(s) and PostScript files should be submitted only by FTP or on diskette unless they are encoded properly to submit through email. (DVI files are binary and PostScript files tend to be very large.)

Electronically prepared manuscripts can be sent via email to `pub-submit@ams.org` (Internet). The subject line of the message should include the publication code to identify it as a Memoir. TeX source files, DVI files, and PostScript files can be transferred over the Internet by FTP to the Internet node `e-math.ams.org` (130.44.1.100).

Electronic graphics. Comprehensive instructions on preparing graphics are available at `www.ams.org/jourhtml/graphics.html`. A few of the major requirements are given here.

Submit files for graphics as EPS (Encapsulated PostScript) files. This includes graphics originated via a graphics application as well as scanned photographs or other computer-generated images. If this is not possible, TIFF files are acceptable as long as they can be opened in Adobe Photoshop or Illustrator. No matter what method was used to produce the graphic, it is necessary to provide a paper copy to the AMS.

Authors using graphics packages for the creation of electronic art should also avoid the use of any lines thinner than 0.5 points in width. Many graphics packages allow the user to specify a "hairline" for a very thin line. Hairlines often look acceptable when proofed on a typical laser printer. However, when produced on a high-resolution laser imagesetter, hairlines become nearly invisible and will be lost entirely in the final printing process.

Screens should be set to values between 15% and 85%. Screens which fall outside of this range are too light or too dark to print correctly. Variations of screens within a graphic should be no less than 10%.

Inquiries. Any inquiries concerning a paper that has been accepted for publication should be sent directly to the Electronic Prepress Department, American Mathematical Society, 201 Charles St., Providence, RI 02904, USA.

Editors

This journal is designed particularly for long research papers, normally at least 80 pages in length, and groups of cognate papers in pure and applied mathematics. Papers intended for publication in the *Memoirs* should be addressed to one of the following editors. In principle the Memoirs welcomes electronic submissions, and some of the editors, those whose names appear below with an asterisk (*), have indicated that they prefer them. However, editors reserve the right to request hard copies after papers have been submitted electronically. Authors are advised to make preliminary email inquiries to editors about whether they are likely to be able to handle submissions in a particular electronic form.

*__Algebra__ to ROBERT GURALNICK, Department of Mathematics, University of Southern California, Los Angeles, CA 90089-1113; email: `guralnic@math.usc.edu`

Algebraic geometry to DAN ABRAMOVICH, Department of Mathematics, Boston University, 111 Cummington St., Boston, MA 02215; email: `abramovic@bu.edu`

*__Algebraic number theory__ to V. KUMAR MURTY, Department of Mathematics, University of Toronto, 100 St. George Street, Toronto, ON M5S 1A1, Canada; email: `murty@math.toronto.edu`

Algebraic topology and cohomology of groups to STEWART PRIDDY, Department of Mathematics, Northwestern University, 2033 Sheridan Road, Evanston, IL 60208-2730; email: `priddy@math.nwu.edu`

Combinatorics and Lie theory to SERGEY FOMIN, Department of Mathematics, University of Michigan, Ann Arbor, Michigan 48109-1109; email: `fomin@umich.edu`

Complex analysis and complex geometry to DUONG H. PHONG, Department of Mathematics, Columbia University, 2990 Broadway, New York, NY 10027-0029; email: `phong@math.columbia.edu`

*__Differential geometry and global analysis__ to LISA C. JEFFREY, Department of Mathematics, University of Toronto, 100 St. George St., Toronto, ON Canada M5S 3G3; email: `jeffrey@math.toronto.edu`

Dynamical systems and ergodic theory to ROBERT F. WILLIAMS, Department of Mathematics, University of Texas, Austin, Texas 78712-1082; email: `bob@math.utexas.edu`

*__Functional analysis and operator algebras__ to MARIUS DADARLAT, Department of Mathematics, Purdue University, 150 N. University St., West Lafayette, IN 47907-2067; email: `mdd@math.purdue.edu`

*__Geometric analysis__ to TOBIAS COLDING, Courant Institute, New York University, 251 Mercer St., New York, NY 10012; email: `colding@cims.nyu.edu`

*__Geometric analysis__ to MLADEN BESTVINA, Department of Mathematics, University of Utah, 155 South 1400 East, JWB 233, Salt Lake City, Utah 84112-0090; email: `bestvina@math.utah.edu`

Harmonic analysis to ALEXANDER NAGEL, Department of Mathematics, University of Wisconsin, 480 Lincoln Drive, Madison, WI 53706-1313; email: `nagel@math.wisc.edu`

Harmonic analysis, representation theory, and Lie theory to ROBERT J. STANTON, Department of Mathematics, The Ohio State University, 231 West 18th Avenue, Columbus, OH 43210-1174; email: `stanton@math.ohio-state.edu`

*__Logic__ to STEFFEN LEMPP, Department of Mathematics, University of Wisconsin, 480 Lincoln Drive, Madison, Wisconsin 53706-1388; email: `lempp@math.wisc.edu`

Number theory to HAROLD G. DIAMOND, Department of Mathematics, University of Illinois, 1409 W. Green St., Urbana, IL 61801-2917; email: `diamond@math.uiuc.edu`

*__Ordinary differential equations, and applied mathematics__ to PETER W. BATES, Department of Mathematics, Michigan State University, East Lansing, MI 48824-1027; email: `peter@math.msu.edu`

*__Partial differential equations__ to PATRICIA E. BAUMAN, Department of Mathematics, Purdue University, West Lafayette, IN 47907-1395; email: `bauman@math.purdue.edu`

*__Probability and statistics__ to KRZYSZTOF BURDZY, Department of Mathematics, University of Washington, Box 354350, Seattle, Washington 98195-4350; email: `burdzy@math.washington.edu`

*__Real analysis and partial differential equations__ to DANIEL TATARU, Department of Mathematics, University of California, Berkeley, Berkeley, CA 94720; email: `tataru@ math.berkeley.edu`

All other communications to the editors should be addressed to the Managing Editor, WILLIAM BECKNER, Department of Mathematics, University of Texas, Austin, TX 78712-1082; email: `beckner@math.utexas.edu`.

Titles in This Series

815 **Martin Bendersky and Donald M. Davis,** V_1-periodic homotopy groups of $SO(n)$, 2004

814 **Johannes Huebschmann,** Kähler spaces, nilpotent orbits, and singular reduction, 2004

813 **Jeff Groah and Blake Temple,** Shock-wave solutions of the Einstein equations with perfect fluid sources: Existence and consistency by a locally inertial Glimm scheme, 2004

812 **Richard D. Canary and Darryl McCullough,** Homotopy equivalences of 3-manifolds and deformation theory of Kleinian groups, 2004

811 **Ottmar Loos and Erhard Neher,** Locally finite root systems, 2004

810 **W. N. Everitt and L. Markus,** Infinite dimensional complex symplectic spaces, 2004

809 **J. T. Cox, D. A. Dawson, and A. Greven,** Mutually catalytic super branching random walks: Large finite systems and renormalization analysis, 2004

808 **Hagen Meltzer,** Exceptional vector bundles, tilting sheaves and tilting complexes for weighted projective lines, 2004

807 **Carlos A. Cabrelli, Christopher Heil, and Ursula M. Molter,** Self-similarity and multiwavelets in higher dimensions, 2004

806 **Spiros A. Argyros and Andreas Tolias,** Methods in the theory of hereditarily indecomposable Banach spaces, 2004

805 **Philip L. Bowers and Kenneth Stephenson,** Uniformizing dessins and Belyĭ maps via circle packing, 2004

804 **A. Yu Ol'shanskii and M. V. Sapir,** The conjugacy problem and Higman embeddings, 2004

803 **Michael Field and Matthew Nicol,** Ergodic theory of equivariant diffeomorphisms: Markov partitions and stable ergodicity, 2004

802 **Martin W. Liebeck and Gary M. Seitz,** The maximal subgroups of positive dimension in exceptional algebraic groups, 2004

801 **Fabio Ancona and Andrea Marson,** Well-posedness for general 2×2 systems of conservation law, 2004

800 **V. Poénaru and C. Tanas,** Equivariant, almost-arborescent representation of open simply-connected 3-manifolds; A finiteness result, 2004

799 **Barry Mazur and Karl Rubin,** Kolyvagin systems, 2004

798 **Benoît Mselati,** Classification and probabilistic representation of the positive solutions of a semilinear elliptic equation, 2004

797 **Ola Bratteli, Palle E. T. Jorgensen, and Vasyl' Ostrovs'kyĭ,** Representation theory and numerical AF-invariants, 2004

796 **Marc A. Rieffel,** Gromov-Hausdorff distance for quantum metric spaces/Matrix algebras converge to the sphere for quantum Gromov-Hausdorff distance, 2004

795 **Adam Nyman,** Points on quantum projectivizations, 2004

794 **Kevin K. Ferland and L. Gaunce Lewis, Jr.,** The $RO(G)$-graded equivariant ordinary homology of G-cell complexes with even-dimensional cells for $G = \mathbb{Z}/p$, 2004

793 **Jindřich Zapletal,** Descriptive set theory and definable forcing, 2004

792 **Inmaculada Baldomá and Ernest Fontich,** Exponentially small splitting of invariant manifolds of parabolic points, 2004

791 **Eva A. Gallardo-Gutiérrez and Alfonso Montes-Rodríguez,** The role of the spectrum in the cyclic behavior of composition operators, 2004

790 **Thierry Lévy,** Yang-Mills measure on compact surfaces, 2003

789 **Helge Glöckner,** Positive definite functions on infinite-dimensional convex cones, 2003

788 **Robert Denk, Matthias Hieber, and Jan Prüss,** $\mathcal{R}$-boundedness, Fourier multipliers and problems of elliptic and parabolic type, 2003

787 **Michael Cwikel, Per G. Nilsson, and Gideon Schechtman,** Interpolation of weighted Banach lattices/A characterization of relatively decomposable Banach lattices, 2003

786 **Arnd Scheel,** Radially symmetric patterns of reaction-diffusion systems, 2003

785 **R. R. Bruner and J. P. C. Greenlees,** The connective K-theory of finite groups, 2003

784 **Desmond Sheiham,** Invariants of boundary link cobordism, 2003

783 **Ethan Akin, Mike Hurley, and Judy A. Kennedy,** Dynamics of topologically generic homeomorphisms, 2003

782 **Masaaki Furusawa and Joseph A. Shalika,** On central critical values of the degree four L-functions for GSp(4): The Fundamental Lemma, 2003

781 **Marcin Bownik,** Anisotropic Hardy spaces and wavelets, 2003

780 **S. Marmi and D. Sauzin,** Quasianalytic monogenic solutions of a cohomological equation, 2003

779 **Hansjörg Geiges,** h-principles and flexibility in geometry, 2003

778 **David B. Massey,** Numerical control over complex analytic singularities, 2003

777 **Robert Lauter,** Pseudodifferential analysis on conformally compact spaces, 2003

776 **U. Haagerup, H. P. Rosenthal, and F. A. Sukochev,** Banach embedding properties of non-commutative L^p-spaces, 2003

775 **P. Lochak, J.-P. Marco, and D. Sauzin,** On the splitting of invariant manifolds in multidimensional near-integrable Hamiltonian systems, 2003

774 **Kai A. Behrend,** Derived ℓ-adic categories for algebraic stacks, 2003

773 **Robert M. Guralnick, Peter Müller, and Jan Saxl,** The rational function analogue of a question of Schur and exceptionality of permutation representations, 2003

772 **Katrina Barron,** The moduli space of $N = 1$ superspheres with tubes and the sewing operation, 2003

771 **Shigenori Matsumoto,** Affine flows on 3-manifolds, 2003

770 **W. N. Everitt and L. Markus,** Elliptic partial differential operators and symplectic algebra, 2003

769 **Jie Wu,** Homotopy theory of the suspensions of the projective plane, 2003

768 **R. Höpfner and E. Löcherbach,** Limit theorems for null recurrent Markov processes, 2003

767 **Po Hu,** S-modules in the category of schemes, 2003

766 **Su Gao and Alexander S. Kechris,** On the classification of Polish metric spaces up to isometry, 2003

765 **Robert Bieri and Ross Geoghegan,** Connectivity properties of group actions on non-positively curved spaces, 2003

764 **J. Spandaw,** Noether-Lefschetz problems for degeneracy loci, 2003

763 **Yasuyuki Kachi and Eiichi Sato,** Segre's reflexivity and an inductive characterization os hyperquadrics, 2002

762 **Leiba Rodman, Ilya M. Spitkovsky, and Hugo Woerdeman,** Abstract band method via factorization, positive and band extensions of multivariable almost periodic matrix functions, and spectral estimation, 2002

761 **Oliver Druet and Emmanuel Hebey,** The AB program in geometric analysis : Sharp Sobolev inequalities and related problems, 2002